Management and Industrial Engineering

Series editor

J. Paulo Davim, Aveiro, Portugal

More information about this series at http://www.springer.com/series/11690

Jorge Luis García-Alcaraz
Aidé Aracely Maldonado-Macías

Just-in-Time Elements and Benefits

Springer

Jorge Luis García-Alcaraz
Autonomous University of Ciudad Juárez
Ciudad Juárez, Chihuahua
Mexico

Aidé Aracely Maldonado-Macías
Autonomous University of Ciudad Juárez
Ciudad Juárez, Chihuahua
Mexico

ISSN 2365-0532 ISSN 2365-0540 (electronic)
Management and Industrial Engineering
ISBN 978-3-319-25917-8 ISBN 978-3-319-25919-2 (eBook)
DOI 10.1007/978-3-319-25919-2

Library of Congress Control Number: 2015953786

Springer Cham Heidelberg New York Dordrecht London

Printed on acid-free paper

Springer International Publishing AG Switzerland is part of Springer Science+Business Media (www.springer.com)

Perhaps writing the book was not as difficult as writing something simple in this dedication. I wish to dedicate this work to the pillars and loves of my life:
To GOD for everything
To my wife, Ana Blanca Rodríguez Rendón, because she supports my projects.
To my son, Jorge Andres García Rodríguez, because he is given me the greatest lessons of my life.
To my daughter Mariana Odette García Rodríguez, who teach me that love is infinite. Just when I thought my life is complete, I realize that is not true, a new angel comes into my life and be born within seven months. For him or her, I dedicate this work. We are waiting for you. We love you.

Jorge Luis García-Alcaraz

To my parents, husband, and sons who have always supported me in all possible ways to pursue my dreams.
To my students to encourage me for being better every day.

Aidé Aracely Maldonado-Macías

Foreword

García-Alcaraz, Ph.D., became part of our research group at Universidad La Rioja, Spain, during a two-month stay in 2012 and also in 2014 during six months. At this time, we used to discuss on several projects that we had in mind with the collaboration of Aidé Aracely Maldonado-Macías, Ph.D., and other fellow researchers and professors. Their team aimed at the creation of a research network focused in industrial processes optimization, which would be financially supported by the National Council of Science and Technology (CONACYT) of Mexico. It is thus with pride that I see this work as the result of that research network that they managed to create and to which I belong.

Unquestionably, this book entitled *Just in time, its elements, benefits and reasons for low implementation: some structural equation models* is a response to the increasing manufacturing industry currently developing in Mexico, especially in Ciudad Juarez, the eighth largest city in the country, located in the border with the USA. These companies are characterized by importing their commodities and exporting the products thereby manufactured to other countries. However, this work also constitutes a guide with a high empirical content about the situation of the just-in-time (JIT) technique in the local industry.

The book is divided into thirteen chapters, which I briefly describe below:

Chapter 1 makes a review of the background of JIT and its origins in Japan; other aspects are defined and discussed such as concepts and evolution, its position within the set of tools of slender manufacturing, and also three cases of success of its application in Western industry.

Chapter 2 describes the JIT elements, which refer to the plans and programs that must be present in order to guarantee its successful implementation. This chapter contains a wide literature review that justifies each one of the above-mentioned elements, which are divided into three categories: the ones associated with human resources, the ones associated with the process and, finally, the ones associated with the product and its characteristics.

In Chap. 3, the benefits obtained from a proper implementation of JIT are discussed. These benefits are divided into six categories: benefits attributed to human resources, benefits attributed to production processes, benefits attributed to engineering, benefits attributed to quality, benefits attributed to management of materials, and finally, economic benefits. Furthermore, as in Chap. 2, a thorough literature review of each of the benefits is presented.

Chapter 4 presents a study of the causes and reasons of the slow implementation of JIT, which are most of the times due to the lack or absence of one of the required elements. This chapter is one of the most important, due mainly to the existing bibliographical bias, where most of the times the success cases are the only ones reported, and the failures and their causes are avoided or not mentioned.

Chapter 5 describes the methodology followed to associate JIT elements to the benefits obtained, as well as the causes for its low implementation. It can be seen that the theoretical foundation lies in the factorial analysis as information reduction technique and on structural equation modeling to generate the causal models. It is interesting to read about the process of creating the questionnaire applied in the local manufacturing industry, which enriches this book as it is full of empiric and real cases of industrial application.

Once the questionnaire was applied in the local maquiladora industry, Chap. 6 makes a description of the research sample. Statistical data of 144 cases are reported, out of which 118 are companies with more than 500 employees; among them, the medical and car industry were the ones that participated the most in the study.

The questionnaire applied had to be answered in a Likert-based scale, based on the experience of those polled, as explained in Chap. 7, which reports the central tendencies and dispersion measures, and presents a description of the JIT elements. It is worth mentioning that the analysis is univariable, and no relations or associations of the variables are made.

In the same way that the elements have been defined, the benefits of JIT in the environment are analyzed in Chap. 8, and the measures of central tendency and deviation are also reported. It can be stated that a better use of resources is one of the most reported benefits, as well as the increase of efficacy of the production process. Chapter 9 is similar to the two previous chapters except that it describes the causes of the low implementation of JIT.

The causal analyses appear from Chap. 10 onwards. They are proposed models, where, by means of the use of structural equations, a series of relations between latent variables is solved with algorithms of partial minimum squares.

Last but not least, just as a brief summary, I sincerely consider that this book will be of great use, not only to those who wish to know how to successfully implement JIT in industry, but also, due to its high empiric content, it can be used as an example in academia and research.

Emilio Jiménez Macías
University of La Rioja, Spain

Preface

Mexico is a developing country and member of the North American Free Trade Agreement (NAFTA) together with United States of America (USA) and Canada. This commercial agreement has favored the establishment of a large number of companies from other countries in the northern border of the Mexican land due to tariff preferences. These companies are usually known as maquiladoras, since they import all raw materials and export the finished product.

The arrival of this type of manufacturing companies has led to the implementation of lean manufacturing techniques, especially just in time (JIT), in order to increase efficiency in their production processes and minimize waste. JIT emerged in Japan, a country with social and cultural characteristics different from those present in Mexico. However, the significance of this technique lies precisely on its impact over the processes of import of raw materials, production, and export of finished products.

Based on cases reported by Kumar (2010); Panchal et al. (2012, 2013) in India, this book researched the elements needed for the implementation of just in time, its benefits, and those factors that cause its slow implementation, and addressed this information for the Mexican environment. To achieve this, a questionnaire was designed and applied to 144 companies located in the northern region of the country. Then, a univariate descriptive analysis was performed in order to propose models to relate these three categories researched.

On the one hand, some models proposed associate every JIT element with the benefit it can bring. These models could allow managers and other personnel responsible for the decision making in companies to identify and focus on critical factors that could guarantee benefits pursued. On the other hand, other models relate the same JIT elements with the causes of its slow implementation in order to identify those factors that prevent a successful operation of the approach.

However, this book is not only targeted at supervisors and managers, it is also dedicated to graduate and undergraduate students in the fields of engineering and business management, since it is hoped that its empirical content will be useful to them.

The book includes thirteen chapters divided into five parts. The first part addresses general information of JIT that describes its timeline, discusses its origins, and details its implementation in Western companies. The second part describes the JIT elements identified, its benefits, and the causes of its slow implementation. This part refers to an extensive literature reviewed, which readers are welcome to consult.

On the other hand, the third part of the book defines the methodology used to generate and propose the casual models previously mentioned. It discusses the design of the questionnaire as well as the processes to debug and analyze data. It is worth mentioning that data analysis is based on factor analysis and structural equation modeling techniques (SEM).

In addition, the fourth part includes the descriptive analyses of the sample, which discusses the manufacturing sectors surveyed and the position of respondents of the survey as well as their seniority. Finally, the fifth part describes the models generated from data gathered where JIT elements of companies surveyed are associated with the benefits that these organizations have obtained.

We sincerely hope this work is useful to readers.

Jorge Luis García-Alcaraz
Aidé Aracely Maldonado-Macías

References

Kumar V (2010) JIT based quality management: concepts and implications in Indian context. Int J Eng Sci Technol 2(1):40–50

Panchal V, Gupta A, Ram S, Rai N (2012) Identification of JIT elements in service sector. Int J Latest Res Sci Tech 1(3):211–214

Panchal V, Amit P, Sachin G (2013) Study of JIT and JIT elements. Int J Indus Engg Tech (IJIET) 2(2):38–41

Acknowledgments

It will be unfair not to recognize the efforts and hard work of all fellow workers involved in the development of this project. We are truthfully thankful to these people and well as those whose names were not mentioned. Thank you.

- Diana Jazmin Prieto Luevano, José Roberto Díaz Reza, and Deysi Márquez Gayosso, Sonia Gabriela Valadez Torres—industrial engineering graduate students of the Autonomous University of Ciudad Juárez, México.
- Valeria Martínez Loya, Edgar Sierra Lomelí, Nallali Olguin, Gerardo Arredondo, Karla Castro and Evelyn Trejo—industrial engineering undergraduate students of the Autonomous University of Ciudad Juárez, México.
- Erick Ortega, mechanical engineering undergraduate student of University of Texas, El Paso.
- María Fernanda Villafuerte Bianchi for her valuable comments concerning the translation and edition of this work in English.

Similarly, authors are sincerely grateful to the National Council of Science and Technology (CONACYT) of Mexico for the financial support provided as well as to Thematic Network in Industrial Processes Optimization with number 260320. Finally, authors thank Autonomous University of Ciudad Juarez for the provision of facilities and databases needed for this research.

Contents

Part I General Overview

1 Concepts of Just-in-Time (JIT) 3
1.1 Background (Historical Evolution of JIT). 3
1.2 Definition of Just-in-Time 4
1.3 JIT and Its Origins 7
1.4 Evolution of JIT in History 8
1.5 Importing JIT to the Occident. 9
1.6 JIT as a Lean Manufacturing Tool 11
1.7 JIT and Supply Chain 13
1.8 The Supply Chain as a Whole 14
1.9 Practical Examples of Successful JIT. 15
1.9.1 Toyota. 15
1.9.2 Dell. 16
1.9.3 Harley Davidson 16
1.10 Why Companies Want to Implement JIT? 17
References. 17

Part II JIT Elements, Its Benefits, and Causes of Its Slow Implementation

2 Elements of JIT 23
2.1 JIT Elements Associated with Human Resources. 23
2.2 JIT Elements Associated with Production Processes 29
2.3 JIT Elements Associated with the Product 38
References. 45

3 Benefits of JIT 53
3.1 Benefits Associated with Human Resources 53
3.2 Benefits Associated with Production Processes 56
3.3 JIT Benefits Associated with Engineering 60
3.4 Benefits Associated with Quality. 62

3.5 Benefits Associated with Materials Handling 63
3.6 Economic Benefits Associated with JIT 66
References 69

4 Causes of Slow Implementation of JIT 75
References 82

Part III Methodology

5 Research Methodology 89
5.1 Stage 1: Questionnaire Development 89
5.1.1 Section 1: Demographics 89
5.1.2 Section 2: JIT Elements 90
5.1.3 Section 3: JIT Benefits 90
5.1.4 Section 4: Causes of a Slow JIT Implementation 91
5.2 Measurement Scale 91
5.3 Administration of the Questionnaire 91
5.4 Information Capture and Database Debugging 92
5.4.1 Missing Values 92
5.4.2 Extreme Values or Outliers 93
5.4.3 Zero Variance 93
5.4.4 Normality Test 94
5.4.5 Homoscedasticity Test 94
5.4.6 Multicollinearity Test 95
5.5 Descriptive Analysis of the Sample and Information 95
5.5.1 The Median as a Measure of Central Tendency 95
5.5.2 Interquartile Range as a Measure of Dispersion 96
5.6 Questionnaire Validation 97
5.6.1 Cronbach's Alpha Internal Consistency 97
5.6.2 Average Variance Extracted (AVE), Discriminant Validity 98
5.6.3 Correlation Coefficient, Predictive Validity 98
5.6.4 Dillon-Goldstein Rho Index, Composite Reliability . . . 99
5.7 Integrating the Latent Variables (Factor Analysis) 99
5.8 Hypothesis Formulation, Construction of Structural Equations Models 100
5.8.1 Effects Between Variables 101
5.8.2 Model Efficiency Ratios 104
References 106

Part IV Descriptive Analysis

6 Descriptive Analysis of the Sample 115
6.1 The Industrial Sectors Analyzed 115
6.2 The Size of Companies 116
6.3 Analysis of Size and Industrial Sector 117

6.4 Seniority and Position of Respondents . . . 118
6.5 Gender of Respondents . . . 118
6.6 Chapter Summary . . . 119

7 Descriptive Analysis of the Elements of JIT . . . 121
7.1 JIT Elements Associated with Human Resources . . . 122
7.2 Elements of JIT Associated with the Production Process . . . 125
7.3 Elements Associated with the Product . . . 131
7.4 Chapter Conclusions . . . 135
References . . . 136

8 Descriptive Analysis of JIT Benefits . . . 145
8.1 Benefits Associated with Human Resources . . . 146
8.2 Benefits Associated with Production Process . . . 148
8.3 Benefits Associated with Engineering . . . 152
8.4 Benefits Associated with Quality . . . 153
8.5 Benefits Associated with Material Handling . . . 155
8.6 Economic Benefits . . . 158
8.7 Conclusions . . . 160
References . . . 162

9 Descriptive Analysis of the Causes of Slow JIT Implementation . . . 169
9.1 Descriptive Analysis . . . 169
9.2 Conclusions . . . 176
References . . . 176

Part V Structural Equation Models

10 Causal Models of JIT Elements Associated with Human Resources and Obtained Benefits . . . 183
10.1 Factor Analysis of JIT Elements Associated with Human Resources . . . 183
10.2 Model 1 Elements: Human Resources-JIT Benefits: Human Resources and Production Process . . . 184
10.2.1 Hypotheses of Proposed Model 1 . . . 185
10.2.2 Evaluation of Model 1 . . . 187
10.2.3 Conclusions of Model 1 . . . 191
10.3 Model 2 Elements: Human Resources-JIT Benefits: Quality and Economic Benefits . . . 192
10.3.1 Hypotheses . . . 193
10.3.2 Evaluation of Model 2 . . . 196
10.3.3 Conclusions of Model 2 . . . 202

10.4 Model 3 Elements: Human Resources-JIT Benefits: Materials Handling and Engineering Process 202
10.4.1 Hypotheses. 203
10.4.2 Evaluation of Model 3. 205
10.4.3 Conclusions of Model 3. 210
References. 211

11 Causal Models of JIT Elements Associated with Production Process and the Obtained Benefits 217
11.1 Factor Analysis of JIT Elements Associated with Production Process. 217
11.2 Model 1. JIT Elements: Lean Techniques and Production Organization and Flow of Materials. Benefits: Production Process and Process Engineering. 219
11.2.1 Hypotheses Proposed of Model 1 220
11.2.2 Results of Model 1 222
11.3 Model 2. Elements: Lean Techniques and Production Organization and Material Flow. Benefits: Human Resources and Quality 229
11.3.1 Hypotheses Proposed of Model 2 230
11.3.2 Results of Model 2 232
11.4 Model 3: JIT Elements: Production Organization and Material Flow and Production Capacity and Inventory Management. Benefits: Material Handling and Economic Performance 238
11.4.1 Proposed Model 3. 239
11.4.2 Results of Model 3 242
References. 247

12 Causal Models JIT Elements Associated with Product and Obtained Benefits. 255
12.1 Factor Analysis of JIT Elements Associated with the Product 255
12.2 Model 1. JIT Elements: Quality Planning and Quality Management. Benefits: Human Resources, Quality, and Economic Benefits 257
12.2.1 Hypotheses of Model 1 257
12.2.2 Results of Model 1 261
12.2.3 Conclusions of Model 1. 267
12.3 Model 2: JIT Elements: Quality Planning and Quality Management. Benefits: Engineering Process, Production Process, and Material Handling. 268
12.3.1 Hypotheses of Model 2 268
12.3.2 Results of Model 2 272
References. 280

13 Causal Models of JIT Elements and Causes of Slow JIT Implementation 285
13.1 Factor Analysis. 285
13.2 Model 1: JIT Elements of Human Resources and Causes of Slow JIT Implementation 287
13.2.1 Proposed Working Hypotheses of Model 1. 287
13.2.2 Evaluation of Model 1 290
13.3 Model 2: JIT Elements of Production Process and Causes of Slow JIT Implementation 295
13.3.1 Proposed Working Hypotheses of Model 2. 295
13.3.2 Evaluation of Model 2. 298
13.4 Model 3: Product Characteristics and Causes of Slow Implementation of JIT 303
13.4.1 Proposed Working Hypotheses of Model 3. 303
13.4.2 Evaluation of Model 3. 305
13.4.3 Sum of Indirect Effects of Model 3 308
References. 309

Part I
General Overview

Chapter 1
Concepts of Just-in-Time (JIT)

Just-in-Time (JIT) is a technique widely used in lean manufacturing environments, and since its conception a few decades ago, it has been extensively applied to support lean objectives. The aim of this chapter is to introduce readers to the historical evolution of JIT, its main concepts and principles, and how it has been successfully used with other techniques and methodologies to provide effective solutions for waste and lack of productivity problems.

1.1 Background (Historical Evolution of JIT)

Due to increasing global competition, manufacturing entities and a vast variety of companies around the world have addressed more attention to customer satisfaction and competitive advantage. Because of the challenges of global competition, companies are forced to reduce costs, improve quality, and meet the needs of their fickle consumers (Gupta 2011).

Nowadays, business environments are characterized by three aspects: competitiveness, readiness to confront and adapt to unpredictable changes and fluctuations in demand, and strict customer requirements seeking for high quality products and the fulfillment of specific needs (Marín and Delgado 2000). Thus, the stability and survival of manufacturing companies in an increasingly competitive world certainly depends on their ability to produce with the highest quality, the lowest cost, and the shortest delivery time permissible (Kumar 2010).

Moreover, manufacturing entities should become competitive to stay in business and remain among preferences of customers. Also, these companies must constantly supply high quality products and offer timely and reliable deliveries of their goods all through the supply chain (Singh and Garg 2011a). One of the most commonly used techniques to accomplish these aims is the implementation of JIT, which actually makes it possible for companies to obtain the competitive advantage required. Moreover, its elements are effective to respond to rapid changes and efficiencies demanded today (Machuca 2002).

The philosophy of JIT is mainly based on both reducing waste and improving and maintaining the excellence of quality in products or services, which are only

J.L. García-Alcaraz and A.A. Maldonado-Macías, *Just-in-Time Elements and Benefits*, Management and Industrial Engineering, DOI 10.1007/978-3-319-25919-2_1

possible with the commitment (loyalty) of every member of the organization involved, and with a strong focus of these members on their particular tasks (involvement at work). These should lead to greater productivity of the production system, lower costs per unit product, quality, increased customer satisfaction, increased sales, but most of all, higher profits for the company, which is the ultimate goal (Fullerton and McWatters 2001).

The traditional concept of JIT, mostly associated with delivery times, is not completely accurate, since as a philosophy it goes way beyond (Fiedler et al. 1993). In fact, JIT as a technique applied to production systems is not isolated; it is integrated with many other techniques. What is more, some authors place JIT within the main tools of Lean Manufacturing (LM) (Fullerton and McWatters 2001).

1.2 Definition of Just-in-Time

It seems convenient to define the concept before going into the topic in order to understand the environment and scope of JIT. However, it is worth mentioning that its definition has changed over time.

JIT applied to a production system is a manufacturing philosophy that eliminates waste associated with time, work, and storage space. Its foundations are the fact that the company produces only what is needed, when needed, and in the amount needed. That is, a company produces only what the customer requests as actual orders, not to forecast. JIT can also be defined as the production of the necessary units, with the required quality in the quantities needed at the last moment safe. This means that the company can easily manage and allocate its own resources (Arnold and Bernard 1989).

JIT philosophy is directed toward the elimination of waste by streamlining production processes, reducing setup times, controlling flow of materials, and providing preventive maintenance to equipment and machinery. Through these activities, inventory and resources can be reduced and used more efficiently (Kannan and Tan 2005).

However, the abovementioned definitions are only some of many that exist, and all of them have been proposed from different perspectives, from production management to general management for customer satisfaction, and also as a tool to gain competitive advantage in the market (Zipkin 1991). The list below shows the definitions of JIT proposed by various authors.

- JIT is defined as the act of having the right parts at the right time and at the right amount (Ohno 1982). This is perhaps one of the first definitions of JIT in the Eastern world.
- JIT is a way of visualizing the physical operations of the company, from the purchase of raw materials to the customer delivery of a final product duly transformed with useful features or attributes (Hall 1983).

- JIT is a system to produce and deliver finished products on time to sellers. Required materials are just-in-time purchased to be transformed into parts and subassemblies just-in-time manufactured that are assembled into finished products (Schonberger and Gilbert 1984). This is one of the first Western interpretations of the JIT philosophy.
- JIT is the most important technique for improving productivity and innovating since the turn of the century (Ebrahtmpour and Schonberger 1984). This definition is essential since it actually considers JIT as a means or technique that helps obtain better rates of productivity in production systems. That is, in their results obtained with the implementation of JIT, companies are favored by a number of benefits reflected on their internal indices, which indicate the efficiency and effectiveness of JIT.
- JIT is a philosophy of industrial production that applies to manufacturing systems where commitment and involvement of human resources at all levels of the organization is highly needed to achieve the transformation of raw materials into finished products (Ansari 1986). This is the first definition listed here that explicitly addresses JIT as a philosophy, which thus, depends on human resources, the most precious capital for the company.
- JIT is an approach ensuring that quantities produced of a product are purchased and carried out on time and with the appropriate quality (Voss 1986). This definition is much more specific than the previous ones, since it associates JIT directly with the flow of materials throughout the production system.
- JIT is a production methodology that aims at improving overall productivity by eliminating waste, which leads to improving the quality of waste material (Voss and Robinson 1987). This definition is similar to the one proposed by Ebrahtmpour and Schonberger (1984). However, it shows a more holistic approach since it mentions an overall productivity, although it focuses more on the disposal of waste in the production system and product quality.
- JIT is a production system to produce the exact type of units required in the necessary time and quantities (Moore 1988). This definition considers JIT as a whole system, and agrees with authors (Voss and Robinson 1987), who mention that the impact of JIT is global.
- JIT is a system whose objective is to minimize inventories of raw materials and work in process, control defects, consolidate production, and simplify and create work in a way that is flexible (Calvasina et al. 1989). The word "system" appears again in this definition. Thus, by 1989, the benefits of JIT were undoubtedly identified among enterprises. Moreover, this definition emphasizes on the handling of materials throughout the production system in a flexible way.
- JIT refers to a manufacturing system to achieve excellence through continuous improvements in productivity and eliminating waste (Crawford and Coax 1990). It seems that JIT was still considered as a production system in 1990.
- JIT is an approach in which waste from the production process is identified and systematically eliminated to reduce costs, delivery times, and improve quality (Miltenbirg 1990). This definition is significant since the notion of approach involves people, with a particular and personal way of thinking. Therefore, JIT

here could be understood as a production philosophy applied to production systems, whose purpose is to deliver a high quality end product. Moreover, this definition associates JIT with quality at the end of its implementation. Therefore, as a result of JIT implementation, it is expected to have a quality product.

- JIT is a philosophy of Japanese manufacturing that represents a natural state of simplicity of production efficiency (Zipkin 1991). This concept views JIT as a philosophy again that seeks easiness, a characteristic of the Japanese culture.
- JIT can be defined as a philosophy where employees identify and solve problems and inefficiencies in the workplace (Billesbach 1991). The notion of philosophy appears once more, which implies the presence of human resources of the company. Also, this definition does not address waste but emphasizes on inefficiencies, which can be understood as any problem that may reduce the productivity of a company.
- JIT is described as a way of producing the smallest amount possible of a product at the earliest time possible with the minimum use of resources, and aiming to eliminate waste in the manufacturing process (Hong et al. 1992). This concept seems to be one of the most complete, since it infers that JIT is a technique seeking to optimize the resources available in a company; and in fact, all companies seek to employ their resources better in order to survive, especially in such a globalized market.
- JIT is the system that favors the flow of production and makes smooth continuous improvements in processes and products (Baker et al. 1994). This definition refers to continuous improvements and smooth flows applied to materials in the production system, which makes JIT different from the traditional concept of Kaizen.
- JIT is a philosophy that defines the way of working. It is best defined as an approach to achieve excellence in manufacturing companies (Garg 1997).
- JIT is a philosophy that seeks to produce only what is required and minimize the total manufacturing costs through continuous improvement and commitment to total quality with the participation of all human resources (Power and Sohal 2000).
- JIT has as general objective the continuous improvement of organizational productivity, quality, and flexibility (White and Prybutok 2001).
- JIT is described both as a material flow management in industries to reduce inventory levels at each stage and a measure to simplify the manufacturing system to quickly identify problems and solutions. This definition recognizes that the main goal of the JIT philosophy is to expose hidden problems and involve all employees in their elimination (Singh and Garg 2011b).
- JIT is a philosophy applied to materials and waste management in a company and is widely used in lean manufacturing (García-Alcaraz 2014).

It seems that most of the previous definitions focus on JIT as a philosophy, a way of thinking, since its success depends on human factors, although machines are also a part of the production system even though they do not think. Likewise, it is concluded that the goal of JIT is the reduction of waste, not only in production lines but also in the entire production system. Similarly, definitions agreed on the fact that the ultimate aim of JIT is product quality. It would be worth asking why quality is the ultimate goal. The answer is simple; one characteristic of quality is the delivery of products in a timely manner, which is the primary responsibility of JIT.

Therefore, in order to survive as global, highly competitive businesses, companies must be willing to make strategic adjustments by developing and using innovative manufacturing methods—such as Just in Time, Advanced Manufacturing Technologies, and Total Quality—in response to demands and in order to increase their efficiency, effectiveness, and responsiveness to their customers (Yasin et al. 2003). Just-in-Time methodology is currently recognized as an orderly perspective to achieve competitiveness and excellence that companies and the market demand (Inman et al. 2011).

1.3 JIT and Its Origins

JIT emerged in Japan. It was developed in the decade of the 1950s by Taiichi Ohno and was applied by the car manufacturer Toyota. The company adopted it in the beginning of the decade of the 1970s, and its main purpose was to eliminate all unnecessary elements in the production area (from procurement to customer service departments, including human resources and finances department, among others). Nowadays, it is also used to achieve cost reductions and meet the needs of customers at the lowest possible costs.

In a country of small size such as Japan, the greatest asset is undoubtedly the physical space (Amasaka 2002). Therefore, two of the pillars of new philosophy (JIT) were precisely space reduction and the elimination of waste; that is, removing the burden of the existence of inventory.

However, during the decade of 1950, technological progress and industrial development were almost exclusive to the United States, due mainly to its victory in World War II that greatly hurt the Japanese nation. However, 30 years later this trend was reversed after a great amount of dedication and effort based on education. Japan experienced great advancement in electronics and other major industrial sectors related to the most prosperous industries of the time. These were favored by the positive economic and labor conditions of Japan. Afterwards, the birth of a large number of companies, almost all related to the same technological sectors, raised fierce competition. The struggle for world supremacy then started to focus on aspects that had never been so important, such as product innovation (Yasin et al. 1997; Jarrar and Smith 2014).

This large number of companies from a same sector caused the almost simultaneous appearance of similar products manufactured by different companies, which reduced the market share and the economic benefits for all. Thus, companies sought to be higher than their competitors, especially in those aspects that nobody before had thought of; they had to be innovative. Japanese companies were the first to focus their products on innovation (Lee et al. 2014; Sanidas 2004).

Thus, companies of Japan had to outstand in product innovation among competition, but they also had to be the fastest to avoid those competitors that reduced their profit margin. In this way they could achieve the greatest possible market. However, with rapid technological advancements, there was not much time from the moment a company released a new product to the one competitors "reproduced" or "copied" that product, if the word can be used—although other authors refer to these reproductions as "adapted innovations." As a consequence, companies had to think of a new production method to continue innovating but also to increase profit margins; and this is precisely the production philosophy addressed in this book: Just-in-Time.

Due to its impact on efficiency indexes, JIT was quickly adopted and implemented in Japanese companies, which addressed two issues at once: they solved problems related to the lack of physical space (characteristic of Japanese land) and they maximized profit that was reflected on inventory reduction and the elimination of waste activities (Maiga and Jacobs 2009).

The first company to implant this production philosophy was Toyota, which allowed it to quickly become a global leader in the automobile sector. The effectiveness of JIT quickly led those companies that applied its principles to improve and refine their philosophies, which had an impact on all areas of the organization (personnel management, direction, decision making, economic analysis, etc.), not only in production processes (Fullerton and McWatters 2002; Fullerton et al. 2003).

Thus, the implementation of the JIT philosophy can explain the success of many Japanese companies in recent years, which little by little lead in the sector where they operate. However, many companies have not yet implemented JIT in their production processes, and most of them seem to be Western companies, including American and European enterprises, although it has been stated that efforts have actually been made without reaching the results obtained in Japan (Fullerton and McWatters 2001; White and Prybutok 2001; Monden 2002).

1.4 Evolution of JIT in History

Just-in-Time has recently become prominent in the research of manufacturing systems and many works address it as their central topic of discussion. For instance, Fig. 1.1 graphs the amount of publications found in the ScienceDirect database referring to JIT only by the title. According to the adjustment line generated by a least square regression, there is a change in the intensity of 244.87 new items every year. That is, the number of works referring to JIT seems to increase. In order to adjust the line presented, items found for year 2015 were excluded since it could generate bias to the graph.

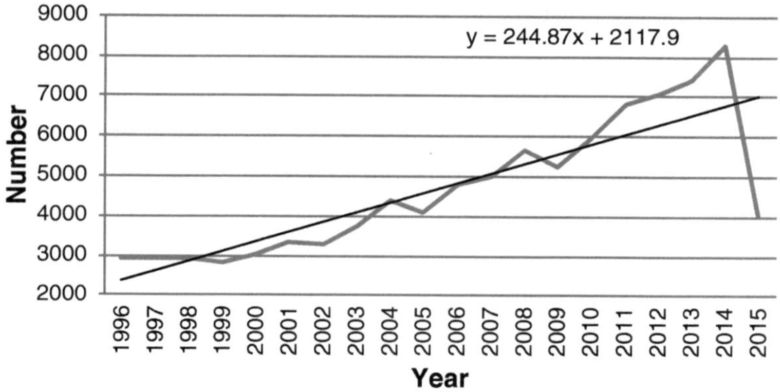

Fig. 1.1 Publications of JIT per year

1.5 Importing JIT to the Occident

The benefits of JIT have encouraged many Western companies to import it. However, although "adopt" can be considered more suitable, it implies that JIT is implemented without making appropriate changes according to the environment and atmosphere in which it operates. Thus, it would be preferable to state that the JIT philosophy must be adapted. These adaptation processes in the West have always been a topic of discussion.

One of the reasons why JIT has not been fully installed in the West may lie in the different lifestyles of these two economic blocks. Japanese culture, for instance, may often involve a more methodic lifestyle; however, what should matter is how companies are perceived (Chhikara and Weiss 1995). On the one hand, companies in America and Europe seem to be the only workplace for employers, which lose all contact with it after working hours. On the other hand, companies are a very important part in the life of Japanese workers who consider problems in these organizations as theirs, too. Thus, they try to solve these issues for the benefit of the whole company rather than for private gain (Aksoy and Öztürk 2011).

Also, it must not be assumed that JIT is not only a productive method but also a philosophy that should not be implemented but taught. Its virtues and drawbacks should thus be presented so that workers learn JIT by their own initiative rather than by imposition. It must be remembered that at least five of the definitions presented in the previous pages refer to JIT as a philosophy, which therefore relies mainly on the people who implement and work with it.

Moreover, advertising JIT has not reached its details in-depth; it has remained only on the surface, often limited to vague and fuzzy concepts. As a result, companies see the outermost layer, which often makes them feel suspicious and reluctant to embrace this new approach. They may see JIT as a method to increase the rate of return on investment of a company or to reduce costs, but they do not have the opportunity to appreciate it and experience it as a philosophy. In these cases, JIT does not reach the required levels of maturity (Yasin et al. 1997, 2003).

However, the adoption of JIT brings a radical change in the way a company is seen and understood, and this is pure philosophy. With JIT implementation, rules and routines already established in a company become obsolete since JIT forces to eliminate overspending, a characteristic often present in large Western enterprises. This actually becomes a factor of the rejection of JIT practices, since not all companies consider themselves flexible enough to make the required changes. In conclusion, there are several reasons why Western organizations still reject JIT, but they all have a coherent explanation.

Another difference between Western and Japanese companies is the way profits are approached. While the former focus on short- and medium-term profits, Japanese enterprises concentrate on longer-term benefits (Green Jr et al. 2014). A Western manager tends to feel enough pressure to deliver profits to shareholders, since its position depends on it. In spite of all, companies from both blocks that implement it visualize JIT as a means of subsistence in global areas due to the benefits it brings. Companies can rely on statistical control to monitor, prevent errors in and improve JIT after its implementation to be informed of its real status.

Similarly, dissimilarity between Western businesses, more specifically American and Japanese, is the role of staff in quality awareness. Eastern companies tend to emphasize more on the added value that every employee can bring to the enterprise when he/she contributes with ideas and points of view to enhance product quality. On the other hand, although the American perspective does seek the involvement of staff in quality awareness, it does not seem to encourage it enough to make suggestions, since it considers that decisions, and therefore JIT, is a primary responsibility of managers.

From the experience of the authors of the present book as university professors and consultants in industries with both focuses, it may be possible that Japanese companies create methodologies whose implementations do not only apply to the company. They can also be seen as forms of life with the purpose of people make their best effort possible; and this is a factor that favors culture from different perspectives, not only in business but also in society. It facilitates the collective involvement of all employees for a common purpose that, in this case, concerns performing tasks in the best way possible and for the benefit of all members (Brox and Fader 1997).

Finally, in this same attempt to explain differences between Western and Japanese firms, a study presented by the Japanese William Ouchi compared Japanese and American companies to obtain a methodology that could take the best aspects of both parts (Ouchi and Price 1993). Authors proposed what may be referred to as Theory Z, as the synergy of both approaches. Authors determined that companies should seek to provide long-term positions and grant promotions to workers to follow the processes of continuous improvement. Mid-term promotions should be avoided since they create vices in the formation of corporate culture. Similarly, authors suggested that decision making be a consensus, since it should focus on giving individual responsibilities and because individuals had to be perceived as an integral part of the company (Flynn 2015).

1.6 JIT as a Lean Manufacturing Tool

There are several tools in Lean Manufacturing (LM) that aim at eliminating all operations that do not add value to products, services, and processes. In this way companies seek to increase the value of each action and eliminate what is not required (Sundar et al. 2014). The main goal of these tools is to reduce waste and improve operations always based on respect for the worker who performs them. At this stage, one may think that this definition of LM is to some extent similar to the concept of JIT. It is partly right since JIT is simply one of the tools of lean manufacturing, and perhaps the most important in these globalized times, when companies purchase components from different suppliers located in remote locations, and thus, make essential the coordinating of the flow of materials.

According to Melton (2005), lean manufacturing system has been defined as a philosophy of excellence in manufacturing based on:

- The planned elimination of all types of waste. This is where JIT makes an appearance as the essential tool of lean manufacturing.
- Respect for the worker, which is achieved with Kaizen.
- The consistent improvement of productivity of the company and product quality delivered to the customer.

Thus, lean manufacturing aims at continuous improvement to allow companies to reduce costs, improve processes, and eliminate waste (with JIT) to enhance customer satisfaction and maintain profit margin. This will also enable them to survive in a global market that demands higher quality of a product, at a faster delivery and lower price (with JIT), and in the required amount.

As a conclusion, lean manufacturing looks for the following goals, most of them through the proper implementation of JIT (Rivera and Frank Chen 2007). Readers may be able to identify these goals that concern JIT, which perhaps will not be a challenge.

- Reduce waste stream dramatically
- Reduce inventory and floor space production
- Create more robust production systems
- Create suitable systems for materials delivery
- Improve production layouts to increase flexibility

However, since lean manufacturing involves a group of techniques or tools often applied together, the benefits of its application are as follows (Shah and Ward 2003). Again, readers can easily identify the JIT benefits in this list.

- 50 % reduction in production costs
- Reduction of inventories
- Reduced lead time
- Improved quality
- Less workforce
- Increased teamwork efficiency
- Reduced waste
  - Overproduction
  - Timeout (delays)
  - Transport
  - The process
  - Inventories
  - Movements
  - Poor quality

As mentioned earlier, JIT is undoubtedly one of the tools of lean manufacturing (LM) that can bring outstanding benefits to companies. Figure 1.2 illustrates the JIT position among the set of tools in companies within an LM environment.

Note that LM has five basic tools that are the foundation of lean thinking in business. These basic tools are standardized process, leveled production, 5S and visual management (usually known as Andon), deployment guidelines, and flow analysis value. Techniques such as JIT and Jidoka are developed over these tools and become the pillars of lean thinking. Therefore, JIT is a technique of second generation in LM, since certain bases must exist in order to be applied. However, JIT also owns tools, among which are:

1. Tack time
2. Continuous flow

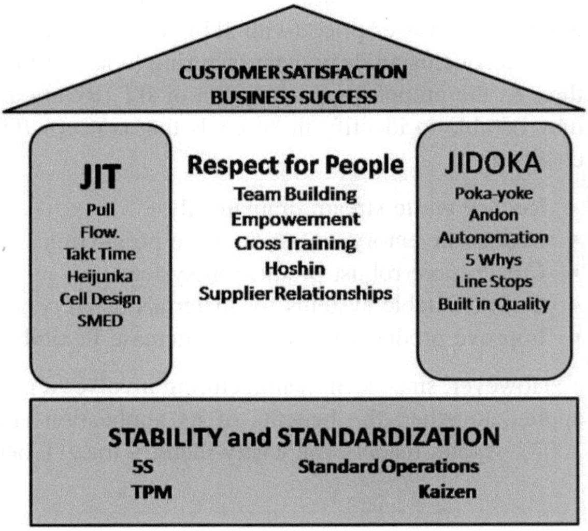

Fig. 1.2 JIT as a tool in lean manufacturing

3. Kanban Pull systems
4. Rapid changes or SMED
5. Manufacturing cells

JIT is not a single technique that generates business success, it is the application of other production techniques, such as Kanban, reduced lot sizes of manufacturing, reduced delivery times, and increased quality of programs required to implement flexible manufacturing processes. Thus, JIT is described as a manufacturing system that strives for excellence through continuous improvement in productivity and the elimination of waste (Dong et al. 2001).

There is no doubt that when JIT is implemented, it seeks to increase operational efficiency and customer satisfaction, improve quality, and gain a strategic competitive advantage (Yasin et al. 2003). This process of implementation is determined by the evolution of the culture of organizations, the review of the working methods and procedures, and the implementation of new administrative processes (Gélinas 1999; Fiedler et al. 1993). Hence, it would be worth wondering about the necessary elements for the implementation of JIT and the benefits obtained from its implementation in the industrial sector, since several authors mention that the same results are not always obtained. These elements of JIT are discussed in the later chapters.

1.7 JIT and Supply Chain

Since JIT is associated with the flow of materials, it is worth mentioning its role in it. In fact customers expect a better product quality, increased responsiveness, short lead times, and lower costs. A number of terms about flow of material have recently emerged, such as supply chain and supply chain management, which often make up JIT.

JIT philosophy by nature tends to eliminate production waste and materials and improve communication, both internal (in the organization) and external (between the organization and its customers and suppliers). Moreover, it has the potential to reduce purchasing costs and is essential to reduce delivery and performance times, improve production quality, and increase productivity and responsiveness to the customer. In addition, JIT tends to encourage the organizational discipline and involvement in management, and seeks to integrate different functions and areas of the organization. Here is where it integrates to the supply chain (Banerjee et al. 2007; Dong et al. 2001; Kojima et al. 2008).

The Supply Chain Management (SCM) provides the integration of buyers and suppliers in the decision-making process in order to improve the flow of material through the supply chain. SCM is seen as the driver of reductions in delivery times, costs, and responsiveness. Therefore, JIT plays an important role in it (González-Benito et al. 2000). According to these authors, the successful implementation of JIT depends on two aspects: the coordination of production schedules with supplier deliveries and high levels of service from providers in terms of product quality. This requires excellent relations with the suppliers.

1.8 The Supply Chain as a Whole

According to Blanchard (2010), the Administration Supply Chain has reached importance as one of the most important aspects of the twenty-first century to improve organizational competitiveness through integration between suppliers and customer companies in order to improve responsiveness and flexibility of these organizations. Similarly, the author mentions that Japanese companies have achieved a competitive advantage by implementing the JIT technique that contributes to the development of suppliers in order to reduce congestion along the supply chain and obtain higher productivity and quality. Nevertheless, one may wonder why JIT is integrated with the supply chain. To answer these questions, one must be familiar with the definitions of SCM in order to compare them with those from JIT. Some definitions are:

- SCM refers to the comprehensive approach to deal with the planning and control of materials from suppliers to end users (Lu and Swaminathan 2015).
- SCM is a network made up of companies that interact to deliver the product or customer service. It connects flows from the supply of raw materials to the delivery of the final product (Diabat et al. 2014).
- SCM refers to the network of organizations involved through the connections in the processes and activities that add value to products and services in the hands of end users (Diabat et al. 2014).
- SCM are those networks of production and distribution that supply raw materials, transform them into subassemblies and finished products, and distribute them to customers (Wu and Wu 2015).
- SCM aims at building trust, exchanging information on market needs, developing new products, and reducing the number of suppliers (Guritno et al. 2015).
- SCM is the chain of trade from the original source of the raw material through those involved in its extraction and processing, manufacturing, distribution, and marketing to end users (Martínez-Jurado and Moyano-Fuentes 2014).
- SCM includes the entire set of entities (suppliers, manufacturers, distributors, and resellers) through which materials, products, and information flow (Caniëls et al. 2013; Miocevic and Crnjak-Karanovic 2012).
- SCM is a network of organizations that begins with suppliers and ends with customers of goods and services (Adobor and McMullen 2007).
- SCM covers materials related to the administration of the offer (from raw materials to finished products, including possible recycling and reuse), the use of processes and technology, and the ability to enhance competitive advantage. Its management aims at optimization and efficiency (Madenas et al. 2014; Wuttke et al. 2013).

The SCM includes manufacturers, suppliers, transporters, warehouses, retailers, and customers in a dynamic flow. After being implemented, it helps reduce costs, increase technological innovation, profitability, and productivity, and improve competitiveness in organizations (David and Eben-Chaime 2003; Aksoy and Öztürk 2011). SCM relies on JIT to achieve this; however, it faces some obstacles such as the fact that the range of products constantly increases, the product life decreases, customers are more demanding, and globalization continues to expand. Thus, the implementation of JIT is not an easy task. Nevertheless, companies have adopted different strategies and practices to create and effectively manage their supply chain. Also, they have focused on integrating processes and forming more efficient alliances with suppliers (Benton and Shin 1998).

1.9 Practical Examples of Successful JIT

The application of JIT in the industrial environment has brought a number of benefits to businesses, many of which are reported in the literature. The purpose of this section is to illustrate some of the most known examples of successful JIT implementation.

1.9.1 Toyota

Toyota is considered by many as the company where JIT was born and became a philosophy applied to production systems. The production strategy of Toyota stands out because raw materials are not brought to the production plant until an order is received and the product is ready to be built. This means that raw material is not largely stored; it often arrives directly in production lines and the integration between the company and its suppliers is very close, since they are partners of the end product. This philosophy has allowed Toyota to maintain a minimum amount of inventory, which translates into lower costs. This also means that the company can quickly adapt to changes in demand without worrying about the elimination of expensive inventory.

Some of the success factors of JIT in Toyota are:

- Small amounts of raw material inventory must be kept in each node of production. They are replenished after being used.
- The accuracy of the prediction of demand is important. It facilitates the proper supply of raw materials.

1.9.2 Dell

Dell is one of the Western companies that have implemented JIT in its production lines that contain a large amount of components, since it is engaged in manufacturing electronics, such as computers, monitors, and projectors.

The benefits for Dell with the implementation of JIT were quickly perceived since the company relies on numerous suppliers. However, the approach of JIT in Dell is different from in Toyota. Dell benefits from its suppliers to achieve the JIT goals, which are often unique and highly specialized. This has allowed it to offer exceptionally fast deliveries to its customers, since suppliers take inventory that Dell does not. Similarly, the company requires and receives the components of their products at a short delivery time, since many of its suppliers are exclusive to them.

Some important factors of success of JIT implementation in the production lines of Dell are:

- Reliable suppliers with the ability to meet time requirements demanded by Dell.
- A unified system of raw materials supply that allows Dell to convey its needs for components to its suppliers. In this way, materials arrive in the company on time to meet the deadlines given to customers.
- The availability of suppliers to keep inventory in their own facilities. This frees Dell from the task, which generates costs, responsibilities, and a great amount of administrative work.

1.9.3 Harley Davidson

Harley Davidson is another company that has adopted Western JIT after it emerged in Japan. Although now it is renowned, after World War II the motorcycle manufacturing company experienced efficiency problems in its production systems and had to make changes. Perhaps the best decision that the company made was becoming an agile manufacturer able to meet demand and provide short deliveries, for which it implemented JIT.

Some of the results of the implementation of JIT in Harley Davidson are:

- Inventory levels decreased by 75 %, and their suppliers were responsible for a great amount of it.
- Productivity increased in the production lines.
- The success of Harley Davidson with the implementation of JIT had much to do with the fact that when JIT was implemented, process problems could no longer be hidden by expensive inventory that helped meet delivery dates.

Hence, process inefficiencies were identified and resolved quickly. This is important to note, since it may be easier to reduce large problems by half, than small problems by half.

1.10 Why Companies Want to Implement JIT?

Many Western companies in the decade of 1980 wished to embrace the JIT philosophy to improve their production systems, since the approach could offer them significant benefits. As a result, companies now invest in its implementation and train human resources.

Thus, many works have reported these benefits and some authors have comprehensively conducted these studies. These works can be consulted in (Singh and Garg 2011a; Melton 2005; Fullerton and McWatters 2001), since they are not the objective of this book. These works analyze JIT both as part of lean manufacturing and in isolation. However, JIT benefits are the subject of study in the third chapter of this book. They are analyzed and discussed in detail.

A question that may arise at this point is whether all companies obtain the same benefits and why some of them abandon certain techniques and philosophies applied to production systems. The truth is that not all companies are equal and there should be a number of elements that guarantee these benefits.

References

Adobor H, McMullen R (2007) Supplier diversity and supply chain management: a strategic approach. Bus Horiz 50(3):219–229. doi:10.1016/j.bushor.2006.10.003

Aksoy A, Öztürk N (2011) Supplier selection and performance evaluation in just-in-time production environments. Expert Syst Appl 38(5):6351–6359. doi:10.1016/j.eswa.2010.11.104

Amasaka K (2002) New JIT: a new management technology principle at Toyota. Int J Prod Econ 80(2):135–144. doi:10.1016/S0925-5273(02)00313-4

Ansari A (1986) Survey identifies critical factors in successful implementation of just-in-time purchasing techniques. Ind Eng 18(10):46–50

Arnold U, Bernard KN (1989) Just-in-time: some marketing issues raised by a popular concept in production and distribution. Technovation 9(5):401–430. doi:10.1016/0166-4972(89)90013-8

Baker RC, Chang RE, Chang IC (1994) Switching rules for JIT purchasing. Prod Inventory Manage J 35(3):13–17

Banerjee A, Kim SL, Burton J (2007) Supply chain coordination through effective multi-stage inventory linkages in a JIT environment. Int J Prod Econ 108(1–2):271–280. doi:10.1016/j.ijpe.2006.12.015

Benton WC, Shin H (1998) Manufacturing planning and control: the evolution of MRP and JIT integration. Eur J Oper Res 110(3):411–440. doi:10.1016/S0377-2217(98)00080-0

Billesbach TJ (1991) A study of the implementation of just-in-time in the United States. Prod Inventory Manage J 32(3):1–4

Blanchard D (2010) Supply chain management best practices, 2nd edn. Wiley, New York

Brox JA, Fader C (1997) Assessing the impact of JIT using economic theory. J Oper Manage 15 (4):371–388. doi:10.1016/S0272-6963(97)00005-3

Calvasina RV, Calvasina EJ, Calvasina GE (1989) Beware the new accounting myths. Manage Acc 12

Caniëls MCJ, Gehrsitz Matthias H, Semeijn J (2013) Participation of suppliers in greening supply chains: an empirical analysis of German automotive suppliers. J Purchasing Supply Manage 19 (3):134–143. doi:10.1016/j.pursup.2013.02.005

Chhikara J, Weiss EN (1995) JIT savings—myth or reality? Bus Horiz 38(3):73–78. doi:10.1016/0007-6813(95)90026-8

Crawford KM, Coax JF (1990) Designing performance measurement systems for just-in-time operations. Int J Prod Res 28(11):2025–2036

David I, Eben-Chaime M (2003) How far should JIT vendor–buyer relationships go? Int J Prod Econ 81–82:361–368. doi:10.1016/S0925-5273(02)00274-8

Diabat A, Kannan D, Mathiyazhagan K (2014) Analysis of enablers for implementation of sustainable supply chain management—a textile case. J Cleaner Prod 83:391–403. doi:10.1016/j.jclepro.2014.06.081

Dong Y, Carter CR, Dresner ME (2001) JIT purchasing and performance: an exploratory analysis of buyer and supplier perspectives. J Oper Manage 19(4):471–483. doi:10.1016/S0272-6963(00)00066-8

Ebrahtmpour M, Schonberger RJ (1984) The Japanese just-in-time/total quality control production system: potential for developing countries. Int J Prod Res 22(3):421–430. doi:10.1080/00207548408942463

Fiedler K, Galletly JE, Bicheno J (1993) Expert advice for JIT implementation. Int J Oper Prod Manage 13(6):23–30. doi:10.1108/01443579310038994

Flynn DN (2015) Building a better model: a novel approach for mapping organisational and functional structure. Procedia Comput Sci 44:194–203. doi:10.1016/j.procs.2015.03.003

Fullerton RR, McWatters CS (2001) The production performance benefits from JIT implementation. J Oper Manage 19(1):81–96. doi:10.1016/S0272-6963(00)00051-6

Fullerton RR, McWatters CS (2002) The role of performance measures and incentive systems in relation to the degree of JIT implementation. Acc Organ Soc 27(8):711–735. doi:10.1016/S0361-3682(02)00012-0

Fullerton RR, McWatters CS, Fawson C (2003) An examination of the relationships between JIT and financial performance. J Oper Manage 21(4):383–404. doi:10.1016/S0272-6963(03)00002-0

García-Alcaraz JL, Maldonado AA, Iniesta AA, Robles GC, Alor-Hernández G (2014) A systematic review/survey for JIT implementation: Mexican maquiladoras as case study. Comput Ind 65(4):761–773. doi:http://dx.doi.org/10.1016/j.compind.2014.02.013

Garg D (1997) Relevance of JIT purchasing in Indian industries. Kurukshetra University, India

Gélinas R (1999) The just-in-time implementation project. Int J Proj Manage 17(3):171–179. doi:10.1016/S0263-7863(98)00029-5

González-Benito J, Suárez-González I, Spring M (2000) Complementarities between JIT purchasing practices: an economic analysis based on transaction costs. Int J Prod Econ 67(3):279–293. doi:10.1016/S0925-5273(00)00032-3

Green Jr, Kenneth W, Inman RA, Birou LM, Whitten D (2014) Total JIT (T-JIT) and its impact on supply chain competency and organizational performance. Int J Prod Econ 147 Part A (0):125–135. doi:http://dx.doi.org/10.1016/j.ijpe.2013.08.026

Gupta A (2011) A conceptual JIT model of service quality. Int J Eng Sci Technol 3(3):2414–2427

Guritno AD, Fujianti R, Kusumasari D (2015) Assessment of the supply chain factors and classification of inventory management in suppliers' level of fresh vegetables. Agric Agric Sci Procedia 3:51–55. doi:10.1016/j.aaspro.2015.01.012

Hall RW (1983) Zero Inventories. Dow Jones-Irwin, Homewood, IL

Hong JD, Hayya JC, Kim SL (1992) JIT purchasing and setup reduction in an integrated inventory model. Int J Prod Res 30(2):255–266. doi:10.1080/00207549208942893

Inman R, Anthony RS, Sale KW, Jr G, Whitten D (2011) Agile manufacturing: relation to JIT, operational performance and firm performance. J Oper Manage 29(4):343–355. doi:10.1016/j.jom.2010.06.001

Jarrar NS, Smith M (2014) Innovation in entrepreneurial organisations: a platform for contemporary management change and a value creator. Br Acc Rev 46(1):60–76. doi:10.1016/j.bar.2013.07.001

Kannan VR, Tan KC (2005) Just in time, total quality management, and supply chain management: understanding their linkages and impact on business performance. Omega 33 (2):153–162. doi:10.1016/j.omega.2004.03.012

Kojima M, Nakashima K, Ohno K (2008) Performance evaluation of SCM in JIT environment. Int J Prod Econ 115(2):439–443. doi:10.1016/j.ijpe.2007.11.017

Kumar V (2010) JIT based quality management: concepts and implications in Indian context. Int J Eng Sci Technol 2(1):40–50

Lee HH, Zhou J, Hsu PH (2014). The role of innovation in inventory turnover performance. Decis Support Syst (0). doi:http://dx.doi.org/10.1016/j.dss.2015.02.010

Lu LX, Swaminathan JM (2015) Supply chain management. In: Wright JD (ed) International encyclopedia of the social and behavioral sciences, 2nd edn. Elsevier, Oxford, pp 709–713

Machuca JAD (2002) JIT facing the new millennium. Int J Prod Econ 80(2):131–134. doi:10.1016/S0925-5273(02)00312-2

Madenas N, Tiwari A, Turner CJ, Woodward J (2014) Information flow in supply chain management: a review across the product lifecycle. CIRP J Manuf Sci Technol 7(4):335–346. doi:10.1016/j.cirpj.2014.07.002

Maiga AS, Jacobs FA (2009) JIT performance effects: a research note. Adv Acc 25(2):183–189. doi:10.1016/j.adiac.2009.06.003

Marín F, Delgado J (2000) Las técnicas justo a tiempo y su repercusión en los sistemas de producción. Economía Ind 331:35–41

Martínez-Jurado PJ, Moyano-Fuentes J (2014) Lean management, supply chain management and sustainability: a literature review. J Cleaner Prod 85:134–150. doi:10.1016/j.jclepro.2013.09.042

Melton T (2005) The benefits of lean manufacturing: what lean thinking has to offer the process industries. Chem Eng Res Des 83(6):662–673. doi:10.1205/cherd.04351

Miltenbirg GJ (1990) Changing MRP's costing procedure to suit JIT. Prod Inventory Manage J 31 (2):77–83

Miocevic D, Crnjak-Karanovic B (2012) The mediating role of key supplier relationship management practices on supply chain orientation—the organizational buying effectiveness link. Ind Mark Manage 41(1):115–124. doi:10.1016/j.indmarman.2011.11.015

Monden Y (2002) The relationship between mini profit-center and JIT system. Int J Prod Econ 80 (2):145–154. doi:10.1016/S0925-5273(02)00314-6

Moore WS (1988) The evolution of a just-in-time environment at Northern Telecom Inc.'s customer service centre. Ind Eng 66–63

Ohno T (1982) How the Toyota production system was created. Japan Econ Stud 10(4):83–101

Ouchi WG, Price RL (1993) Hierarchies, clans, and theory Z: a new perspective on organization development. Organ Dyn 21(4):62–70. doi:10.1016/0090-2616(93)90034-X

Power D, Sohal AS (2000) Human resource management strategies and practices in just-in-time environments: Australian case study evidence. Technovation 20(7):373–387. doi:10.1016/S0166-4972(99)00151-0

Rivera L, Frank Chen F (2007) Measuring the impact of Lean tools on the cost–time investment of a product using cost–time profiles. Robot Comput Integr Manuf 23(6):684–689. doi:10.1016/j.rcim.2007.02.013

Sanidas E (2004) Technology, technical and organizational innovations, economic and societal growth. Technol Soc 26(1):67–84. doi:10.1016/j.techsoc.2003.10.006

Schonberger R, Gilbert J (1984) JIT purchasing: a challenge for US industry. Calif Manage Rev 26 (1):54–68

Shah R, Ward PT (2003) Lean manufacturing: context, practice bundles, and performance. J Oper Manage 21(2):129–149. doi:10.1016/S0272-6963(02)00108-0

Singh S, Garg D (2011a) JIT system: concepts, benefits and motivations in Indian industries. Int J Manage Bus Stud 1(1):26–30

Singh S, Garg D (2011b) JIT system: concepts, benefits and motivation in Indian industries. Int J Manage Bus Stud 1(1):26–30

Sundar R, Balaji AN, Satheesh Kumar RM (2014) A review on lean manufacturing implementation techniques. Procedia Eng 97:1875–1885. doi:10.1016/j.proeng.2014.12.341

Voss CA (1986) Just-in-time manufacture. IFS Limited. Springer-Verlag, London, UK

Voss CA, Robinson SJ (1987) Application of just in-time manufacturing techniques in United Kingdom. Int J Oper Prod Manage 7(4):46–52

White RE, Prybutok V (2001) The relationship between JIT practices and type of production system. Omega 29(2):113–124. doi:10.1016/S0305-0483(00)00033-5

Wu LC, LH Wu (2015) Improving the global supply chain through service engineering: a services science, management, and engineering-based framework. Asia Pac Manage Rev 20(1):24–31. doi:10.1016/j.apmrv.2014.12.002

Wuttke DA, Blome C, Henke M (2013) Focusing the financial flow of supply chains: an empirical investigation of financial supply chain management. Int J Prod Econ 145(2):773–789. doi:10.1016/j.ijpe.2013.05.031

Yasin MM, Small M, Wafa MA (1997) An empirical investigation of JIT effectiveness: an organizational perspective. Omega 25(4):461–471. doi:10.1016/S0305-0483(97)00005-4

Yasin MM, Small MH, Wafa MA (2003) Organizational modifications to support JIT implementation in manufacturing and service operations. Omega 31(3):213–226. doi:10.1016/S0305-0483(03)00024-0

Zipkin PH (1991) Does manufacturing need a JIT revolution? Harvard Bus Rev 1(2):40–50

Part II
JIT Elements, Its Benefits, and Causes of Its Slow Implementation

Chapter 2
Elements of JIT

The aim of this book is to identify the key elements that ensure the success of JIT, the benefits that companies obtain, and the causes of its slow implementation or abandonment. In addition, this book integrates these elements in causal models and associates those JIT elements identified with the benefits that companies obtain and the causes of a slow progress in its implementation. These models could allow managers and administrators to identify the elements that are essential for the success of JIT and those that cause great delay of the expected benefits. Thus, to achieve this, all crucial elements of JIT philosophy are first defined, along with the causes of a slow JIT implementation and the benefits. Once these elements were identified and discussed, the structural models were proposed.

This chapter mostly relies on Kumar et al. (2004), Singh and Garg (2011), Kumar (2010) to discuss the elements that ensure a successful JIT implementation. Authors identified a total of 34 elements, which were divided into three groups for the purpose of this research:

1. Elements associated with human resources
2. Elements associated with the production process
3. Elements associated with the product

2.1 JIT Elements Associated with Human Resources

If JIT is considered a philosophy of production, some of its elements must be associated with human resources or factors, since any philosophy is inherent of humans. The 12 elements associated to these human resources and proposed by Kumar et al. (2004), Singh and Garg (2011), Kumar (2010) were categorized and listed in Table 2.1. The first column stands for the names of every element while the second column includes the names of authors who have referred to every element. Finally, the third column contains the number of authors in total who referred to each element.

Note that in Table 2.1 elements are arranged in a descending order according to the number of references reported for every one. The first five elements have a total

J.L. García-Alcaraz and A.A. Maldonado-Macías, *Just-in-Time Elements and Benefits*, Management and Industrial Engineering, DOI 10.1007/978-3-319-25919-2_2

Table 2.1 JIT elements associated with human resources

Name of item	Authors	Quote
Flexible workforce	(Baker et al. 1994; Hall 1983; Garg et al. 1996b; Hong et al. 1992; Yasin et al. 2003; Prodipto 1999; Pyane 1993; Saxena and Sohay 1999)	8
Zero deviation in production programs	(Garg et al. 1994, 1996a; Singh 1989; Ebrahimpour and Schonberger 1984; Garg 1997; Garg and Deshmukh 1999b; Voss and Robinson 1987; Yasin et al. 2003)	8
Multifunctional workers	(Bonito 1990; Delbridge 1995; Garg 1997; Garg and Deshmukh 1999b; Garg et al. 1994; Sewell and Wilkinson 1992; Singhvi 1992; Yasin et al. 2003)	8
Workers motivation	(Bonito 1990; Delbridge 1995; Fiedler et al. 1993; Garg et al. 1994, 1996a; Sewell and Wilkinson 1992; Vrat et al. 1993; Kumar et al. 2001)	8
Short delivery times	(Singh 1989; Chong and Rundus 2000; Ebrahimpour and Schonberger 1984; Garg and Deshmukh 1999a; Singhvi 1992; Vrat et al. 1993; Voss 1986; Kumar and Garg 2000)	8
Error prevention	(Chong and Rundus 2000; Garg 1997; Garg and Deshmukh 1999b; Hall 1983; Vrat et al. 1993; Kumar and Garg 2000)	6
Long-term contracts	(Kumar and Garg 2000; Vrat et al. 1993; Garg and Deshmukh 1999b; Garg et al. 1996a; Garg 1997)	5
Self-correction of errors	(Voss 1986; Garg and Deshmukh 1999b; Garg et al. 1994; Garg 1997; Ebrahimpour and Schonberger 1984)	5
Employee empowerment in QC	(Singh 1989; Delbridge 1995; Garg and Deshmukh 1999a; Hall 1983; Priestman 1985)	5
Supplier quality certification	(Garg et al. 1996a; Garg 1997; Garg and Deshmukh 1999b; Kumar et al. 2001)	5
Evaluation and selection of suppliers	(Garg et al. 1996a; Garg 1997; Garg and Deshmukh 1999b)	3
Effective communication	(Gilbert 1990; Yasin et al. 2003)	2

of eight references, there is only one element with six references, four elements with five, one element with three, and finally, one element with two references is reported.

A definition of each element is provided in order to identify why they are part of this group.

1. *Flexible workforce*: This element has several dimensions, but the most important concerns availability of operators and the opportunity for them to hold more than one position, which is discussed in the third element related to multifunctional workers (Asendorf and Schultz-Wild 1984). As for flexible hours, its importance can be understood by analyzing the following scenario: Let us suppose that a certain company receives an urgent request or production order at 2:30 p.m., and the departure time of all workers is at 3:00 p.m. Due to

the urgency of this order, workers will be required to work overtime; otherwise, the production order would probably be delivered late to the customer. Flexible workforce is created through education and training, along with the establishment of clear policies for its implementation, since there are often Constitutional or Federal rights in labor laws that are clear in terms of the payment of overtime work, the holidays granted, and more. Authors Kohler and Schultz-Wild (1985) discussed the issues and policies that may be found when companies seek to implement a flexible workforce. Similarly, Toikka (1987) analyzed the procedures that companies should follow in order to develop their workforce in an environment of visible manufacturing.

2. *Zero deviation in the production program*: JIT programs are important to monitor the work plans generated, since deviations could cause delays in other production orders. To avoid any deviation in production programs, some managers stated that there must be some certainty in demand. However, at the same time they can complain that many customers continuously ask for urgent orders. Thus, in order to meet the requests, companies require making changes in their production plans already set. Fortunately, the task-scheduling problem has been widely studied from an academic point of view, since deviations in production plans are very common. Several models can be found nowadays based on optimization aimed at minimizing deviation. For instance, Crama (1997) refers to a model of combinatorial optimization for task scheduling, while Greene and Sadowski (1986), Sawik (2007) mentioned a model based on integer programming. Similarly, models in Li et al. (2015) are multi-target, since the proper programming of activities does not only seek to minimize the cost, but also to make appropriate use of available resources.
3. *Multifunctional workers*: This element refers to the availability of workers or people responsible for the production process to perform various activities in the production line. Hence, if any member were missing or stopped attending, any other worker could replace him/her in the position or job. That means that the production line would not be affected and would always have continuous flow thanks to the training and flexibility of workers. This could ensure a just-in-time production system without delay. Some authors recommend drawing upon the rotation of employees around the different positions so they can perform and train in different skills. This way, an employer can make adjustments with greater fluidity whenever he/she requires (Corominas et al. 2006) and without the obstacles due to attendance problems or unfortunate accidents. Although this element is ranked third in this category, authors such as Martínez-Jurado et al. (2014) consider it the most important element for the success of lean manufacturing tools, including the JIT methodology or philosophy.
4. *Workers motivation*: Many authors discuss this element and it is also one of the most important associated with human resources. Although it is difficult to actually measure the motivation of employees or operators online, JIT considers that this element is crucial, since an unmotivated employee will not seek the

achievement of the goals of the company. Therefore, companies should seek by all means to motivate their employees and make them feel part of the company with the purpose of creating a corporate identity. Thus, companies must seek that their employees feel proud of their employment, the product that they manufacture, and the organization to which they belong. In other words, employers want their employees to put their shirt on. Only motivated employees will seek to achieve the objectives of the company. Several works in the literature address the role of motivation in the industry and many of them are recent. For instance, Diefendorff and Seaton (2015) analyzed in detail of the importance of motivation in the workplace. Their analysis focused primarily on operational levels, while Barron and Hulleman (2015) rather emphasized on the analysis of financial costs for a company when it obtained an adequate program of work motivation. Finally, authors Anderman and Gray (2015) discussed the relationship between motivation and the educational levels of operators.

5. *Short delivery times*: In this element, managers must seek short delivery rates in order to have constant flow of deliveries in the production system. Some managers mentioned that companies held a much closer relationship with customers who asked for short delivery times than with those customers whose deliveries took longer periods of time. In these cases, managers stated that sometimes companies did not know the customers. Thus, short delivery times with customers strengthen relations between the two parties in which people maintain contact.
6. *Error prevention*: JIT philosophy is based on human resources, but as part of our human nature, many mistakes can occur both in the production line and in the administration. Hence, it is important that companies have a system of error prevention to avoid problems within the production system, since mistakes can delay a production order. This often happens because defective pieces must be reprocessed when possible, since sometimes an entire piece is wasted and must be thus rebuilt from its beginning. This implies a delay in just-in-time delivery. Error prevention is also aimed at preventing fatalities, such as accidents of the operators that may result in the loss of a limb. Thus, researchers have also discussed its importance. For instance, Liu et al. (2009) proposed an economic analysis of the mistakes made by the employee in a manufacturing environment. Similarly, Ruckart and Burgess (2007) proposed an analysis of the cost of errors in the mining and manufacturing sectors.

   Clear examples of the efforts made by many companies to avoid mistakes are training simulators, such as those offered to pilots in many airlines. In these simulators, workers are introduced to specific situations to measure their responsiveness and quality of decisions. The study of Di Pasquale et al. (2015) discussed a simulator to analyze human errors, a concept that many works on manufacturing recently name reliability (Baraldi et al. 2015; Mkrtchyan et al. 2015).
7. *Long-term contracts*: These types of contracts generally happen when demand has little uncertainty with the customer and when the company is familiarized

with the client and vice versa. Thus, the client knows the processing capacity of the producer and the producer consequently knows the needs of the customer. This means that with long-term contracts and a reduction of uncertainty from both parties, deliveries of production orders tend to be on time.
Long-term contracts with suppliers have been extensively studied, but as noted in Cannon et al. (2010), these relationships are not always possible, especially because of factors related to the culture where companies operate. In these cases, it is recommended that buyers frequently perform a series of audits to their suppliers in order to maintain their relationship (Chen and Jeter 2008). Also, other authors go beyond seeing suppliers as sellers of raw materials, they rather consider suppliers as partners in the production system (Chicksand). However, these long-term relationships that consolidate that much, could pose a number of problems. However, Costello (2013) proposed a series of strategies to resolve such conflicts that may arise, and Frascatore and Mahmoodi (2008) defined a number of possibilities related to the sanctions that could be applied when any of the parties fails in the purchasing and selling process. Similarly, Xu et al. (2015) addressed batch sizes that could be ordered due to unforeseen changes in the demand.

8. *Self-correction*: This concerns the ability of workers to correct their mistakes without the need of supervisors. It is related to the degree of authority that supervisors grant to their employees, since if there were no authority, there would not be responsibility either. However, to ensure self-correction of defects, employees or workers in the production line must have high education, since they must be able to identify the error and provide a solution. Thus, if a certain employee does not have the authority to self-correct defects, the supervisor will likely detect it and decide how to correct it. Then, the process to solve that problem would certainly require much time, since the supervisor would assign its solution to another employee, and this represents high administrative costs. Granting authority to an employee is commonly known as empowerment and this concept involves many cultural and psychological aspects, since many people do not wish to take responsibility for their actions (Chu 2003; Paul et al. 2000) in Western companies and which are very natural. As Cheung et al. (2012) stated, the administration should be based on giving authority and responsibility to employees. Recently, Tong et al. (2015) expressed that strong leadership in work teams of the manufacturing process is required in order to delegate responsibilities, since these are an immediate reflection or mirror of supervisors or senior managers.
9. *Authority for quality control to workers*: This element is closely related to the previous one, since it means that every worker has the control of quality over every processes carried out. This means that workers can take appropriate actions to remedy quality defects that they can detect, which implies an ability of self-correcting defects. However, a worker can be able to correct the defects he found either in his/her process or in previous ones, but if he/she does not

have the authority to correct them, it is likely that their education and training will not be enough.

Empowerment of workers must be related to the responsibilities they hold in the production process. Workers must be trained people so they can be granted with authority but also responsibility. Authority given to workers over quality control is essential in JIT because product quality largely depends on the skills and abilities of employees. Thus, a product or order can be delivered on time if employees are able to determine, identify, and correct errors without waiting for a supervisor to detect them. This type of authority conferred to workers in relation to the quality of products has called for the attention of many researchers and academics. In de Macedo-Soares and Lucas (1995), authors carried out a study of this phenomenon in which they compared Brazilian and American companies by identifying cultural and educational aspects. Four years later, Howard and Foster (1999) discussed the influence of human factors and their perspective on the final quality of a product. More recently, Tong et al. (2015) performed an analysis about leadership and delegation of authority and responsibility to employees. They emphasized on confidence when performing manufacturing activities. They also concluded that employees were to some extent responsible for the safety performance of activities, and that companies must provide proper methods to perform them.

10. *Supplier quality certification*: A supply chain begins with raw materials from suppliers; hence, the production process starts with them, which will be later transformed into finished products. It is well known that if waste enters the production process, waste will come out as a final product. Therefore, raw material suppliers most count on quality certifications. This suggests that they can be an integral part of the production system, not just as a supplier but also as a partner. Let us suppose that a company requests an order to a supplier and the supplier delivers the wrong part or component, or a correct but defective component. Obtaining the appropriate piece would mean, from an administrative point of view, to return the material to be replaced, which would imply waste of time and costs. Hence, a quality accreditation or certification is also a guarantee of the administrative capacities of suppliers. The vendor certification process and its impact on the quality of the final product have been studied by Park et al. (1996), who argued that the best way to hold a good relationship with a supplier is having few of them, although, in times of uncertainty, having a single supplier for a particular part or component could probably pose a number of problems of shortage (Larson and Kulchitsky 1998). With recent advances in certification, many providers already count on ISO regulations and its guidelines (Terlaak and King 2006) and some of them include environmental or ecological aspects (Ullah et al. 2014; Nguyen and Hens 2015).
11. *Seller rating*: Every company should conduct an evaluation of its suppliers in order to know if they meet the expected requirements, which are often properly set by their quality systems and certifications (Kohlbeck 2011). Low valuations of suppliers indicate that they do not meet the expected metrics for the

company. Therefore, these suppliers can be easily identified by companies, which could then look for and identify alternative suppliers. It is worth mentioning that all suppliers have the right and must know the metrics under which they were evaluated, so companies have the obligation to make these public. Some other companies offer a monthly report to their suppliers with statistics of the relationships they maintain. The report also informs of delays in deliveries, incomplete deliveries, and percentage of defective products, among others. Suppliers with high valuations metrics ensure that products are delivered on time since the supplier system is guaranteed.

12. *Effective communication*: Now that market globalization forces companies to have suppliers from around the world, effective communication to the outside must be effective. However, communication must also be successful within the company between the different hierarchical levels. It must be wide and ensure the flow of appropriate information to avoid mistakes and misunderstandings. If there is no communication between departments in the production line, they will surely have issues with information and material flow, and this can heavily compromise just-in-time deliveries. In Japanese companies, the following process in a production line is always considered as a customer; this perspective ensures that all departments establish open communication. This way they are guaranteeing the satisfaction of their clients. From a social point of view, there are many studies that have analyzed the phenomenon of communication in companies; some of them also discussed the protocols to be followed to ensure it (Huang et al. 2009). Undoubtedly, information and communication technologies today play an important role, since information is rapidly shared between suppliers and buyers. As a result, production systems are often linked and exchange this information in real time (Aliu and Halili 2013), minimizing supply chains globally. As authors Jiang and Liu (2015) stated, who conducted a study in the coal sector.

2.2 JIT Elements Associated with Production Processes

JIT is a philosophy that largely depends on human resources. However, once raw materials arrive in warehouse, they enter the production process, and this process also plays an essential role in just-in-time deliveries of finished products and the elimination of waste.

Table 2.2 of this section discusses those JIT elements associated with the production process. The first column contains the name of the elements, while the second stands for the name of authors who identified each element and considered it important to illustrate JIT. Finally, the third column includes the total number of authors who referred to that element. The elements were arranged in descending order according to the total number of authors who mentioned them.

Table 2.2 JIT elements associated with production processes

JIT elements—process	Authors	Total
Kanban system	(Singh 1989; Hall 1983; Saxena and Sohay 1999; Pyane 1993; Prodipto 1999; Pan and Liao 1989; Muralidharan et al. 2001; Yasin et al. 2003; Hong et al. 1992; Kim and Ha 2003; Kumar and Garg 2000; Voss and Robinson 1987; Vrat et al. 1993; Padukone and Subba 1993; Garg and Deshmukh 1999b; Garg et al. 1996a; Garg 1997)	17
Preparation time reduction (SMED)	(Singh 1989; Hall 1983; Saxena and Sohay 1999; Pyane 1993; Prodipto 1999; Vrat et al. 1993; Hong et al. 1992; Baker et al. 1994; Kumar and Garg 2000; Singhvi 1992; Padukone and Subba 1993; Daesung and Seung-Lae 1997)	17
Small lot sizes	(Dutton 1990; Ebrahimpour and Schonberger 1984; Hall 1983; Singhvi 1992; Vrat et al. 1993; Baker et al. 1994; Bose and Rao 1988; Golhar and Stamm 1991; Hong et al. 1992; Pan and Liao 1989; Prodipto 1999; Garg et al. 1996a; Garg and Deshmukh 1999b; Daesung and Seung-Lae 1997)	14
Cellular manufacturing	(Singh 1989; Hall 1983; Saxena and Sohay 1999; Prodipto 1999; Vrat et al. 1993; Kumar and Garg 2000; Yasin et al. 2001; Voss and Robinson 1987; Garg and Deshmukh 1999b; Garg et al. 1996b; Garg 1997)	12
Safety stock	(Singh 1989; Hall 1983; Kumar and Garg 2000; Vuppalapati et al. 1995; Vrat et al. 1993; Padukone and Subba 1993; Dutton 1990; Chong and Rundus 2000; Ebrahimpour and Schonberger 1984; Garg 1997; Garg and Deshmukh 1999b; Garg et al. 1996a)	12
Improved plant layout	(Hall 1983; Pyane 1993; Prodipto 1999; Baker et al. 1994; Padukone and Subba 1993; Yasin et al. 2001; Voss and Robinson 1987; Garg and Deshmukh 1999b; Garg et al. 1996b; Garg 1997)	10
Just-in-time purchases	(Hall 1983; Saxena and Sohay 1999; Pyane 1993; Prodipto 1999; Vrat et al. 1993; Mohan and Singh 1995; Yasin et al. 2001; Goyal and Deshmukh 1992)	8
Process control	(Singh 1989; Prodipto 1999; Hong et al. 1992; Ha and Kim 1997; Kumar et al. 2001; Priestman 1985; Padukone and Subba 1993; Garg and Deshmukh 1999b)	8
Standardized containers	(Hall 1983; Prodipto 1999; Kumar and Garg 2000; Ebrahimpour and Schonberger 1984; Yasin et al. 2001; Voss and Robinson 1987; Garg and Deshmukh 1999b)	7
Group technology	(Singh 1989; Kumar and Garg 2000; Vuppalapati et al. 1995; Padukone and Subba 1993; Chong and Rundus 2000; Voss and Robinson 1987; Garg and Deshmukh 1999b)	7
Process flexibility	(Hall 1983; Vrat et al. 1993; Hong et al. 1992; Daesung and Seung-Lae 1997; Baker et al. 1994)	5
Reduction of work in process	(Ebrahimpour and Schonberger 1984; Vrat et al. 1993; Kumar and Garg 2000; Prodipto 1999; Garg and Deshmukh 1999b)	5

(continued)

Table 2.2 (continued)

JIT elements—process	Authors	Total
Continuous improvement	(Hall 1983; Prodipto 1999; Vrat et al. 1993; Garg et al. 1996b)	4
Specialized factories	(Saxena and Sohay 1999; Pyane 1993; Yasin et al. 2001)	3
Kaizen	(Ebrahimpour and Schonberger 1984; Garg 1997; Voss and Robinson 1987)	3
Scheduling below installed capacity	(Yasin et al. 2001; Garg et al. 1996b)	2
Use of robots	(Conzalez and Suarez-Gonzalez 2001; Garg et al. 1996b)	2
Pull system	(Baker et al. 1994; Conzalez and Suarez-Gonzalez 2001)	2

The following paragraphs briefly discuss each element in order to explain their importance in this group:

1. *Kanban system*: We cannot speak of a Kanban system without defining it. Kanban is an information system traditionally used to harmoniously control and indicate what must be produced in a production system. In other words, it controls and indicates the amount and timing required to carry out an activity (Panayiotou and Cassandras 1999). A Kanban is usually represented by a simple card indicating the characteristics of the product-to-be. This information card can be used inside and outside the company. It can be used in the company as the production system itself, and outside as a method to exchange information between companies (Chan 2001). A typical example could be in an automotive assembly line. A card could be placed in the car at the beginning of the production system, when only the chassis or the main support of the car is placed. This card would indicate the characteristics of the car-to-be, such as the type, its color, the type of built-in speakers, whether it would have electric windows and buttons, or whether it would include fog lights. This way, operators in the assembly line could always know the kind of components to integrate into the chassis without a supervisor indicating. It would ensure continuous production flows, which would favor just-in-time deliveries and waste disposal of assemblies of unwanted components.
   Kanban as a system or tool applied in a production line might be the most important tool to ensure the flow of materials; which is why it was ranked as the first element (17 references identified in total). One could also consider that Kanban is a very efficient tool to ensure proper delivery of information in the production line and, thus, avoid mistakes. For a more thorough review of the application of Kanban in the industry, readers are welcome to consult (Lage Junior and Godinho Filho 2010).
2. *Preparation Time Reduction (SMED)*: Maintaining the continuous flow of materials through the production system is not an easy task, since machines and equipment used to transform the materials frequently break down or must be

adjusted when companies make changes in their product (Ferradás and Salonitis 2013). This is why Henry Ford said that he could paint a car in any color that customers asked, as long as it was black. Henry Ford knew that making changes to the paint line in a process costs money, and as a result, his product could become more expensive. However, it must be kept in mind that a batch size for a client is only one product; this ensures that the product is fully customized.

The SMED program was proposed to minimize time required to put the machinery and equipment to point (Almomani et al. 2013). This technique emerged in the automotive sector and focused mainly on changing tools from lathes, milling machines, and material grinding equipment. Thus, in a production system, the programs to reduce setup times for machines and equipment are based both on the analysis of unnecessary movements by operators and machines and the design of these machines. It is important to highlight that in order to meet short preparation times, people in the company should be highly trained and skilled to identify all unnecessary movements.

3. *Small lot sizes*: As it was previously mentioned, large batch is desirable for suppliers, since they will not need to make changes in their production line. This will reduce production costs and, thus, favor a greater competitiveness. However, from the point of view of customer companies, small batch sizes or low volume production are rather preferable since they allow these organizations to own much more customized products with very specific characteristics. Thus, there must be a balance between the needs of the customer firm and those from the suppliers, although in an environment of JIT, it is not an easy task (Khan and Sarker 2002; Kim and Ha 2003). Moreover, small batch sizes represent less expenses of inventory maintenance to manufacturing companies, who will always look for that. However, having little inventory implies a mature and trustful relationship with the supplier, which does not always happen. If the supplier does deliver raw materials on time and the producer does not have sufficient inventory, the production process might stop due to a lack of raw materials (Lovell 2003). Therefore, a company that has begun applying just-in-time manufacturing must have certified suppliers with high valuations and which are fully integrated into the production system and have become reliable. This will save inventory costs for companies, as it did for Toyota, Dell, and many others that have reported success in implementing JIT in their production systems.
4. *Cellular manufacturing*: Companies usually have a representative product and always seek to be leaders in it, which means that they have many other products in their production line (Erenay et al. 2015). In order to achieve high levels of expertise, some firms generate a mini-factory within the large factory where they bring together all machinery and equipment to develop a certain product or process. As it was already mentioned, the gain is that the company becomes specialist in that product, suggesting high reliability of its production process and high training of its human resources (Deep and Singh 2015). With these

resources, the company can always face rapid changes in the product and avoid human mistakes and mistakes in the process, which assures just-in-time deliveries. Many approaches have been proposed for the integration of manufacturing cells, which are often based on binary optimization that always aims at minimizing costs and delivery times. Some examples can be found in Aalaei and Davoudpour (2015), Brown (2014), Mahdavi et al. (2012).

5. *Safety stock*: The uncertainty in demand forces companies to have a safety stock inventory. The most suitable scenario would be one where producers know with absolute certainty what will happen with the demand; they could order the raw materials only when needed without experiencing storage costs. Companies do not want to experience marginal costs due to a lack of product in inventory. Thus, a security stock may be a necessary evil until the emergence of models that integrate all variables to predict demand. A safety stock can be defined as the amount of raw material or finished product that the company has in arms to meet any unforeseen event (Sarkar and Tripathy 2002). For instance, if a company receives a production order for any material or product and one component is missing since it is out of both the traditional and the safety stock inventories, it is likely that the product will suffer from late deliveries even with a just-in-time philosophy. This is why having a safety stock is considered as one of the most important elements of JIT associated with production processes. Other typical examples of safety stocks are food stocks (Lee et al. 2014) and blood banks from the government. They exist in order to be used when required.
6. *Improved plant layout*: There is one scenario that I usually give my students to demonstrate the importance of material handling. The diagram of a traditional process includes operations, inspections, transport, storage, and some delays. If the process removed all transport, delays, and stores under the premise that they do not add value to the product, the process diagram would consist only of those activities associated with operations and inspections. This is most likely to be a paradox, or fictitious, since the storage of raw material and safety stocks are needed, as it was mentioned above. However, engineers can focus on minimizing these values, especially the ones related to the transport of raw materials and finished products. This is achieved with an appropriate distribution of the process.

   The impact plant layout in companies and the improvements in efficiency ratios in a just-in-time program have been extensively studied. For example, while Yasin et al. (1997) introduced an analysis of the efficiency of JIT and found that one of its reasons was precisely distribution. Similarly, White and Prybutok (2001) analyzed the good practices of the JIT philosophy and concluded that distribution of the production process was perhaps the most important. Similarly, Inman et al. (2011), Alcaraz et al. (2014), authors described structural equation models which demonstrated that a good distribution of the production process was the source of agility, since it avoided unnecessary movements of raw materials and finished products that added no value whatsoever.

7. *Just-in-time purchases*: It is traditionally thought that JIT philosophy is only applied in the production system; however JIT starts from the process of procurement of raw materials and components. Thus, as mentioned above, implementing JIT involves the commitment from the purchasing department early in the production process and the distribution department at the end of it. JIT Purchasing ensures that companies meet the needs of their production systems, since delays in raw materials could cause delays all throughout the production process. However, this element requires certified and reliable suppliers that are part of the production system, which would imply great level of dependence between the buyer and seller (Handfield 1993) and could bring a number of advantages for the manufacturing company (Giunipero and O'Neal 1988).
   The obstacles that may arise when reaching a purchasing program with JIT tend to be quickly resolved. Furthermore, the program proposes many benefits for business (Dong et al. 2001), although two questions will always be under discussion: how far the relationship between buyer and seller can go, and to what extent the seller should be involved in the decision-making process of the buyer (David and Eben-Chaime 2003).
8. *Process control*: The supply chain of a company could be divided into three stages: the supply of raw materials, the production process, and the distribution of the finished product. How would it be if any of these stages were not under control? What would happen if there was a lack of control in the production process? Could JIT be applied in an uncontrolled production process?
   If any of the stages of the supply chain lacked of control, companies would be unable to know the kind of materials in stock, the products that are currently in the production line, or those that are being delivered to customers. This is a situation that managers would certainly wish to avoid. Similarly, if the production process lacked of control companies may not know which product or orders have initiated the production process, their stage in it, or when they would be finished, which may be catastrophic. Thus, it would be impossible to implement JIT philosophy without any control on the production process, which demonstrates that control plays a key role in the success of the JIT program. In fact, Benton and Shin (1998) suggested that companies needed a proper planning process, since it was recently shown that the relationship between process control and performance indices of the supply chain was statistically significant and noticeably high (Alcaraz et al. 2014).
   In Huq (1999) the author mentions a series of procedures that seek to approach JIT philosophy from a production process based on projects. This may be interesting to many, since JIT philosophy is generally applied to mass production systems.
9. *Standardized containers*: In Chap. 1 of this book mentioned, lean manufacturing (LM) is based, among others, on a series of tools that are essential for its

success. One of these tools is the standardization of processes, which begins by standardizing parts, one of which are the containers in which raw materials are placed in order to be assembled or turned into finished products to distribute.

10. *Group technology*: If a company has been or must be a specialist in a given operation or activity, its equipment and machinery are likely divided into different technology groups. This means that similar operations are performed in the same place, which requires the formation of product families. This allows companies to take advantage of the similarities of machinery and processes for greater fluidity in the material. This consequently improves the efficiency of just-in-time deliveries. It has been found that group technology for production offers several advantages such as agility (Inman et al. 2011), higher levels of product quality and processing, and specialized care provided to machinery and equipment (Cua et al. 2001). However, these changes in production organization also require a series of organizational modifications, which are often neglected in businesses but should never be forgotten in order to ensure on-time deliveries (Yasin et al. 2003).
11. *Process flexibility*: Flexibility can be defined as the ability of a company to make quick adjustments to the products (Gupta et al. 1992). It must not be confused with agility, which is the capacity of quick response to fluctuations in demand. A flexible process is where machinery and equipment can perform different activities and are not specialized in one. For instance, if a very specialized machine in a company broke down, the production manager could be in a serious problem, since the company may not count on another machine to replace the damaged one and perform its activities. If instead of being specialized, the two were multifunctional pieces of equipment, one could surely replace the other and adapt to its new activities through rapid reprogramming. Indeed, the specialization process definitely provides many advantages, but it also implies drawbacks, such as the one described above. The advantages of the specialized equipment are high quality and speed in its processes; nonetheless, it may be challenging to adapt a specialized machine so it can carry out a different activity. This becomes a particular disadvantage since purchasing industrial machinery is usually expensive. Thus, as Moattar Husseini et al. (2006), Inman et al. (2011) stated, there must exist a balance between flexibility in the machinery and equipment and the cost that companies are willing to experience for the agility that it requires.
12. *Reduction of work in process*: Having piles of raw material in the process must be an issue for production managers, since they are not being processed at the moment and are just stored anywhere. Besides, they are not receiving any added value whatsoever. This is why process inventory should be reduced to a minimum, although it is usually a challenging task because many operations and activities throughout the production process have different execution times and, therefore, grouping activities and balancing the production line is a permanent issue (Houghton and Portougal 1997). A reduced work in process means that the production line is balanced and, thus, the flow of materials throughout it is smooth and continuous. This suggests that a considerable

amount of waste in the production process has been eliminated and companies have a greater opportunity to meet deliveries on time.

13. *Continuous improvement*: Kaizen or continuous improvement has been widely researched in production systems and represents a philosophy by itself. The main idea in continuous improvement is that there is not such thing as a perfect process. That is, the process can be improved. Companies that have applied this philosophy have gained great economic benefits. Indeed, accepting that a whole process can be improved makes organizations think that it can always be upgraded, simplified or, expedited. This would lead to faster production cycles and, therefore, timely delivery of processes and products to customers.
    It is considered that the implementation of Kaizen largely depends on operators, since they can suggest most of the changes and improvements because they are the most acquainted of the machinery and the equipment they operate as well as the process for which they are responsible. Thus, implementing a system of suggestions from them would mean listening to the voice of those who run the operations. Similarly, it is often stated that in quality systems it is important to listen to the voice of the customer; however, in this case, it is imperative to listen to the voice of operators.
    The implementation of Kaizen in Western companies is not an easy task, since some working conditions are different from those of Eastern companies, which is why Western business might face major challenges for its implementation (Machuca 2002). These challenges have become the topic of discussion for several pieces of research that seek to identify the critical success factors of Kaizen. For instance, Farris et al. (2009) reported a list of factors found in manufacturing systems in the United States, while structural equations in Mexico have reported a series of relationships that ensure the success of this technique (García et al. 2013, 2014). It shall not be forgotten that people are responsible for the improvement of any process, and that people and processes together produce quality products. Therefore, improvements always come from human resources.
14. *Specialized factories*: Another tool to ensure on-time deliveries is to rely on specialized factories. In fact, a minor change to a new product could lead to big changes in equipment, since these tend to be less flexible. Thus, from the viewpoint of JIT, a specialized factory is more suitable due to the advantages that it provides, which are: the process is usually quickly executed, the personnel who operates it is highly skilled and trained in quality processes, preventive maintenance of equipment is possible, and the process has high fluidity of materials since operations are properly balanced. However, it is always recommended to find a balance in the flexibility expected from these specialized factories, since they require a very standardized product that would suffer few changes in its design.
15. *Kaizen*: Although number 13 refers to continuous improvement and this one discusses Kaizen, it must be emphasized that both concepts stand for the same element. While Eastern companies commonly employ the term Kaizen,

Western industries are much more familiar with the expression "continuous improvement."

The list of elements associated with production processes provided was based on the results found by Kumar (2010), Kumar and Garg (2000), Kumar et al. (2001, 2004), who actually discussed and analyzed the two concepts separately. As a personal conclusion regarding both concepts, it would be suitable, as Recht and Wilderom (1998) stated, to focus on analyzing the cultures of the companies that implement JIT and Kaizen. Thus, it is vital for the success of both JIT and organizations that they implement a system of opinions through which employees make suggestions to improve the production process. A production process that constantly improves is certainly a highly efficient process in all its aspects, starting with its workforce, who makes these suggestions and gives opinions.

16. *Scheduling below installed capacity*: Although not frequently mentioned, scheduling production below the installed capacity guarantees orders delivered on time. However, when companies schedule above their installed production capacity it usually results in problems, such as delayed deliveries, poor-quality products and, consequently, loss of customer trust.

    Many businesses accept orders above its production capacity so they do not lose their opportunity to make business. However, it is also true that these companies can outsource to other companies to perform some of their production. These techniques are traditionally called outsourcing. However, the company that accepts the production order is responsible for the quality and on-time delivery of products of the company that hired it to manufacture part of the amount committed. Some typical examples of outsourcing are found in the automotive sector (Collins et al. 1997). It has also been observed that sometimes this process of outsourcing has lead to issues of strong reliance on external companies (Yilmaz and Bedük 2014), which is why control systems must be extremely rigorous. Others authors, however, have wondered whether a service provider is actually required when outsourcing activities are performed (Dabhilkar et al. 2009).
17. *Use of robots*: It is traditionally believed that robots do not make mistakes but people who fabricate, design, and program them do. However, robots are machines that can be disqualified and lose their ability and accuracy over time and with their uses in production systems. As for their roles inside JIT production systems, robots perform tasks that are difficult, require high precision, or are dangerous, guarantee that the production process will not stop in case of accidents or human errors. Moreover, in a program of preventive and corrective maintenance, robots continue working without the constraints of a possible strike for salary increase. Thus, the flow of materials through the production process is kept continuous and steady, which favors just-in-time deliveries. However, it must also be mentioned that robots are highly specialized equipment that require highly trained personnel, not only to operate and maintain them, but also to program them or reprogram them when they are taken from

one line of production to another, or when changes are made to the design specifications and activities of the robot.

18. *Pull system*: A pull system can be defined as a system that only produces what the client requests. That means that a production order is generated solely when the company already relies on the customer and a sale contract. It does not occur to store products or involve workers in the production process, because if it did, the product could remain long time in the warehouses of the company with its own storage costs. Inventories along the production process or at the end of it as finished products do not exist in pull systems, since the product is manufactured and delivered to the customer at the moment. Readers who wish to further learn about this production system may consult (Ohno 2011), which discusses control levels in pull systems with JIT.

2.3 JIT Elements Associated with the Product

Table 2.3 lists the elements for the success of JIT that are associated with the product. As in prior tables, the first column includes the name of the element, while the second stands for the authors who have referred to that element. The third column corresponds to the total number of authors who have quoted each element.

All elements in this category relate to quality, which is one characteristic of products. Some readers may think that these elements really belong to the category of production process and not product, which may be correct since quality is generated in the process. However, since all of the elements stood for a product characteristic (quality), they were included in this category.

1. *Quality Circles*: Quality circles were a huge phenomenon in the decade of 1980. Their purpose was to know the voice of the employees or workers who were, and still are, the most familiarized with the production system. A quality circle is integrated by a group of coworkers seeking to solve problems in production lines. QCs have nowadays become rare, since the popularity of Kaizen has replaced them, although they essentially remain the same. Participating in quality circles implies a great commitment to the company, as it sought and seeks to provide solutions to everyday problems. Quality circles were more present in Japan, although the West quickly adopted them due to their operational benefits (Blair and Ramsing 1983). One of the first nations to integrate quality circles in the West was apparently the United Kingdom, (Hayward and Dale 1984; Frazer and Dale 1986); yet, the rush for its adoption caused many mistakes in their application and most of the time the approach was abandoned (Pascarella 1985).
   Some authors even expressed that there were many cultural factors associated with quality circles, since they required collaborative work from a group of employees, which is often difficult to achieve (Manson and Dale 1989).

Table 2.3 JIT elements associated with the product

JIT elements—product	Authors	Total
Quality circles	(Saxena and Sohay 1999; Maass et al. 1989; Singhvi 1992; Vrat et al. 1993; Priestman 1985; Padukone and Subba 1993; Nandi 1988; Bonito 1990; Garg et al. 1994; Chong and Rundus 2000; Garg 1997; Singh 1989; Yasin et al. 2001; Garg and Deshmukh 1999b)	14
Total quality control	(Hall 1983; Saxena and Sohay 1999; Vrat et al. 1993; Padukone and Subba 1993; Dutton 1990; Ebrahimpour and Schonberger 1984; Garg 1997; Flynn et al. 1995; Yasin et al. 2001)	9
Statistical quality control	(Kumar et al. 2001; Priestman 1985; Padukone and Subba 1993; Dutton 1990; Singh 1989; Voss and Robinson 1987; Garg and Deshmukh 1999b; Kumar and Garg 2000)	8
Quality development programs	(Hall 1983; Macbeth et al. 1988; Baker et al. 1994; Chong and Rundus 2000; Yasin et al. 2001; Garg 1997; Voss and Robinson 1987)	7
Continuous quality improvement	(Hall 1983; Singhvi 1992; Chong and Rundus 2000; Ebrahimpour and Schonberger 1984; Garg and Deshmukh 1999b; Garg 1997)	6
Zero defects	(Prodipto 1999; Kumar and Garg 2000; Ebrahimpour and Schonberger 1984; Garg and Deshmukh 1999b; Garg 1997)	5
Quality-oriented training	(Singhvi 1992; Vrat et al. 1993; Garg et al. 1994, 1996b; Singh 1989)	5
High visibility of QC	(Hall 1983; Kumar et al. 2001; Priestman 1985; Dutton 1990; Ebrahimpour and Schonberger 1984)	5
Long-term QC commitment	(Hall 1983; Kumar et al. 2001; Priestman 1985; Dutton 1990)	4
Regulate quality and reliability of audits	(Kumar et al. 2001; Nandi 1988; Garg and Deshmukh 1999b)	3
Quality culture	(Kumar et al. 2001; Singhvi 1992; Flynn et al. 1995)	3
100 % quality inspection	(Kumar et al. 2001; Kumar and Garg 2000; Singh 1989)	3
Simplifying the process of total quality	(Kumar et al. 2001; Bonito 1990; Ebrahimpour and Schonberger 1984)	3

However, it is a consensus that quality circles really improve production processes and bring quality benefits (Brennan 1990; Rosenfeld et al. 1991). As far as JIT philosophy is concerned, quality circles have a great impact due to their objectives, the people who integrate them, and the leadership required. For instance, if issues with the movement and handling of materials arise, they may be solved or minimized by a quality circle.

2. *Total Quality Control*: Quality is an imputable feature of the product and has several attributes, which means that it is measured by various properties of the product. These properties must be controlled throughout the entire production

process. If a product does not meet any of these characteristics or attributes, it can be considered defective and will not be accepted in the market by the customer. Quality must be first analyzed in the production process, since this process concentrates all the activities that add value to a product or focus on obtaining the desired characteristics. However mistakes can occur during any stage of the production; thus, the piece of product affected would be unable to continue with the regular flow and would undergo additional activities to correct the defects and provide the characteristics desired for the product. As a result, the continuous flow of material would be interrupted and the goals of JIT philosophy would not be achieved.

The impact of quality programs on JIT has previously been studied. For instance, Withers et al. (1997) associated total quality management (TQM) and JIT with companies holding or having held quality certifications according to ISO standards. However, Vuppalapati et al. (1995) extended the research and stated that these two tools could not be applied independently, since they are interdependent. More recently, Cua et al. (2001) added a third variable to the relation and associated TQM, JIT, and programs of Total Productive Maintenance (TPM), which were considered essential to obtain proper performance and efficiency indicators in companies. Finally, a more general study was provided by Ahmad et al. (2012); it addressed several lean manufacturing tools and discussed their association with total quality programs and JIT.

3. *Statistical quality control*: In a quality system, numbers and statistics generated by the same system are the real indicators of the status of companies. Companies may rely on sophisticated quality control programs, but the interpretation of their results can still be inadequate. Numbers is the natural language of business and, in any company, this language is represented with the results shown by the production process. Statistical quality control in companies is a snapshot of their health. Thus, if the quality is not adequate, products or subassemblies cannot continue their normal flow, and the goals of JIT may be compromised.

   To implement a program of statistical quality control, companies must be familiar with all the characteristics desired for the product. This can be achieved through proper planning (Sengupta et al. 1993). Thus, only when the objectives to be achieved are known, it is easy to associate them with just-in-time programs (Yasin et al. 1997). For example, Fullerton and McWatters (2002) discussed different techniques of lean manufacturing, as well as the impact of these techniques over the level of maturity of a just-in-time implementation. Authors mentioned that appropriate plans and programs of quality control along with their application in the production process by statistical quality control are essential to the continuous flow throughout the supply chain and one of the key success factors of JIT.

4. *Quality development program*: JIT philosophy is part of the quality development program of a company, since it seeks to eliminate waste throughout the production process. Therefore, meeting these objectives is also part of the

development of the company as a profit organization. It seems impossible to apply without quality and business development programs. This means that JIT programs are a means to easily obtain the quality sought. Moreover, the idea of generating quality development plans is not a recent topic of research, since they are clearly linked with the performance indicators of a company (Valmohammadi and Roshanzamir 2015; Cua et al. 2001). However, it is accepted that product quality depends not only on machinery and equipment, but also on the education and training of managers who develop plans and workers who execute these plans (Sila 2007; Ooi 2014). Hence, a quality program must be closely linked to education and training programs.

5. *Continuous quality improvement*: As mentioned above, the quality of a product depends on the production process, but above all, it relies on the ability and skills of human resources. A quality plan or program should include strategies for continuous quality improvement, since no product is totally perfect; all products can be improved. Continuous quality improvement is mainly supported by two concepts, quality circles and Kaizen events (Knechtges and Decker 2014), where it is intended that workers or operators solve problems arising in the production lines. However, these improvements in quality depend to a great extent on the culture of these workers and their knowledge of the activities carried out in the production process (Choo et al. 2015). In other words, continuous improvement can only be understood after a suitable education program. Accepting that a product meets all specifications is being sure that there is no possibility to improve it. If these programs of continuous quality improvement are properly executed, they will always target at the elimination of waste throughout the supply chain and, therefore, companies will benefit from a better group of materials that facilitates the success of a JIT program. Recent studies reported economic aspects related the human factor and its impact on continuous improvement (Helander and Burri 1995).
6. *Zero defects*: When Crosby mentioned a program of zero-defect production in the arms industry in the United States, he knew that defects had to be corrected if the product could be saved or because discarding them would imply great economic costs. Thus, he believed that it was best not to make mistakes. However, reaching the goal of zero defects is almost impossible (Stroebel 1993), and many programs nowadays, such as such as Six Sigma philosophy (Love et al. 1995), focus primarily on obtaining acceptable quality rather than flawless quality. To achieve acceptable levels of quality and, as far as possible with zero defects, improvement programs should be based on education for all operators and the entire organizational structure (Westkämper and Warnecke 1994). Nevertheless, companies should also make use of their equipment and technologies to test quality to keep an adequate monitoring system throughout the entire production process. Fortunately, information and communication technologies today are able to achieve this. Similarly, producers must be very clear and specific when establishing guarantees, since the products can fail not just because of a mistake made during the manufacturing process, but also because a certain component may have completed its life cycle

(Myklebust 2013). Thus, companies could benefit from a continuous flow of materials throughout the production process when they focus on achieving zero defects in the production process. As a result, all products will be just-in-time delivered.

7. *Quality-oriented training*: This book has often mentioned the importance of JIT training and the impact of education and skills from operators on the production line and managers and supervisors on the senior and middle management. Quality-oriented training emphasizes on the fact that companies must guide training and education not only toward quality of both the product and production process, but also toward the people and their quality of life. A program of education and training oriented to quality would improve the work of employees as well as their everyday life. They would feel proud of the company they work for.
   The role of training and education in many of the techniques and tools of lean manufacturing has been extensively studied (Chryssolouris et al. 2013); however, companies nowadays require specific knowledge of advanced technologies for training and education due to their rapid change and modernization (Mital et al. 1999). Some universities and colleges now propose a number of courses that students must pursue in industries. That is, students must develop real projects in companies, which in turn grade them in the project. In some countries, such as Mexico, the government finances postgraduate programs to be pursued both inside the University and within the industry. These mixed programs seek to ensure the formation of people with high academic and scientific levels, but also practical skills. These educational programs always focus on the quality of the products and services offered. Thus, quality-oriented training ensures the fulfillment of the objectives of JIT.
8. *High visibility of quality control*: Accomplishments related to quality and all indices and metrics that represent the efficiency of the company, must be highly visible so that all operators can be familiar with the status of their workplace. This is very important because in many companies senior managers are usually the only ones acquainted with quality and efficiency statistics. This may be a mistake, because operators are not at least informed. However, some other companies tend to place neon signs indicating the number of defects detected in a shift. The sign is visible to all employees, which allows operators and supervisors to make an effort to reduce these defects.
   Similarly, not only bad news must be reported; the achievements of a company ought to be informed through different media. For instance, announcing the completion of a shift with zero defects can surely motivate employees to continue working toward this goal. When this kind of information is shared with workers in the production line, it helps keep them informed in and make decisions, both in real time. Some authors recommend planning the system to inform results from the moment facilities of the workplace are designed, since this would guarantee the expected results. When unplanned, the announcements are usually hung around the facilities or put in a place that was not designed for

that purpose at all (Colledani et al. 2014). It shall not be forgotten that not only quality issues must be available to workers. Instructions and the status of the production system, such as a product delay, must be visual to all. This improves the flow of raw materials throughout the production system and helps identify the biggest wastes, which is why it is considered as an element of JIT. Companies would be unable to find a production target or goal if workers are not familiar with the system at any time.

9. *Long-term commitment to QC*: Product quality is not achieved overnight; it is the result of hard work over many quality plans, which agrees with the principles of JIT. Both JIT and quality are philosophies applied to production systems, and philosophies are not implemented within a short time since they must be integrated into the lifestyle of workers. Therefore, quality must be the product of plans previously established by the top management, since it is responsible for setting the parameters and desirable attributes in the products or services offered. This form of ensuring quality is not new (Shturtz 1992). Some authors even recommend that companies have their own quality laboratory, so supervisors and workers themselves are responsible for quality control (Roderick and English 1990). Similarly, a more recent element has been discussed, the fact that quality planning ought to integrate environmental aspects to demonstrate how companies could get rid of products with polluting components (Madu et al. 1995). Surely, then, a long-term quality plan can help generate strategies to eliminate waste in the production system, which would increase fluidity of materials since the defective subassemblies would not remain in it.
10. *Regulate quality and reliability of audits*: A successful quality program must be actually executed, instead of being a beautiful book in a file cabinet. Quality should be a philosophy of production, but this is not always achieved. Therefore, companies should continuously perform a series of audits to analyze and detect deviations related to the plans established. A quality program without follow-up will most likely have a number of deviations in a short time, and these deviations tend to be difficult to correct and could consequently imply economic cost for the company, and companies are set to generate capital.
    It should be mentioned that quality audits do not generate quality; they merely indicate the status of a company in relation to a product or process. On the one hand, it is recommended that, in order to maintain standards and high-quality levels, audits must be conducted by experts who are fully acquainted with the desired characteristics of the product (Power and Terziovski 2007). On the other hand, other authors recommend that audits be performed by personnel outside the company, which preferably have no relation to the heads of the quality department to avoid complicity and concealment of errors (Kouaib and Jarboui 2014). Finally, it should be mentioned that no audit is sufficient to ensure product quality (Taggart 2013; Powell et al. 2013), and they always ought to be performed, especially when there are critical points or high risks. Therefore, an audit would help quickly detect errors and waste in the production process so they can be addressed as soon as possible to prevent

defective products on the market or the stoppage of production system due to internal faults, which could generate late deliveries to customers.

11. *Quality culture*: As previously stated, quality is obtained as a result of education and culture; yet, reaching it demands great conceptualization and awareness. Indeed, while manufacturing quality products costs nothing, poor quality or no quality at all implies high costs, and one of them is failure to achieve just-in-time deliveries (Janipha and Ismail 2013). This failure may contribute to a bad reputation for the company.
    An important work addressing the culture of quality emerged in Detert et al. (2003), which proposed to generate an index point to indicate the degree or level of quality culture in a company; yet, this may be further studied, since there seems to be a direct relationship between quality culture and operational and financial performance indexes (Roldán et al. 2012). However, it should be stressed that a culture of quality is the responsibility of top managers and operators, who are the primary reflection of the culture that exists in the company. As a summary, while senior management is responsible for the development, monitoring, supervision, and control of quality plans and programs, employees are responsible for the successful execution of these.
12. *100 % quality inspection*: The inspection of all raw materials arriving in the warehouses may not be a proper approach, since it is assumed that companies only order acceptable and high-quality components. The quality of these components is the responsibility of the manufacturer. The cost of a 100 % inspection is very high, and it does not guarantee first quality products, although it helps detect those who do not meet specifications. However, it may be more suitable if every operator performed an inspection of his/her corresponding raw materials and became responsible for the operations that he/she must execute. Moreover, it may not be desirable to count on an inspection section within a production line. It might be more appropriate if each activity could be checked by the person in charge. This could ensure a more effective and rapid 100 % inspection, which guarantees that the components or raw materials entering the production process are free from defects and problems related to product quality will not emerge in subsequent activities. This ensures a continuous flow of materials along the system and just-in-time deliveries.
    Fortunately, a number of technologies nowadays help conduct 100 % quality inspections within the production process, and they are all supported by sophisticated computer equipment (Xie et al. 1998). As a result, ensuring quality can sometimes be a process that is fully automated by vision systems (Lee and Park 2000).
13. *Simplify the process of total quality*: Product quality is not warranted with a sophisticated program. It is frequently mentioned that quality programs must be straightforward, which indicates that a company has fully accepted quality as a production philosophy. A quality management system is often mistaken with the quality philosophy itself, since quality control is not an easy task and workers usually know little of it since they fear their complexity (Dooley and

Flor 1998). The previous page referred to an index to measure quality culture in companies, and perhaps simplification of the quality process is a valid index. It is true that during the production process, companies must draw upon documentation and survey statistics to identify areas of opportunity for improvement. Nevertheless, if a product is incorrectly processed at some point of the production process, and it can be corrected by the operator who detected the issue, the easiest and perhaps most logical decision would be to correct it so the product does not continue in the production line. However, it is required for administrative purposes to count and record such errors, since they may indicate the need of a training program for the person in charge of the activity (Forza and Filippini 1998). This proposes two scenarios, one in which quality assurance is simplified by correcting an identified effect, and another in which companies issue an administrative act to take corrective actions. It is likely that both be necessary to ensure product quality in the future. If the report is not issued, the problem may not be solved from its roots, and the error may continue to appear.

References

Aalaei A, Davoudpour H (2015) Revised multi-choice goal programming for incorporated dynamic virtual cellular manufacturing into supply chain management: a case study. Eng Appl Artif Intell (0). doi:10.1016/j.engappai.2015.04.005

Ahmad MF, Zakuan N, Jusoh A, Takala J (2012) Relationship of TQM and business performance with mediators of SPC, lean production and TPM. Procedia—Soc Behav Sci 65:186–191. doi:10.1016/j.sbspro.2012.11.109

Alcaraz JLG, Maldonado AA, Iniesta AA, Robles GC, Hernández GA (2014) A systematic review/survey for JIT implementation: Mexican maquiladoras as case study. Comput Ind 65 (4):761–773. doi:10.1016/j.compind.2014.02.013

Aliu A, Halili A (2013) The impact of information and communication technologies as a tool to facilitate growth in the manufacturing sector in Republic of Kosovo. Procedia Technol 8:465–470. doi:10.1016/j.protcy.2013.11.060

Almomani MA, Aladeemy M, Abdelhadi A, Mumani A (2013) A proposed approach for setup time reduction through integrating conventional SMED method with multiple criteria decision-making techniques. Comput Ind Eng 66(2):461–469. doi:10.1016/j.cie.2013.07.011

Anderman EM, Gray D (2015) Motivation learning and instruction. In: Wright JD (ed) International encyclopedia of the social and behavioral sciences, 2nd edn. Elsevier, Oxford, pp 928–935. doi:10.1016/B978-0-08-097086-8.26041-8

Asendorf I, Schultz-Wild R (1984) Work organization and training in a flexible manufacturing system—an alternative approach. In: Martin T (ed) Design of work in automated manufacturing systems. Pergamon, pp 45–49. doi:10.1016/B978-0-08-031118-0.50015-4

Baker RC, Chang RE, Chang IC (1994) Switching rules for JIT purchasing. Prod Inventory Manag J 35(3):13–17

Baraldi P, Podofillini L, Mkrtchyan L, Zio E, Dang VN (2015) Comparing the treatment of uncertainty in Bayesian networks and fuzzy expert systems used for a human reliability analysis application. Reliab Eng Syst Saf 138:176–193. doi:10.1016/j.ress.2015.01.016

Barron KE, Hulleman CS (2015) Expectancy-value-cost model of motivation. In: Wright JD (ed) International encyclopedia of the social and behavioral sciences, 2nd edn. Elsevier, Oxford, pp 503–509. doi:10.1016/B978-0-08-097086-8.26099-6

Benton WC, Shin H (1998) Manufacturing planning and control: the evolution of MRP and JIT integration. Eur J Oper Res 110(3):411–440. doi:10.1016/S0377-2217(98)00080-0

Blair JD, Ramsing KD (1983) Quality circles and production/operations management: concerns and caveats. J Oper Manage 4(1):1–10. doi:10.1016/0272-6963(83)90021-9

Bonito J (1990) Motivating employees for continuous improvement efforts. Prod Inventory Manage Rev APICS News 8:225–236

Bose GJ, Rao A (1988) Implementing JIT with MRP-II creates hybrid manufacturing environment. Ind Eng (September):49–53

Brennan M (1990) Management and quality circles: case studies in direct participation. J Oper Manage 9(3):426–427. doi:10.1016/0272-6963(90)90171-9

Brown JR (2014) A capacity constrained mathematical programming model for cellular manufacturing with exceptional elements. J Manuf Syst (0). doi:10.1016/j.jmsy.2014.09.005

Cannon JP, Doney PM, Mullen MR, Petersen KJ (2010) Building long-term orientation in buyer–supplier relationships: the moderating role of culture. J Oper Manage 28(6):506–521. doi:10.1016/j.jom.2010.02.002

Chan FTS (2001) Effect of kanban size on just-in-time manufacturing systems. J Mater Process Technol 116(2–3):146–160. doi:10.1016/S0924-0136(01)01022-6

Chen H, Jeter D (2008) The role of auditing in buyer-supplier relations. J Contemp Account Econ 4(1):1–17. doi:10.1016/S1815-5669(10)70027-8

Cheung C, Baum T, Wong A (2012) Relocating empowerment as a management concept for Asia. J Bus Res 65(1):36–41. doi:10.1016/j.jbusres.2011.07.012

Chong V, Rundus M (2000) The effect of new manufacturing practices and intensity of market competition on organizational performance: an empirical investigation. Paper presented at the international conference on management research accounting research and case symposium, Mesa, Arizona, USA, 6–8 Jan 2000

Choo AS, Nag R, Xia Y (2015) The role of executive problem solving in knowledge accumulation and manufacturing improvements. J Oper Manage 36:63–74. doi:10.1016/j.jom.2015.03.001

Chryssolouris G, Mavrikios D, Mourtzis D (2013) Manufacturing systems: skills and competencies for the future. Procedia CIRP 7:17–24. doi:10.1016/j.procir.2013.05.004

Chu KF (2003) An organizational culture and the empowerment for change in SMEs in the Hong Kong manufacturing industry. J Mater Process Technol 139(1–3):505–509. doi:10.1016/S0924-0136(03)00527-2

Colledani M, Tolio T, Fischer A, Iung B, Lanza G, Schmitt R, Váncza J (2014) Design and management of manufacturing systems for production quality. CIRP Ann—Manuf Technol 63(2):773–796. doi:10.1016/j.cirp.2014.05.002

Collins R, Bechler K, Pires S (1997) Outsourcing in the automotive industry: from JIT to modular consortia. Eur Manage J 15(5):498–508. doi:10.1016/S0263-2373(97)00030-3

Conzalez B, Suarez-Gonzalez J (2001) Effect of organizational variables in JIT purchasing implementation. Int J Prod Res 39(10):2231–2259

Corominas A, Pastor R, Rodríguez E (2006) Rotational allocation of tasks to multifunctional workers in a service industry. Int J Prod Econ 103(1):3–9. doi:10.1016/j.ijpe.2005.05.015

Costello AM (2013) Mitigating incentive conflicts in inter-firm relationships: evidence from long-term supply contracts. J Account Econ 56(1):19–39. doi:10.1016/j.jacceco.2013.02.001

Crama Y (1997) Combinatorial optimization models for production scheduling in automated manufacturing systems. Eur J Oper Res 99(1):136–153. doi:10.1016/S0377-2217(96)00388-8

Cua KO, McKone KE, Schroeder RG (2001) Relationships between implementation of TQM, JIT, and TPM and manufacturing performance. J Oper Manage 19(6):675–694. doi:10.1016/S0272-6963(01)00066-3

Dabhilkar M, Bengtsson L, von Haartman R, Åhlström P (2009) Supplier selection or collaboration? Determining factors of performance improvement when outsourcing manufacturing. J Purchasing Supply Manage 15(3):143–153. doi:10.1016/j.pursup.2009.05.005

Daesung HA, Seung-Lae K (1997) Implementation of JIT purchasing: an integrated approach. Prod Plan Control 8(2):152–157. doi:10.1080/095372897235415

David I, Eben-Chaime M (2003) How far should JIT vendor–buyer relationships go? Int J Prod Econ 81–82:361–368. doi:10.1016/S0925-5273(02)00274-8

de Macedo-Soares TDLA, Lucas DC (1995) Empowerment and total quality: comparing research findings in the USA and Brazil. Technovation 15(8):475–495. doi:10.1016/0166-4972(95)96587-J

Deep K, Singh PK (2015) Design of robust cellular manufacturing system for dynamic part population considering multiple processing routes using genetic algorithm. J Manuf Syst 35:155–163. doi:10.1016/j.jmsy.2014.09.008

Delbridge R (1995) Surviving JIT: control and resistance in a Japanese transplant. J Manage Stud 32:803–817

Detert JR, Schroeder RG, Cudeck R (2003) The measurement of quality management culture in schools: development and validation of the SQMCS. J Oper Manage 21(3):307–328. doi:10.1016/S0272-6963(02)00130-4

Di Pasquale V, Miranda S, Iannone R, Riemma S (2015) A simulator for human error probability analysis (SHERPA). Reliab Eng Syst Saf 139:17–32. doi:10.1016/j.ress.2015.02.003

Diefendorff JM, Seaton GA (2015) Work motivation. In: Wright JD (ed) International encyclopedia of the social and behavioral sciences, 2nd edn. Elsevier, Oxford, pp 680–686. doi:10.1016/B978-0-08-097086-8.22036-9

Dong Y, Carter CR, Dresner ME (2001) JIT purchasing and performance: an exploratory analysis of buyer and supplier perspectives. J Oper Manage 19(4):471–483. doi:10.1016/S0272-6963(00)00066-8

Dooley KJ, Flor RF (1998) Perceptions of success and failure in TQM initiatives. J Qual Manage 3(2):157–174. doi:10.1016/S1084-8568(99)80111-4

Dutton B (1990) Switching to quality excellence. Manuf Syst 8:245–256

Ebrahimpour M, Schonberger RJ (1984) The Japanese just-in-time/total quality control production system: potential for developing countries. Int J Prod Res 22:421–430

Erenay B, Suer GA, Huang J, Maddisetty S (2015) Comparison of layered cellular manufacturing system design approaches. Comput Ind Eng 85:346–358. doi:10.1016/j.cie.2015.02.021

Farris JA, Van Aken EM, Doolen TL, Worley J (2009) Critical success factors for human resource outcomes in Kaizen events: an empirical study. Int J Prod Econ 117(1):42–65. doi:10.1016/j.ijpe.2008.08.051

Ferradás PG, Salonitis K (2013) Improving changeover time: a tailored SMED approach for welding cells. Procedia CIRP 7:598–603. doi:10.1016/j.procir.2013.06.039

Fiedler K, Galletly JE, Bicheno J (1993) Expert advice for JIT implementation. Int J Oper Prod Manage 13(6):23–30. doi:10.1108/01443579310038994

Flynn B, Salakibara S, Schroeder R (1995) Relationship between JIT and TQM practices and performance. Acad Manage J 38(5):1325–1360

Forza C, Filippini R (1998) TQM impact on quality conformance and customer satisfaction: a causal model. Int J Prod Econ 55(1):1–20. doi:10.1016/S0925-5273(98)00007-3

Frascatore MR, Mahmoodi F (2008) Long-term and penalty contracts in a two-stage supply chain with stochastic demand. Eur J Oper Res 184(1):147–156. doi:10.1016/j.ejor.2006.10.056

Frazer VCM, Dale BG (1986) UK quality circle failures—the latest picture. Omega 14(1):23–33. doi:10.1016/0305-0483(86)90005-8

Fullerton RR, McWatters CS (2002) The role of performance measures and incentive systems in relation to the degree of JIT implementation. Acc Organ Soc 27(8):711–735. doi:10.1016/S0361-3682(02)00012-0

García J, Rivera D, Iniesta A (2013) Critical success factors for Kaizen implementation in manufacturing industries in Mexico. Int J Adv Manuf Technol 68(1–4):537–545. doi:10.1007/s00170-013-4750-2

García J, Maldonado A, Alvarado A, Rivera D (2014) Human critical success factors for kaizen and its impacts in industrial performance. Int J Adv Manuf Technol 70(9–12):2187–2198. doi:10.1007/s00170-013-5445-4

Garg D (1997) Relevance of JIT purchasing in Indian industries. Kurukshetra University, India

Garg D, Deshmukh SG (1999a) JIT concepts and relevance in Indian context. Ind Eng J 28 (1):14–18

Garg D, Deshmukh SG (1999b) JIT purchasing: literature review and implications for Indian industry. Prod Plan Control 10(3):276–285. doi:10.1080/095372899233235

Garg S, Vart P, Kanda A (1994) Work culture in JIT environment. Prod J 35(3):463–466

Garg D, Deshmukh S, Kaul O (1996a) Critical analysis in JIT purchasing in Indian context. Prod J 37:271–279

Garg D, Deshmukh SG, Kaul ON (1996b) Parametric analysis of inventory models under JIT purchasing analysis of inventory models under JIT purchasing agreements. Paper presented at the national conference operational research in modern technology, REC Kurukshetra, 8–9 Mar 1996

Gilbert JP (1990) The state of JIT implementation and development in the USA. Int J Prod Res 28 (6):1099–1109

Giunipero LC, O'Neal C (1988) Obstacles to JIT procurement. Ind Mark Manage 17(1):35–41. doi:10.1016/0019-8501(88)90024-7

Golhar DY, Stamm CL (1991) The just-in-time philosophy: a literature review. Int J Prod Res 29 (4):657–676

Goyal SK, Deshmukh SG (1992) A critique of the literature on just-in-time manufacturing. Int J Oper Prod Manage 12(1):18–28

Greene TJ, Sadowski RP (1986) A mixed integer program for loading and scheduling multiple flexible manufacturing cells. Eur J Oper Res 24(3):379–386. doi:10.1016/0377-2217(86) 90031-7

Gupta YP, Lonial SC, Glynn Mangold W (1992) Empirical assessment of the strategic orientation of JIT manufacturers versus non-JIT manufacturers. Comput Integr Manuf Syst 5(3):181–189. doi:10.1016/0951-5240(92)90029-C

Hall R (1983) Zero inventories. Dow Jones-Irwin, Homewood, IL

Handfield RB (1993) A resource dependence perspective of just-in-time purchasing. J Oper Manage 11(3):289–311. doi:10.1016/0272-6963(93)90005-A

Hayward SG, Dale BG (1984) Quality circle failures in UK manufacturing companies—a study. Part II. Omega 12(6):557–568. doi:10.1016/0305-0483(84)90058-6

Helander MG, Burri GJ (1995) Cost effectiveness of ergonomics and quality improvements in electronics manufacturing. Int J Ind Ergon 15(2):137–151. doi:10.1016/0169-8141(94)00028-2

Hong J-D, Hayya JC, Kim S-L (1992) JIT purchasing and setup reduction in an integrated inventory model. Int J Prod Res 30(2):255–266. doi:10.1080/00207549208942893

Houghton E, Portougal V (1997) Trade-offs in JIT production planning for multi-stage systems: balancing work-load variations and WIP inventories. Int Trans Oper Res 4(5–6):315–326. doi:10.1016/S0969-6016(97)00032-4

Howard LW, Foster ST (1999) The influence of human resource practices on empowerment and employee perceptions of management commitment to quality. J Qual Manage 4(1):5–22. doi:10.1016/S1084-8568(99)80093-5

Huang C-Y, Huang T-S, Chen W-L (2009) Communication protocols for order management in collaborative manufacturing. Int J Prod Econ 122(1):257–268. doi:10.1016/j.ijpe.2009.06.004

Huq Z (1999) Conventional shop control procedures to approximate JIT inventory performance in a job shop. J Manuf Syst 18(3):161–174. doi:10.1016/S0278-6125(99)80029-6

Inman RA, Sale RS, Green KW Jr, Whitten D (2011) Agile manufacturing: relation to JIT, operational performance and firm performance. J Oper Manage 29(4):343–355. doi:10.1016/j.jom.2010.06.001

Janipha NAI, Ismail F (2013) Conceptualisation of quality issues in Malaysian construction environment. Procedia—Soc Behav Sci 101:53–61. doi:10.1016/j.sbspro.2013.07.178

Jiang X, Liu Y (2015) Global value chain, trade and carbon: case of information and communication technology manufacturing sector. Energy Sustain Dev 25:1–7. doi:10.1016/j.esd.2014.12.001

Junior ML, Godinho Filho M (2010) Variations of the Kanban system: literature review and classification. Int J Prod Econ 125(1):13–21. doi:10.1016/j.ijpe.2010.01.009

Khan LR, Sarker RA (2002) An optimal batch size for a JIT manufacturing system. Comput Ind Eng 42(2–4):127–136. doi:10.1016/S0360-8352(02)00009-8

Kim S-L, Ha D (2003) A JIT lot-splitting model for supply chain management: enhancing buyer–supplier linkage. Int J Prod Econ 86(1):1–10. doi:10.1016/S0925-5273(03)00006-9

Knechtges P, Decker MC (2014) Application of Kaizen methodology to foster departmental engagement in quality improvement. J Am Coll Radiol 11(12, Part A):1126–1130. doi:10.1016/j.jacr.2014.08.027

Kohlbeck M (2011) Investor valuations of suppliers' major customer disclosures. Adv Acc 27 (2):278–285. doi:10.1016/j.adiac.2011.07.002

Kohler C, Schultz-Wild R (1985) Flexible manufacturing systems—manpower problems and policies. J Manuf Syst 4(2):135–146. doi:10.1016/0278-6125(85)90020-2

Kouaib A, Jarboui A (2014) External audit quality and ownership structure: interaction and impact on earnings management of industrial and commercial Tunisian sectors. J Econ Financ Adm Sci 19(37):78–89. doi:10.1016/j.jefas.2014.10.001

Kumar V (2010) JIT based quality management: concepts and implications in Indian context. Int J Eng Sci Technol 2(1):40–50

Kumar V, Garg D (2000) JIT elements in Indian context: an analysis. Prod J 41(2):217–222

Kumar V, Garg D, Mehta N (2001) JIT based quality management in Indian industries: prospectus and future directions. Paper presented at the national conference supply chain management, Institute of Public Enterprise, Hyderabad, India, 19–20 Jan 2001

Kumar V, Garg D, Mehta N (2004) JIT elements in Indian context: an analysis. Prod J 63 (August):662–665

Larson PD, Kulchitsky JD (1998) Single sourcing and supplier certification: performance and relationship implications. Ind Mark Manage 27(1):73–81. doi:10.1016/S0019-8501(97)00039-4

Lee KH, Park H (2000) Automated inspection planning of free-form shape parts by laser scanning. Rob Comput Integr Manuf 16(4):201–210. doi:10.1016/S0736-5845(99)00060-5

Lee Y-M, Mu S, Shen Z, Dessouky M (2014) Issuing for perishable inventory management with a minimum inventory volume constraint. Comput Ind Eng 76:280–291. doi:10.1016/j.cie.2014.08.007

Li X, Chen L, Xu H, Gupta JND (2015) Trajectory Scheduling Methods for minimizing total tardiness in a flowshop. Oper Res Perspect 2:13–23. doi:http://dx.doi.org/10.1016/j.orp.2014.12.001

Liu H, Hwang S-L, Liu T-H (2009) Economic assessment of human errors in manufacturing environment. Saf Sci 47(2):170–182. doi:10.1016/j.ssci.2008.04.006

Love CE, Guo R, Irwin KH (1995) Acceptable quality level versus zero-defects: some empirical evidence. Comput Oper Res 22(4):403–417. doi:10.1016/0305-0548(94)00049-E

Lovell MC (2003) Optimal lot size, inventories, prices and JIT under monopolistic competition. Int J Prod Econ 81–82:59–66. doi:10.1016/S0925-5273(02)00349-3

Maass RA, Brown JA, Bossert JL (1989) Supplier certification—a continuous improvement strategy. Am Soc Qual Control

Macbeth DK, Baxter L, Farguson N, Neil G (1988) Buyer-vendor relationship with just-in-time: lessons from US multinationals. Ind Eng 20(9):38–41

Machuca JAD (2002) JIT facing the new millennium. Int J Prod Econ 80(2):131–134. doi:10.1016/S0925-5273(02)00312-2

Madu CN, C-h Kuei, Winokur D (1995) Environmental quality planning: a strategic total quality management (STQM) approach. Futures 27(8):839–856. doi:10.1016/0016-3287(95)00048-2

Mahdavi I, Aalaei A, Paydar MM, Solimanpur M (2012) A new mathematical model for integrating all incidence matrices in multi-dimensional cellular manufacturing system. J Manuf Syst 31(2):214–223. doi:10.1016/j.jmsy.2011.07.007

Manson M, Dale B (1989) The operating characteristics of quality circles and yield improvement teams: a case study comparison. Eur Manag J 7(3):287–295. doi:10.1016/0263-2373(89)90122-9

Martínez-Jurado PJ, Moyano-Fuentes J, Jerez-Gómez P (2014) Human resource management in lean production adoption and implementation processes: success factors in the aeronautics industry. BRQ Bus Res Q 17(1):47–68. doi:10.1016/j.cede.2013.06.004

Mital A, Pennathur A, Huston RL, Thompson D, Pittman M, Markle G, Kaber DB, Crumpton L, Bishu RR, Rajurkar KP, Rajan V, Fernandez JE, McMulkin M, Deivanayagam S, Ray PS, Sule D (1999) The need for worker training in advanced manufacturing technology (AMT) environments: a white paper. Int J Ind Ergon 24(2):173–184. doi:10.1016/S0169-8141(98)00024-9

Mkrtchyan L, Podofillini L, Dang VN (2015) Bayesian belief networks for human reliability analysis: a review of applications and gaps. Reliab Eng Syst Saf 139:1–16. doi:10.1016/j.ress.2015.02.006

Moattar Husseini SM, O'Brien C, Hosseini ST (2006) A method to enhance volume flexibility in JIT production control. Int J Prod Econ 104(2):653–665. doi:10.1016/j.ijpe.2005.05.025

Mohan MTR, Singh NP (1995) Buyer-supplier relationships: the case from Indian diary co-operatives. Prod J 35(4):643–647

Muralidharan C, Anantharaman N, Deshmukh SG (2001) Vendor rating in purchasing scenario: a confidence interval approach. Int J Oper Prod Manage 21(10):1305–1326. doi:10.1108/01443570110404736

Myklebust O (2013) Zero defect manufacturing: a product and plant oriented lifecycle approach. Procedia CIRP 12:246–251. doi:10.1016/j.procir.2013.09.043

Nandi SN (1988) Implementing quality control circles in India. Prod J 28(4):419–424

Nguyen QA, Hens L (2015) Environmental performance of the cement industry in Vietnam: the influence of ISO 14001 certification. J Clean Prod 96:362–378. doi:10.1016/j.jclepro.2013.09.032

Ohno K (2011) The optimal control of just-in-time-based production and distribution systems and performance comparisons with optimized pull systems. Eur J Oper Res 213(1):124–133. doi:10.1016/j.ejor.2011.03.005

Ooi K-B (2014) TQM: a facilitator to enhance knowledge management? A structural analysis. Expert Syst Appl 41(11):5167–5179. doi:10.1016/j.eswa.2014.03.013

Padukone H, Subba R (1993) Global status of JIT—implication for developing countries. Prod J 34(3):419–429

Pan AC, Liao CJ (1989) An inventory model under just-in time purchasing agreements. Prod Inventory Manage J 30(1):49–52

Panayiotou CG, Cassandras CG (1999) Optimization of kanban-based manufacturing systems. Automatica 35(9):1521–1533. doi:10.1016/S0005-1098(99)00074-6

Park HY, Reddy CS, Shin G-C, Eckerle C (1996) Impact of the supplier certification program in US firms. Eur J Purchasing Supply Manage 2(2–3):107–118. doi:10.1016/0969-7012(95)00025-9

Pascarella P (1985) 19—Quality circles: just another management headache? In: Hodgetts JWGM (ed) Readings and exercises in organizational behavior. Academic Press, USA, pp 171–177. doi:10.1016/B978-0-12-054752-4.50027-0

Paul RJ, Niehoff BP, Turnley WH (2000) Empowerment, expectations, and the psychological contract—managing the dilemmas and gaining the advantages. J Socio-Econ 29(5):471–485. doi:10.1016/S1053-5357(00)00083-4

Powell DA, Erdozain S, Dodd C, Costa R, Morley K, Chapman BJ (2013) Audits and inspections are never enough: a critique to enhance food safety. Food Control 30(2):686–691. doi:10.1016/j.foodcont.2012.07.044

Power D, Terziovski M (2007) Quality audit roles and skills: perceptions of non-financial auditors and their clients. J Oper Manage 25(1):126–147. doi:10.1016/j.jom.2006.02.005

Priestman S (1985) SQC and JIT: partnership in quality, does culture make a difference? Qual Prog 18(5):31–35

Prodipto R (1999) Data warehousing for textiles, Texinfotech-99. Paper presented at the Techno park (I) Ltd., New Delhi, India

Pyane TE (1993) Acme manufacturing: a case study in JIT implementation. Prod Inventory Manage J 34(2):82–86

Recht R, Wilderom C (1998) Kaizen and culture: on the transferability of Japanese suggestion systems. Int Bus Rev 7(1):7–22. doi:10.1016/S0969-5931(97)00048-6

Roderick LM, English JR (1990) Laboratory for quality planning and control. Comput Ind Eng 19 (1–4):165–169. doi:10.1016/0360-8352(90)90098-7

Roldán JL, Leal-Rodríguez AL, Leal AG (2012) The influence of organisational culture on the total quality management programme performance. Investigaciones Europeas de Dirección y Economía de la Empresa 18(3):183–189. doi:10.1016/j.iedee.2012.05.005

Rosenfeld Y, Warszawski A, Laufer A (1991) Quality circles in temporary organizations: lessons from construction projects. Int J Project Manage 9(1):21–27. doi:10.1016/0263-7863(91) 90051-V

Ruckart PZ, Burgess PA (2007) Human error and time of occurrence in hazardous material events in mining and manufacturing. J Hazard Mater 142(3):747–753. doi:10.1016/j.jhazmat. 2006.06.117

Sarkar SK, Tripathy N (2002) An empirical analysis of the impact of stock index futures trading on securities dealers' inventory risk in the NASDAQ market. Rev Financ Econ 11(1):1–17. doi:http://dx.doi.org/10.1016/S1058-3300(01)00036-2

Sawik T (2007) Integer programming approach to reactive scheduling in make-to-order manufacturing. Math Comput Model 46(11–12):1373–1387. doi:10.1016/j.mcm.2007.01.010

Saxena KBC, Sohay BS (1999) World-class manufacturing and global competitiveness. Prod J 40 (1):88–96

Sengupta S, Davis RP, Ferrell WG (1993) Production planning and control in a JIT environment. Appl Math Model 17(1):41–46. doi:10.1016/0307-904X(93)90126-2

Sewell G, Wilkinson B (1992) Someone to watch over me: surveillance, discipline and just-in-time process. Sociology 26:271–289

Shturtz GT (1992) Quality planning. Met Powder Rep 47(10):57. doi:10.1016/0026-0657(92) 91976-Q

Sila I (2007) Examining the effects of contextual factors on TQM and performance through the lens of organizational theories: an empirical study. J Oper Manage 25(1):83–109. doi:10.1016/ j.jom.2006.02.003

Singh A (1989) Just-in-time system: an integrated system. Prod J 30(3):309–314

Singh S, Garg D (2011) JIT system: concepts, benefits and motivations in Indian industries. Int J Manage Bus Stud 1(1):26–30

Singhvi S (1992) Employee involvement in JIT success: Eicher experience. Prod J 33:366–369

Stroebel T (1993) Targeting zero-defect manufacturing: test and measurement world. Microelectron Reliab 33(13):2060. doi:10.1016/0026-2714(93)90396-G

Taggart D (2013) 12—Quality assurance and audits in the nuclear industry in the USA. In: Devgun J (ed) Managing nuclear projects. Woodhead Publishing, Cambridge, pp 237–254. doi:10.1533/9780857097262.4.237

Terlaak A, King AA (2006) The effect of certification with the ISO 9000 quality management standard: a signaling approach. J Econ Behav Organ 60(4):579–602. doi:10.1016/j.jebo.2004. 09.012

Toikka K (1987) Development of work in FMS—case study on new manpower strategy. In: BrÖDner P (ed) Skill based automated manufacturing. Pergamon, Oxford, pp 7–12. doi:10. 1016/B978-0-08-034800-1.50007-0

Tong DYK, Rasiah D, Tong XF, Lai KP (2015) Leadership empowerment behaviour on safety officer and safety teamwork in manufacturing industry. Saf Sci 72:190–198. doi:10.1016/j.ssci. 2014.09.009

Ullah B, Wei Z, Xie F (2014) ISO certification, financial constraints, and firm performance in Latin American and Caribbean countries. Glob Financ J 25(3):203–228. doi:10.1016/j.gfj. 2014.10.003

Valmohammadi C, Roshanzamir S (2015) The guidelines of improvement: relations among organizational culture, TQM and performance. Int J Prod Econ 164:167–178. doi:10.1016/j.ijpe.2014.12.028

Voss CA (1986) Just-in-time manufacture. IFS Limited. Springer, London, UK

Voss C, Robinson S (1987) Application of just-in-time manufacturing techniques in United Kingdom. Int J Oper Prod Manage 7(4):46–52

Vrat P, Mittal S, Tyagi K (1993) Implementation of JIT in Indian environment: a Delhi study. Prod J 34:251–256

Vuppalapati K, Ahire SL, Gupta T (1995) JIT and TQM: a case for joint implementation. Int J Oper Prod Manage 15(5):84–94. doi:10.1108/01443579510083686

Westkämper E, Warnecke HJ (1994) Zero-defect manufacturing by means of a learning supervision of process chains. CIRP Ann—Manuf Technol 43(1):405–408. doi:10.1016/S0007-8506(07)62240-9

White RE, Prybutok V (2001) The relationship between JIT practices and type of production system. Omega 29(2):113–124. doi:10.1016/S0305-0483(00)00033-5

Withers BE, Ebrahimpour M, Hikmet N (1997) An exploration of the impact of TQM and JIT on ISO 9000 registered companies. Int J Prod Econ 53(2):209–216. doi:10.1016/S0925-5273(97)00118-7

Xie M, Goh TN, Lu XS (1998) Computer-aided statistical monitoring of automated manufacturing processes. Comput Ind Eng 35(1–2):189–192. doi:10.1016/S0360-8352(98)00058-8

Xu J, Feng G, Jiang W, Wang S (2015) Optimal procurement of long-term contracts in the presence of imperfect spot market. Omega 52:42–52. doi:10.1016/j.omega.2014.10.003

Yasin MM, Small M, Wafa MA (1997) An empirical investigation of JIT effectiveness: an organizational perspective. Omega 25(4):461–471. doi:10.1016/S0305-0483(97)00005-4

Yasin MM, Wafa MA, Small MH (2001) Just-in-time implementation in the public sector, an empirical examination. Int J Oper Prod Manage 21(9):1195–1204

Yasin MM, Small MH, Wafa MA (2003) Organizational modifications to support JIT implementation in manufacturing and service operations. Omega 31(3):213–226. doi:10.1016/S0305-0483(03)00024-0

Yilmaz A, Bedük A (2014) Evaluation of the effect of the outsourcing on resource dependency and transaction cost approach: a research in Konya Oiz, Turkey. Procedia—Soc Behav Sci 109:737–752. doi:10.1016/j.sbspro.2013.12.538

Chapter 3
Benefits of JIT

Companies implement JIT philosophy in order to obtain something in return, since these manufacturing and service companies aim to generate economic benefits. Therefore, if JIT cannot offer any of these benefits, they simply suspend its implementation. This chapter describes the benefits ensured to companies with the implementation of JIT. They were classified into the following categories:

- Benefits associated with Human Resources;
- Benefits associated with Production Processes;
- Benefits associated with Engineering;
- Benefits associated with Quality;
- Benefits associated with Materials Handling; and
- Economic Benefits.

The following sections identify and describe these categories.

3.1 Benefits Associated with Human Resources

Human resources are responsible for implementing JIT philosophy and they are also the main beneficiaries. Table 3.1 lists a total of five benefits perceived from JIT implementation concerning human resources according to authors of literature reviewed.

1. *Improved worker motivation*. The average number of on-time deliveries to customers increases after the implementation of JIT (Kuo-Ting et al. 2009; Hill and Vollmann 1986). This should be one of the many sources of pride and motivation for the employees, although senior managers must make sure this motivation is well transmitted through different actions. For instance, employees should be informed of their achievements either visually or orally. Similarly, as it has been stated before, quality and its parameters must be completely visible; thus, senior management must encourage the employees to feel the achievements of the company as theirs (Dwivedula and Bredillet 2010).

J.L. García-Alcaraz and A.A. Maldonado-Macías, *Just-in-Time Elements and Benefits*,
Management and Industrial Engineering, DOI 10.1007/978-3-319-25919-2_3

Table 3.1 JIT benefits associated with human resources

JIT benefits	Authors	Total
Improved worker motivation	(Balakrishnan et al. 1996; Hall 1983; Padukone and Subba 1993; Hong et al. 1992; Miltenbirg 1990; Daesung and Seung-Lae 1997; Bartezzaghi et al. 1992; Voss and Robinson 1987; Singh and Bhandarkar 1996; Garg et al. 1996b; Dutton 1990)	11
Increased teamwork	(Balakrishnan et al. 1996; Hall 1983; Bonito 1990; Singhvi 1992; Hong et al. 1992; Daesung and Seung-Lae 1997; Garg 1997; Garg and Deshmukh 1999b; Fiedler et al. 1993)	8
Reduced classification of positions	(Bonito 1990; Kumar et al. 2001; Chong and Rundus 2000; Ebrahimpour and Schonberger 1984)	4
Increased resource utilization	(Kumar and Garg 2000; Garg 1997; Flynn et al. 1995)	4
Increased communication	(Voss 1986; Garg and Deshmukh 1999b)	2

Unmotivated workers may be a challenging situation to handle inside any organization, since their involvement in quality activities will be likely at a minimum level. On the other hand, motivated and proud workers would certainly be involved in programs and activities of continuous improvement. In addition, they would be looking forward to hearing new ideas that could generate better efficiency levels for the company. Large research has been dedicated to motivation and the way it should be instilled in employees of companies, and enough evidence has demonstrated its importance. Unfortunately, topics such as this one have not been considered an essential part for the training of engineers. They have rather been left exclusive for administrators and social sciences researchers. However, engineers today interact with people who perform important tasks and functions; therefore, they are, to a great extent, responsible for maintaining their co-workers motivated and transferring professional commitment. Therefore, it is recommended that engineers be trained and educated in such forms that they could improve skills related to personnel management, since they will likely hold administrative and managerial positions in the industry.

Furthermore, significant motivation of employees must be enforced in a company, and it is a major point to be attended by managers, since workers are the most acquainted with the production processes and will often provide most of the suggestions and actions for improvement (Landherr and Constantinescu 2012). However, excessive workload might cause employees feel stressed and, consequently, production efficiency can often decrease. Therefore, companies must find a balance between keeping workers motivated and assigning them a certain amount of workload (Khalatbari et al. 2013). Stressed workers could be psychologically affected and their motivation will likely diminish. Companies must thus identify the barriers that obstruct full motivation in their employees.

Dissatisfied personnel in production lines are commonly found in companies, and a very clear example in the food industry is described in Macheka et al. (2013). Similarly, there are some studies highlighting the existence of an incentive program for employees, which is often associated with a stimulus or bonus program, either financial or nonfinancial (Lau and Roopnarain 2014). However, employees usually need a proper recognition of the work their perform, and not necessarily a monetary compensation or economic reward. As for the implementation of a universal index to measure the level of motivation of employees in a company, this may seem even more difficult, since motivation is purely psychological and everyone can expect different results from their work (Zámečník 2014).

2. *Increased teamwork*. This element is easy to understand, although the following scenario could also depict its importance without problem (McLachlin 1997). Let us suppose that a company wishes to find the ideal size of a team in order to load and unload trucks in one single platform. If one worker can load and unload one truck in one hour, although an average of two trucks per hour is expected, at least two people would be required to perform the loading and unloading tasks of the two trucks in the same hour to obtain a balanced line. However, if the company wished to increase their service rate, the best decision might be to increase the size of the team. In fact, if two operators covered the typical queuing system discussed above, the utilization ratio would be at 100 % without operators taking breaks. On the other hand, if four people constituted the team, companies could offer the loading and unloading capacity of four trucks in one hour. Event though this example is simple due to the didactic purposes pursued here, the solution does not only consist of increasing the number of team members randomly. There must be a balance between the costs of having a line or queue and the opportunity to offer the highest service rate. Companies can offer shorter waiting times to the customer, but also higher costs to provide the service.
3. *Reduced classification of positions*. As it was mentioned earlier, employees must possess certain skills to develop activities within the production process to ensure any position they may need to occupy. This diversification of skills will rapidly bring reduction in the classification of positions and, consequently, companies will benefit from a much flatter organizational structure, in which the relationship between workers and senior management could be significantly closer (Globerson and Korman 2001). The benefits of this kind of organizational structure rely on a rapid and efficient communication among all organizational levels.

   Reducing the classification of positions is common when numerous organizational levels exist, and information may distort when it reaches operators and employees who are in the lowest levels of the structure. Moreover, administrative costs may be lower for companies with a flat organizational structure than for those relying on tall and complex ones, since various people and positions must be justified with the production rate from operators. When the information flow is improved on a flat organizational structure, production orders also flow faster, and a better compliance to delivery times can be achieved. These benefits add value to the implementation of a JIT program.

4. *Increased and better use of resources*. One important advantage in JIT is the easy identification of waste, particularly waste of the workforce from human resources. This seems to be one of the most relevant benefits of JIT implementation and has become the focus of various studies. The increased and better use of resources occurs along the supply chain, as it was stated by Kojima et al. (2008), who also reported the improvement of several performance indices after JIT application in a production system. Similarly, (Sandanayake et al. 2008) addressed a model that sought to simulate the optimization of available resources in a company when JIT had been implemented. More recently, the use of a structural equation approach found that Mexican maquiladora industries had improved its import and export processes of raw materials and finished goods in a JIT environment (García-Alcaraz et al. 2015).
5. *Increased communication*. This benefit refers to a more direct and fluid (both horizontally and vertically) communication among the different organizational levels of a company as a consequence of flat structures obtained from a successful JIT implementation in the production process.
   For over two decades, communication has been considered an essential part in the success of JIT; yet, it is also a benefit that can be obtained from the purchasing process, if it is based on the trust in certified suppliers (Nassimbeni 1996). Furthermore, increased communication has been identified as an essential element to achieve the integration of the JIT philosophy with the material requirements program (MRP). In fact, communication among the different members of the supply chain can be considered not only as a required element, but also as a result of the proper practices that lead to a more competitive business (Ahmad et al. 2012).
   Furthermore, some parts of the organizational structure can be modified as the result of improved communication among the members of the supply chain, since they usually perform control tasks functions. However, with the elimination of these processes, internal entities should be removed and a much flatter organizational structure is achieved (Yasin et al. 2003). Finally, structural equation techniques recently revealed that communication is one of the most valued benefits by managers (Alcaraz et al. 2014; Green et al. 2012).

3.2 Benefits Associated with Production Processes

According to authors found in literature, Table 3.2 lists eight perceived benefits identified and related to the production processes. As in previous tables, these items are sorted in descending order for a better appreciation of its importance, according to the number of references obtained for every one.

Table 3.2 Benefits associated with production process

Benefits of JIT	Authors	Total
Increased productivity	(Singh 1989; Hall 1983; Padukone and Subba 1993; Prodipto 1999; Vrat et al. 1993; Hong et al. 1992; Muralidharan et al. 2001a; Guinipero and Law 1990; Daesung and Seung-Lae 1997; Bartezzaghi et al. 1992; Baker et al. 1994; Kumar and Garg 2000; Priestman 1985; Garg et al. 1994; Ebrahimpour and Schonberger 1984; Flynn et al. 1995; Fiedler et al. 1993; Garg et al. 1996a)	19
Reduced waste and rework	(Hall 1983; Singhvi 1992; Hong et al. 1992; Muralidharan et al. 2001a; Daesung and Seung-Lae 1997; Kumar and Garg 2000; Vrat et al. 1993; Garg 1997; Ebrahimpour and Schonberger 1984; Flynn et al. 1995; Garg et al. 1996a)	11
Increased efficiency	(Hall 1983; Hong et al. 1992; Muralidharan et al. 2001a; Miltenbirg 1990; Daesung and Seung-Lae 1997; Bartezzaghi et al. 1992; Kumar and Garg 2000; Garg and Deshmukh 1999b; Ebrahimpour and Schonberger 1984)	9
Increased process flexibility	(Hall 1983; Prodipto 1999; Vrat et al. 1993; Kumar et al. 2001; Garg and Deshmukh 1999b; Chong and Rundus 2000; Dutton 1990; Garg et al. 1996b)	8
Reduced production lead time	(Lee and Ebrahimpour 1985; Bartezzaghi et al. 1992; Baker et al. 1994)	3
Reduced work in progress	(Vrat et al. 1993; Hong et al. 1992; Daesung and Seung-Lae 1997)	3
Integration of different manufacturing activities	(Roy and KK 1996; Garg 1997)	2

1. *Increased productivity*. Although the word productivity may seem vague and diffuse, it is the most frequent concept reported as a benefit. Indeed, companies can have productivity in different sectors and in various areas of the same company, which is why the word may be fuzzy. They can be productive in human resource management, production processes, or raw material management, among others.
   Technically, a company is productive when the relationship between inputs and outputs is notably large (Banerjee and Golhar 1993; Oral et al. 2003). However, productivity has several connotations, and to say that a company is productive involves measuring various operational, economic, and social aspects. As for the productivity benefits obtained after the implementation of a JIT program, (García-Alcaraz et al. 2015) proposed a detailed analysis using structural equation modeling to evaluate the cause and effect of JIT in some indices of productivity of companies. Similarly, (Green et al. 2014) analyzed the same causal relationship but with the integration of other variables or items in the model.
2. *Reducing waste and rework*. This benefit refers to the reduction of waste of raw materials and reprocessing, and it is notably important since it is purely

operational. Moreover, it refers to the production process from human resources or the production process needed to transform raw materials.

On the one hand, reduced waste of raw materials refers to the fact that operators are more skilled at and accurate when executing their tasks; consequently, less subassemblies or products are wasted. However, these skills are also the result of an appropriate training program. Among all subassemblies or products with defects, some of them may be redeemable and could be saved with certain activities. This is called a rework processes in the field of engineering.

In JIT, the main type of waste is associated with time. Two of the elements that could reduce waste when working on a JIT system are time, indeed, and the production costs since they reduce inventories. In their work, (Miltenburg 1993) explained the relationships between time and production cost and time and design, as well as the relationships that they all had with quality. In addition, (Yasin et al. 1997) provided a list of benefits obtained from the implementation of JIT and found that the reductions of waste materials and delivery times were top benefits. A deeper study on JIT benefits can be found in Fullerton and McWatters (2001).

3. *Increased efficiency*. Efficiency is defined as the ability to adequately perform or complete a function (Bernard 1996). Thus, an efficient company can fulfill the function of generating money or economic dividends for its shareholders. Therefore, being efficient requires education, facilities, management from senior managers, and above all, determination and perseverance. Similarly, any efficient company must apply a series of techniques and tools for its production systems, and JIT is one of these tools. A great amount of research in the last three decades has studied how JIT philosophy can help a company become more efficient. For instance, (Bäck 1985) proposed a diversification and decentralization of businesses so they could become specialized in specific areas. Also, (Monden 2002) proposed a model which associated the financial performance of a company with the implementation of JIT. The model clearly demonstrated that JIT philosophy was beneficial for companies that had applied it. Likewise, years later (Kojima et al. 2008) studied the relationship and partnership of JIT with the performance of the supply chain. In this study, JIT was merely considered as a part or component of the supply chain. Finally, (Inman et al. 2011) mentioned that one of the greatest benefits obtained from applying JIT philosophy was the agility in the movement of raw materials along the entire production system. All these improvements or benefits derived from the implementation of JIT sooner or later translated into an indication of efficiency.
4. *Increased process flexibility*. Process flexibility is not merely due to a successful implementation of JIT, but also to many other techniques, tools, or actions applied simultaneously in the process, such as preventive maintenance programs, systems to prevent errors, and proper training of operators in various activities of the production process. Technically speaking, flexibility refers to the ability of companies to make adjustments to the products by integrating or removing new activities with new features. However, flexibility must not be

confused with agility. In fact, while flexibility refers to the volume and diversified range of products, agility relies on time as its main source.

Authors White and Prybutok (2001) carried out an extended study concerning the best practices in JIT philosophy and the performance indicators of a company. The study considered flexibility as an important factor for both aspects. In addition, (Fullerton and McWatters 2002) identified human resources as a source of flexibility and proposed an incentive plan based on production and results in an environment of JIT. Furthermore, (Maiga and Jacobs 2009) studied the different sources of flexibility in a company. They proposed a number of relationships between JIT and the performance indices of a company and found flexibility as one of the greatest benefits obtained from this relation. More recently, (Inman et al. 2011) proposed a structural equation model in which they cited flexibility as one of the most pursued operational benefits by companies. If businesses are flexible in their processes, they can be able to manufacture a wide range of products using few machines and equipment.

5. *Reduced production lead time*. The delivery time of finished products is primarily a result of time reduction in the production process. In order to reduce production lead time, companies must remove uncertainties along their production processes, starting with the suppliers of raw materials, because they are an external partner in supply chain (Hill and Vollmann 1986). Achieving a reduction in delivery time is one of the most important indications of efficiency and productivity, and often guarantees companies an early entry into new markets (Yasin et al. 1997).

   The most relevant at this time might be to determine which elements promote the reduction of production lead time, which is one objective of this book. However, it is likely that on-time deliveries by suppliers, production process agility, and multiple trainings of operators or human resources be some of these elements, although a statistical proof of this would be highly desirable.

6. *Reduced of work in progress*. If this benefit is achieved throughout the production process, all machinery and equipment used, as well as human resources, would be highly efficient. Moreover, the material would continuously and steadily flow, and it shall not be forgotten that the continuous flow of materials is one of the most important goals of JIT.

   As it was stated by Sewell (1990), some of the elements that can promote the reduction of work in progress relate to the information technologies implemented throughout the production system, since they allow companies to be familiar with the location of the material in the production process at all times, which facilitates management and decision-making.

7. *Integration of different manufacturing activities*. Integrating JIT and synchronizing processes along the production lines are not easy tasks (Bäck 1985). The ideal scenario would be one where production lines were always completely balanced, but this may not happen often. As a result, companies tend to cluster or integrate manufacturing activities to achieve a better balance in production lines (Banerjee and Armouti 1992). The integration of these activities requires

an arduous planning through a logical clustering often supported by information technologies. An example of this is the material requirements planning (MRP) (Benton and Shin 1998), where coordination in the material flow plays an important role.

The integration of these manufacturing activities is frequently based on their similarities, the skills required from the different human resources along the production line, or the production strategies applied to the system (Xiong and Nyberg 2000).

Some authors mention that the integration of manufacturing activities is one of the most important benefits derived from JIT implementation, since companies obtain agility in their supply chains as a result of it. Agility is an efficiency rating since it measures the speed of a company to respond to the demands of customers (Inman et al. 2011).

3.3 JIT Benefits Associated with Engineering

According to authors found in literature, Table 3.3 lists four perceived benefits related to engineering. The elements were arranged according to the number of references obtained for every one of them.

1. *Reduced space requirements*. As it was mentioned in previous chapters, several case studies have reported their success with the implementation of JIT philosophy. These companies have reduced and/or eliminated their inventories of raw material, which consequently lead to a reduction in space requirements.
   Since there is a great flow of raw materials and subproducts along the production line, companies do not require their temporary storage. As a result, they benefit from a decrease in space requirements (Emde and Boysen 2012; Fullerton and McWatters 2001). The technique of reducing space requirements emerged in Japan, where the space available is extremely limited for companies

Table 3.3 JIT benefits obtained in engineering

JIT benefits	Authors	Total
Reduced space requirements	(Prodipto 1999; Vrat et al. 1993; Hong et al. 1992; Muralidharan et al. 2001b; Guinipero and Law 1990; Daesung and Seung-Lae 1997; Bartezzaghi et al. 1992)	7
Reduced movement distances	(Singh 1989; Bonito 1990; Voss 1986; Ebrahimpour and Schonberger 1984; Garg et al. 1996a)	5
Quick responses to engineering changes	(Muralidharan et al. 2001b; Hall 1983; Martel 1993)	3
Increased innovation	(Hall 1983; Kumar et al. 2001)	2

and people. The synchronization and high fluidity of materials was crucial as a benefit of JIT because it represented considerable space saving.
In order to achieve this synchronization of activities, companies now should draw upon the implementation of many information systems, such as the MRP, which is one of the main bases of JIT philosophy in the West, more specifically in the United States and Europe (White and Prybutok 2001).

2. *Reduced movement distances*. This benefit is related to the previous one, since reducing movement distances of the required materials and raw materials demands less space for the production process. A process diagram contains several symbols to represent the production process. One of these symbols is an arrow indicating the transport of the materials. Thus, the goal is to reduce them as much possible.
The objective of reducing the movement of raw materials during the production process is to avoid adding value to the product and only add a cost. Therefore, (O'Neal 1989) stated that a reduction of distances and motions of materials, and the integration of the entire company in JIT philosophy, could produce a rapid approach to the market, and, therefore, a commercial entrance and positioning for the company. Also, (White and Pearson 2001) mentioned that through proper implementation of JIT, movement distances were a natural consequence, which, according to Moattar Husseini et al. (2006), is simply greater flexibility in the production process. However, companies must follow a set of rules to obtain these benefits (Hum and Lee 1998). Based on these rules, a number of proposals can be followed for a proper production scheduling with a minimum motion of materials.
3. *Quick responses to engineering changes*. Preventive maintenance programs are a key element in JIT philosophy, so responses to design changes and product modifications often occur rapidly. Another element that is commonly found as a requirement for the implementation of JIT philosophy is the single-minute exchange of die (SMED), which seeks to make quick modifications in machinery and equipment when changes occur in the production line. One study performed by Gupta et al. (1992) compared the benefits of companies with and without JIT and found that there was significant difference between those that had applied JIT philosophy and those that had not.
Similarly, (Yasin et al. 1997) mentioned that these quick changes in engineering were simple; that is, with no administrative management since supervisors usually trusted their employees, who were proficient at work. More recently, (Amasaka 2007) indicated that the Toyota plant was able to make big changes and innovate in their product lines due to the flexibility offered by JIT philosophy. Changes are often a limiting factor for some companies when making modifications in their production systems, since when these changes occur slowly, companies analyze their cost from an economic point of view. As a result, quick responses to engineering changes may not be purely the result of technological innovation in the production process and flexibility. They also depend on the administrative flexibility of every company (Yasin et al. 2003).

4. *Increased innovation*. When it is possible to make quick changes in the production process to adapt to the new specifications of a product, innovation can become limited (Alcaraz et al. 2014). One might remember the previous example of Henry Ford, who mentioned that he could paint a car of any color as long as it was black. Ford was aware of the fact that changes in the production line were not fast and he wished to manufacture serial products at a low cost. This meant that he could not make innovations in his product because of the high cost of making changes to the production line.

3.4 Benefits Associated with Quality

Quality is traditionally accepted as the ability to deliver a product with the required technical and quantity specifications, when it is required, and at a fair price. The implementation of JIT philosophy can support this concept of quality, since it seeks to achieve on-time deliveries.

According to authors from the literature reviewed, Table 3.4 lists the three benefits identified and directly associated with quality. The first column of the table lists every benefit, while the second includes the name of authors who referred to each one. Finally, the third column specifies the total number of authors who addressed every benefit.

1. *Increased product quality*. Although JIT traditionally emphasizes on materials handling, it also aims at the elimination of waste throughout the production system; and this contributes to the quality of products and processes. Numerous studies have demonstrated the relationship between JIT philosophy and total quality achievement, and some more have sought to associate this philosophy with the success of companies with certain quality certifications (Withers et al. 1997). Similarly, other pieces of research have found not only a relation between JIT and quality programs, but also between JIT and the total preventive

Table 3.4 JIT benefits associated with quality

JIT benefits	Authors	Total
Increased product quality	(Singh 1989; Hall 1983; Padukone and Subba 1993; Prodipto 1999; Vrat et al. 1993; Hong and Hayya 1992; Muralidharan et al. 2001b; Martel 1993; Daesung and Seung-Lae 1997; Kumar and Garg 2000; Bartezzaghi et al. 1992; Kumar et al. 2001; Voss and Robinson 1987; Priestman 1985; Dutton 1990; Flynn et al. 1995)	17
Paperwork reduction	(Bartezzaghi et al. 1992; Kumar and Garg 2000; Ebrahimpour and Schonberger 1984; Garg et al. 1996a)	4
Higher quality of production process	(Singh 1989; Padukone and Subba 1993; Vrat et al. 1993)	3

maintenance programs for machinery and equipment (Cua et al. 2001). More recently, another study found that two of the most important effects of JIT implementation over a production system were product quality and improved processes (Maiga and Jacobs 2009).

In a strict sense, product quality emerges from people and workers in the bottom line of the production and efficient machinery and equipment. Therefore, if companies invest in proper education and training for their staff and implement preventative maintenance programs to increase flexibility, they will likely obtain high-quality products.

2. *Paperwork Reduction.* One might be skeptical about the fact that companies can benefit from less paperwork with a successful implementation of JIT. However, administrative simplification in companies is part of product quality, since many administrative procedures exist due to a lack of trust and proper integration between suppliers and manufacturers. The better the integration between the customer companies and the supplier, the larger the administrative simplification, which thus implies a reduction of paperwork. However, if this integration is not complete, companies may need numerous and tedious administrative procedures.

   In order to achieve high levels of integration between suppliers and manufacturers, some authors have recommended applying theories based on organizational change (Vora 1992). However, a suitable option to achieve a proper level of integration with suppliers is the implementation of shared information systems, such as MRP. This would allow both parts to become familiar with the status of the production system (Benton and Shin 1998).
3. *Production process with higher quality.* A quality product is the result of a quality production process, which means that quality is not actually born. It is created by workers or human resources that interact with machines, equipment, and data during the production process. There are several lean manufacturing tools that can be applied to the production process to improve it, such as total productive maintenance (TPM), card system or Kanban, rapid change programs, and continuous improvement programs. All of them required elements of JIT.

3.5 Benefits Associated with Materials Handling

Although JIT brings many kinds of benefits, it must be recognized that those associated with materials handling seem to be one of the most important. Table 3.5 lists these benefits, which were identified after a literature reviewing process. It must be noted that authors reported here are fewer than those reported in Tables 3.1, 3.2, and 3.3. A maximum of 7 references were found for elements in this new category, while previous categories listed a minimum of 10 references for each

Table 3.5 JIT benefits associated with materials handling

JIT benefits	Authors	Total
Reduced part numbers	(Prodipto 1999; Vrat et al. 1993; Hong and Hayya 1992; Daesung and Seung-Lae 1997; Kumar and Garg 2000; Voss 1986; Garg 1997)	7
Reduced inventory	(Hong et al. 1992; Muralidharan et al. 2001b; Guinipero and Law 1990; Macbeth et al. 1988; Daesung and Seung-Lae 1997; Carter and Narsimhan 1996; Bartezzaghi et al. 1992)	7
Reduced materials handling	(Singh 1989; Vrat et al. 1993; Dutton 1990; Ebrahimpour and Schonberger 1984)	4
Close supplier–customer relations	(Sakuri 1986; Parnaby 1988; Macbeth et al. 1988; Martel 1993).	4
Reduced lot sizes purchased	(Macbeth et al. 1988; Baker et al. 1994)	2
Increased inventory rotation	(Guinipero and Law 1990)	1

benefit. This rejects the traditional assumption that JIT philosophy is merely associated with materials handling.

1. *Reduced part numbers*. When a product is required, a work order must be set. This system requires efficient and opportune communication between suppliers and manufacturers. Such high-quality communication can be achieved through the proper integration of the information technologies and production systems that these two parts share. Proper integration between suppliers and manufacturers helps develop a sense of trust and commitment, leading to certified operations and, consequently, a reduction in the number of parts. This also reduces uncertainty in the two organizations (Hill and Vollmann 1986).
   Uncertainty can be reduced by the efficient flow of information between suppliers and manufacturers. This efficient flow can be supported, again, through the proper management of information and communication technologies and helps both organizations become familiar with the needs and abilities of each other (Sewell 1990; Huq and Huq 1994). Furthermore, a reduced number of parts or components for a product make the management system straightforward for both customers and vendors. Thus, less paperwork is required, which also represents a reduction of costs.
2. *Reduced inventory*. Another of the most desirable benefits of JIT system is the reduction of inventory from the beginning of the production system, along the transformation process, and until the product is finished (Balakrishnan et al. 1996). One may remember at this point the cases of those companies introduced in the first chapter who have succeeded with JIT. Some of them actually eliminated inventories of raw materials to try to reduce the inventory management and high administrative costs (Lovell 2003). Today, companies search for this balance between having an adequate inventory level of raw materials and the marginal costs of not having a product, since the latter may represent losing

customers (Min and Sui Pheng 2006). This means that sometimes it may be suitable to rely on minimum inventories even when they may represent an additional cost.

Several mathematical models nowadays help determine the optimum inventory size for the effectiveness of JIT environments. Some of these models can be found in Cao and Schniederjans (2004), and Min and Sui Pheng (2006). To achieve the optimal reduction of lot sizes, there must exist proper integration between the production systems of both suppliers and buyers. Otherwise, great uncertainty may arise and the two parts could be forced to maintain high volumes of inventory (Cao and Schniederjans 2004; Lovell 2003). Thus, both parts should be highly integrated with each other in order to minimize inventories resulting from the uncertainty in demand that could arise.

3. *Reducing materials handling*. Transportation of materials throughout the production system does not add value to the product, which is the reason that organizations always seek to reduce it to its lowest level. Besides, it shall not be forgotten that a proper distribution plant is one of the most important elements for the success of a JIT program (White and Prybutok 2001; Halim et al. 2012). In their work, (Fullerton and McWatters 2001) stated that one of the economic benefits that companies could obtain from JIT is the reduction of costs due to a better distribution of the production system and the efficient management of materials. Therefore, these two practices should be continuously applied in companies seeking to increase their profits (White and Prybutok 2001).

   Similarly, (Ahmad et al. 2003) stated that in addition to the physical layout, another key element to reduce costs of materials handling was the infrastructure of production systems, which should be as modern and flexible as possible to be quickly adapted to other production systems. Also, some researchers have mentioned that materials handling should make use of forces, such as gravity, to reduce its cost (Halim et al. 2012), and that companies should rely on automated transportation systems since they could improve the efficiency, but specially safety, of production systems (Kesen and Baykoç 2007).

4. *Close relations supplier/customer*. When companies successfully implement JIT in their production systems, it means that their suppliers certainly played an important role and the processes that both parts share are highly integrated, which is chiefly due to their close relationship. Also, these companies with successful JIT implementation often rely on a few selected suppliers and their levels of trust are high.

   Such close relationships between suppliers and manufacturers have been studied over many years. One of the first works approaching the topic was conducted by Imrie and Morris (1992), although (Gules and Burgess 1996) carried out a specific case study applied to advanced manufacturing technologies. Similarly, (Dong et al. 2001) proposed a descriptive analysis of the expectations from suppliers and manufacturers concerning the raw materials purchasing process. They found that manufacturing companies needed reliable suppliers that could also integrate into their production systems. Furthermore, (David and

Eben-Chaime 2003) researched on the suitable degree of closeness in the relationships between suppliers and manufacturers when a JIT system was installed. Finally, (Hald et al. 2009) introduced an analysis of these relationships within the context of globalization, where the geographical distance between suppliers and manufacturers is often large.

5. *Reduced lot sizes purchased.* When high integration in the production system exists between suppliers and manufacturers, and they work to eliminate the uncertainty in demand, lot sizes purchased become smaller. This benefit has been vastly approached in the literature. For instance, (Khan and Sarker 2002) proposed a model to determine the optimal size of a purchase lot for a JIT production system. However, (Kim and Ha 2003) stated that in cases where a production order could not be completely delivered, it might be divided into two or more possible deliveries. Nevertheless, this may happen after a negotiation between the two parts involved (Kelle et al. 2003). Similarly, (Lovell 2003) introduced a specific and special case when studying the optimal size of purchase orders for a manufacturer that had implemented the JIT system but its business environment was totally monopolistic.
6. *Increased inventory rotation.* Inventory rotation has been one of the most important benefits reported from the implementation of JIT (Primrose 1992; Huson and Nanda 1995; Yasin et al. 1997), since its effect can be translated quickly without any complications in purely economic terms (Fullerton et al. 2003).
   Inventory rotation implies that companies count on inventory levels of raw materials for very short periods of time, which indicates a reduction in the administrative cost of these inventories and the implementation of laws of strong financial impact (Cannon 2008). Inventory rotation has been studied by academics through various techniques that aim to diminish the time and cost of storage. In their study, (Demeter and Matyusz 2011) analyzed inventory rotation as a result of the application of various lean manufacturing techniques in the production system. More specifically, (Pong and Mitchell 2012) carried out an analysis regarding the investment levels and inventory control policies that are followed to increase inventory rotation in companies from the United Kingdom. Inventory rotation implies two aspects: the reduction of administrative costs by keeping inventory only for short periods of time and lower investment in raw materials, which could be consumed quickly. In economic terms, the two can demonstrate that companies optimize their economic resources and capital is properly invested.

3.6 Economic Benefits Associated with JIT

Due to their high importance, the economic benefits of JIT are discussed as the last section. These benefits might justify to the greatest extent the implementation of JIT in companies. That is, if companies did not obtain any economic benefit from JIT,

Table 3.6 Economic benefits associated with JIT

JIT benefits	Authors	Total
Reduced production costs	(Singh 1989; Kumar and Garg 2000; Garg and Deshmukh 1999b; Chong and Rundus 2000; Dutton 1990; Garg 1997; Ebrahimpour and Schonberger 1984; Garg et al. 1996a)	8
Reduced labor costs	(Guinipero and Law 1990; Bartezzaghi et al. 1992; Vuppalapati et al. 1995; Priestman 1985; Garg and Deshmukh 1999a; Garg et al. 1994, 1996; Ebrahimpour and Schonberger 1984)	8
Reduced general expenses	(Singh 1989; Hall 1983; Prodipto 1999; Vrat et al. 1993; Kumar and Garg 2000; Garg and Deshmukh 1999b)	6
Reduced product costs	(Hall 1983; Bartezzaghi et al. 1992)	2
Improved competitiveness	(Bartezzaghi et al. 1992)	1
Increased profit margin	(Bartezzaghi et al. 1992)	1

the philosophy may not be implemented, since it would surely represent a cost for these organizations. Table 3.6 lists the economic benefits identified from JIT implementation according to literature reviewed.

1. *Reduced production costs*. Previous sections of this chapter have discussed JIT benefits such as the reduction of materials transportation and materials handling, the improvement of the distribution in plants for the production system, and a high inventory rotation. All these benefits can probably be translated into the reduction of production costs for businesses, although it must be statistically verified. However, the reduction of production costs will reflect on lower prices for the customer and thus a better market positioning for the company. This is why the administration and enforcement of JIT focuses on reducing costs and eliminating waste as much as it can be possible.
   The economic impact of JIT has been widely studied; nevertheless, certain companies have reported operative problems during its implementation. Thus, before implementing JIT philosophy in their processes, these organizations may have to carry out an economic analysis. Then, they should propose the initiative to the senior management and demonstrate JIT efficiency in the production systems (Miltenbirg 1990).
   Authors Brox and Fader (1997) sought to justify the implementation of JIT system in companies by relying on economic theory. However, the work of Fullerton et al. (2003) can be considered another relevant piece of research concerning JIT, since it analyzed its implementation from an economic or financial perspective. Similarly, three years later (Cannon 2008) conducted a study emphasizing on JIT and its economic benefits in inventory management of production systems. Finally, (Maiga and Jacobs 2009) established aspects of and relations between JIT and the operational and financial performance of companies.

2. *Reduced labor costs*. Reduction of labor costs is another form of reducing waste in JIT. It can be achieved by reducing the tasks to perform by the workforce, transporting raw materials, or simply because operations on raw materials become more efficient.
   In their work, (Wisner and Stanley 1994) described new business practices to apply JIT philosophy and their benefits. They noted a significant reduction in labor cost, especially due to the levels of education in companies implementing JIT. Similarly, (Brox and Fader 1997) carried out an analysis of the economic effects of JIT and reported that the reduction of labor cost is one of the benefits gained from JIT implementation.
   Finally, (Yasin et al. 1997) studied the efficiency and effectiveness of JIT within a company whose economic benefits were not clearly visible. However, those obtained from human resources were highly valued and reflected on their efficiency levels, since operators developed the skills to perform more than one task and hold various positions. They had become multifunctional, which is a benefit usually not observed. However, since in a JIT environment low and high levels of the organization build elevated levels of trust between them, companies rely on a very flat organizational structure, indicating that a few people are required to run the business and every worker is familiar with his/her responsibilities. This also represents a reduction of cost in administrative labor (Yasin et al. 2003).
3. *Less general expenses*. Several of the benefits obtained from JIT are reflected economically, which to a great extent can justify its implementation. In the decade of 1980, the difference between companies that applied JIT and those who did not was particularly visible, especially when manufacturing systems were audited (Ball et al. 2015).
   Some researchers directly associate the economic impact of JIT with the performance of the company and its physical growth and ongoing reputation (Alcaraz et al. 2014; Green et al. 2014). Similarly, other studies addressing the economic benefits obtained from the successful implementation of JIT programs can be found in Fullerton and McWatters (2001), Fullerton et al. (2003), Cua et al. (2001), and Inman et al. (2011).
4. *Reduced product cost*. The elimination of waste along a JIT production system can be translated into a lower cost of the finished product and will represent a competitive advantage for the company. Thus, the quality of a product should not be neglected.
   Great part of this reduced cost is the result of lower inventory management in warehouses and production processes, a better plant distribution, and more capable and multifunctional employees, among others. However, cost reduction is not only visible in the production line when the product design was established; it can also be perceived from the phase of design and planning (Banerjee and Armouti 1992) (Meybodi 2003), when companies could set a rigorous plan with the expectations for the product-to-be (Benton and Shin 1998).

5. *Improve competitiveness*. Competitive advantage can be perceived as the increased product quality with the lowest production cost, and the least materials handling by operators (replaced by automated systems that frequently result in a minimization of accidents). Thus if JIT implementation does not offer a competitive advantage, it may be wiser not to implement it. However, it should be noted that the only benefit usually visible for customers is the sale price. Businesses frequently consider JIT program as a failure in early stages of its implementation, since important economic investments must be undertaken and are generally not reflected in short-time profitability (Fullerton and McWatters 2001). In fact, companies may mistakenly assume that competitive advantage and commercial benefits of JIT will be reflected within a short period of time (Oral et al. 2003). However, they should first concentrate on eliminating waste. Afterward, they can move on to a new stage to improve product quality and obtain lower production costs that imply a better market positioning (Rentroia-Bonito et al. 2005).
6. *Profit margin increase*. Higher profit margin with JIT implemented in production systems is only a result of the proper practices applied. The relationship between JIT and the financial performance of the company and the profits earned has been extensively studied. For instance, (Balakrishnan et al. 1996; Inman et al. 2011) explained that pure economic benefits are significantly visible in audits run to the production systems, and these economic benefits entail a higher profit margin for the company. Similarly, (Inman et al. 2011) introduced an economic justification for JIT implementation in production systems for companies that were not fully convinced to apply it due to a lack of economic indications of improved performance. Furthermore, years ago, (Brox and Fader 1997) found that JIT had a clear impact on the financial statements of companies who relied on it. In addition, (Balakrishnan et al. 1996) carried out a detailed analysis of the economic and financial benefits of companies with JIT. They also expanded their research and sought to find a commercial image that companies could reflect to their customers. Finally, (Fullerton et al. 2003) proposed one of the most recent works regarding the economic benefits that companies could obtain from JIT systems.

References

Ahmad S, Schroeder RG, Sinha KK (2003) The role of infrastructure practices in the effectiveness of JIT practices: implications for plant competitiveness. J Eng Technol Manage 20(3):161–191. doi:10.1016/S0923-4748(03)00017-1

Ahmad MF, Zakuan N, Jusoh A, Takala J (2012) Relationship of TQM and business performance with mediators of SPC, lean production and TPM. Proc Soc Behav Sci 65:186–191. doi:10.1016/j.sbspro.2012.11.109

Alcaraz JLG, Maldonado AA, Iniesta AA, Robles GC, Hernández GA (2014) A systematic review/survey for JIT implementation: Mexican maquiladoras as case study. Comput Ind 65 (4):761–773. doi:10.1016/j.compind.2014.02.013

Amasaka K (2007) Applying new JIT—Toyota's global production strategy: epoch-making innovation of the work environment. Robot Comput Integr Manuf 23(3):285–293. doi:10.1016/j.rcim.2006.02.001

Bäck H (1985) Decentralized integration—advantage or disadvantage of logistics and JIT-concepts. Comput Ind 6(6):529–541. doi:10.1016/0166-3615(85)90033-8

Baker RC, Chang RE, Chang IC (1994) Switching rules for JIT purchasing. Prod Inventory Manage J 35(3):13–17

Balakrishnan R, Linsmeier T, Venkatakchalam M (1996) Financial benefits from JIT adoption: effects of customer concentration and cost structure. Acc Rev 71(2):183–205

Ball F, Tyler J, Wells P (2015) Is audit quality impacted by auditor relationships? J Contemp Acc Econ 11(2):166–181. doi:10.1016/j.jcae.2015.05.002

Banerjee P, Armouti HA (1992) JIT approach to integrating production order scheduling and production activity control. Comput Integr Manuf Syst 5(4):283–290. doi:10.1016/0951-5240(92)90045-E

Banerjee S, Golhar DY (1993) EDI implementation in JIT and non-JIT manufacturing firms: a comparative study. Int J Oper Prod Manage 13(3):25–37

Bartezzaghi E, Turco F, Spina G (1992) The impact of the just-in-time approach on production system performance: a survey of Italian industry. Int J Oper Prod Manage 12(1):5–17. doi:10.1108/EUM0000000001292

Benton WC, Shin H (1998) Manufacturing planning and control: the evolution of MRP and JIT integration. Eur J Oper Res 110(3):411–440. doi:10.1016/S0377-2217(98)00080-0

Bernard KN (1996) Just-in-time as a competitive weapon: the significance of functional integration. J Mark Manage 12(6):581–597

Bonito J (1990) Motivating employees for continuous improvement efforts. Prod Inventory Manage Rev APICS News 8:225–236

Brox JA, Fader C (1997) Assessing the impact of JIT using economic theory. J Oper Manage 15(4):371–388. doi:10.1016/S0272-6963(97)00005-3

Cannon AR (2008) Inventory improvement and financial performance. Int J Prod Econ 115(2):581–593. doi:10.1016/j.ijpe.2008.07.006

Cao Q, Schniederjans MJ (2004) A revised EMQ/JIT production-run model: an examination of inventory and production costs. Int J Prod Econ 87(1):83–95. doi:10.1016/S0925-5273(03)00094-X

Carter JR, Narsimhan R (1996) Is price really strategic? Int J Purchasing Mater Manage 32(1):20–28

Chong V, Rundus M (2000) The effect of new manufacturing practices and intensity of market competition on organizational performance: an empirical investigation. Paper presented at the international conference on management research accounting research and case symposium, Mesa, 6–8 Jan 2000

Cua KO, McKone KE, Schroeder RG (2001) Relationships between implementation of TQM, JIT, and TPM and manufacturing performance. J Oper Manage 19(6):675–694. doi:10.1016/S0272-6963(01)00066-3

Daesung HA, Seung-Lae K (1997) Implementation of JIT purchasing: an integrated approach. Prod Plann Control 8(2):152–157

David I, Eben-Chaime M (2003) How far should JIT vendor–buyer relationships go? Int J Prod Econ 81–82:361–368. doi:10.1016/S0925-5273(02)00274-8

Demeter K, Matyusz Z (2011) The impact of lean practices on inventory turnover. Int J Prod Econ 133(1):154–163. doi:10.1016/j.ijpe.2009.10.031

Dong Y, Carter CR, Dresner ME (2001) JIT purchasing and performance: an exploratory analysis of buyer and supplier perspectives. J Oper Manage 19(4):471–483. doi:10.1016/S0272-6963(00)00066-8

Dutton B (1990) Switching to quality excellence. Manuf Syst 8:245–256

Dwivedula R, Bredillet CN (2010) Profiling work motivation of project workers. Int J Proj Manage 28(2):158–165. doi:10.1016/j.ijproman.2009.09.001

Ebrahimpour M, Schonberger RJ (1984) The Japanese just-in-time/total quality control production system: potential for developing countries. Int J Prod Res 22:421–430

Emde S, Boysen N (2012) Optimally routing and scheduling tow trains for JIT-supply of mixed-model assembly lines. Eur J Oper Res 217(2):287–299. doi:10.1016/j.ejor.2011.09.013

Fiedler K, Galletly JE, Bicheno J (1993) Expert advice for JIT implementation. Int J Oper Prod Manage 13(6):23–30. doi:10.1108/01443579310038994

Flynn B, Salakibara S, Schroeder R (1995) Relationship between JIT and TQM practices and performance. Acad Manage J 38(5):1325–1360

Fullerton RR, McWatters CS (2001) The production performance benefits from JIT implementation. J Oper Manage 19(1):81–96. doi:10.1016/S0272-6963(00)00051-6

Fullerton RR, McWatters CS (2002) The role of performance measures and incentive systems in relation to the degree of JIT implementation. Acc Organ Soc 27(8):711–735. doi:10.1016/S0361-3682(02)00012-0

Fullerton RR, McWatters CS, Fawson C (2003) An examination of the relationships between JIT and financial performance. J Oper Manage 21(4):383–404. doi:10.1016/S0272-6963(03)00002-0

García-Alcaraz J, Prieto-Luevano D, Maldonado-Macías A, Blanco-Fernández J, Jiménez-Macías E, Moreno-Jiménez J (2015) Structural equation modeling to identify the human resource value in the JIT implementation: case maquiladora sector. Int J Adv Manuf Technol 77(5–8):1483–1497. doi:10.1007/s00170-014-6561-5

Garg D (1997) Relevance of JIT purchasing in Indian industries. Kurukshetra University, Kurukshetra

Garg D, Deshmukh SG (1999a) JIT concepts and relevance in Indian context. Ind Eng J 28(1):14–18

Garg D, Deshmukh SG (1999b) JIT purchasing: literature review and implications for Indian industry. Prod Plann Control 10(3):276–285. doi:10.1080/095372899233235

Garg S, Vart P, Kanda A (1994) Work culture in JIT environment. Prod J 35(3):463–466

Garg D, Deshmukh S, Kaul O (1996a) Critical analysis in JIT purchasing in Indian context. Prod J 37:271–279

Garg D, Deshmukh SG, Kaul ON (1996b) Parametric analysis of inventory models under JIT purchasing analysis of inventory models under JIT purchasing agreements. Paper presented at the National conference operational research in modern technology, REC Kurukshetra, 8–9 Mar 1996

Globerson S, Korman A (2001) The use of just-in-time training in a project environment. Int J Proj Manage 19(5):279–285. doi:10.1016/S0263-7863(00)00012-0

Green KW Jr, Whitten D, Inman RA (2012) Aligning marketing strategies throughout the supply chain to enhance performance. Ind Mark Manage 41(6):1008–1018. doi:10.1016/j.indmarman.2012.02.003

Green KW Jr, Inman RA, Birou LM, Whitten D (2014) Total JIT (T-JIT) and its impact on supply chain competency and organizational performance. Int J Prod Econ Part A 147:125–135. doi:10.1016/j.ijpe.2013.08.026

Guinipero LC, Law WK (1990) Organizational change and JIT implementation. Prod Inventory Manage J 31(3):71–73

Gules HK, Burgess TF (1996) Manufacturing technology and the supply chain: linking buyer-supplier relationships and advanced manufacturing technology. Eur J Purchasing Supply Manage 2(1):31–38. doi:10.1016/0969-7012(95)00014-3

Gupta YP, Lonial SC, Glynn Mangold W (1992) Empirical assessment of the strategic orientation of JIT manufacturers versus non-JIT manufacturers. Comput Integr Manuf Syst 5(3):181–189. doi:10.1016/0951-5240(92)90029-C

Hald KS, Cordón C, Vollmann TE (2009) Towards an understanding of attraction in buyer–supplier relationships. Ind Mark Manage 38(8):960–970. doi:10.1016/j.indmarman.2008.04.015

Halim NHA, Jaffar A, Yusoff N, Adnan AN (2012) Gravity flow rack's material handling system for just-in-time (JIT) production. Proc Eng 41:1714–1720. doi:10.1016/j.proeng.2012.07.373

Hall R (1983) Zero inventories. Dow Jones-Irwin, Homewood

Hill AV, Vollmann TE (1986) Reducing vendor delivery uncertainties in a JIT environment. J Oper Manage 6(3–4):381–392. doi:10.1016/0272-6963(86)90011-2

Hong J-D, Hayya JC (1992) Just-in-time purchasing: single or multiple sourcing? Int J Prod Econ 27(2):175–181. doi:10.1016/0925-5273(92)90011-U

Hong J-D, Hayya JC, Kim S-L (1992) JIT purchasing and setup reduction in an integrated inventory model. Int J Prod Res 30(2):255–266. doi:10.1080/00207549208942893

Hum S-H, Lee C-K (1998) JIT scheduling rules: a simulation evaluation. Omega 26(3):381–395. doi:10.1016/S0305-0483(97)00066-2

Huq Z, Huq F (1994) Embedding JIT in MRP: the case of job shops. J Manuf Syst 13(3):153–164. doi:10.1016/0278-6125(94)90001-9

Huson M, Nanda D (1995) The impact of just-in-time manufacturing on firm performance in the US. J Oper Manage 12(3–4):297–310. doi:10.1016/0272-6963(95)00011-G

Imrie R, Morris J (1992) A review of recent changes in buyer-supplier relations. Omega 20(5–6):641–652. doi:10.1016/0305-0483(92)90008-U

Inman RA, Sale RS, Green KW Jr, Whitten D (2011) Agile manufacturing: relation to JIT, operational performance and firm performance. J Oper Manage 29(4):343–355. doi:10.1016/j.jom.2010.06.001

Kelle P, Al-khateeb F, Anders Miller P (2003) Partnership and negotiation support by joint optimal ordering/setup policies for JIT. Int J Prod Econ 81–82:431–441. doi:10.1016/S0925-5273(02)00357-2

Kesen SE, Baykoç ÖF (2007) Simulation of automated guided vehicle (AGV) systems based on just-in-time (JIT) philosophy in a job-shop environment. Simul Model Pract Theor 15(3):272–284. doi:10.1016/j.simpat.2006.11.002

Khalatbari J, Ghorbanshiroudi S, Firouzbakhsh M (2013) Correlation of job stress, job satisfaction, job motivation and burnout and feeling stress. Proc Soc Behav Sci 84:860–863. doi:10.1016/j.sbspro.2013.06.662

Khan LR, Sarker RA (2002) An optimal batch size for a JIT manufacturing system. Comput Ind Eng 42(2–4):127–136. doi:10.1016/S0360-8352(02)00009-8

Kim S-L, Ha D (2003) A JIT lot-splitting model for supply chain management: enhancing buyer–supplier linkage. Int J Prod Econ 86(1):1–10. doi:10.1016/S0925-5273(03)00006-9

Kojima M, Nakashima K, Ohno K (2008) Performance evaluation of SCM in JIT environment. Int J Prod Econ 115(2):439–443. doi:10.1016/j.ijpe.2007.11.017

Kumar V, Garg D (2000) JIT elements in Indian context: an analysis. Prod J 41(2):217–222

Kumar V, Garg D, Mehta N (2001) JIT based quality management in Indian industries: prospectus and future directions. Paper presented at the National conference supply chain management. Institute of Public Enterprise, Hyderabad, 19–20 Jan

Kuo-Ting H, Ro YK, Liker JK (2009) Further motivation for continuous improvement in just-in-time logistics. IEEE Trans Eng Manage 56(4):571–583. doi:10.1109/TEM.2009.2030336

Landherr M, Constantinescu C (2012) Intelligent management of manufacturing knowledge: foundations, motivation scenario and roadmap. Proc CIRP 3:269–274. doi:10.1016/j.procir.2012.07.047

Lau CM, Roopnarain K (2014) The effects of nonfinancial and financial measures on employee motivation to participate in target setting. Br Acc Rev 46(3):228–247. doi:10.1016/j.bar.2014.02.006

Lee SM, Ebrahimpour M (1985) Just-in-time production system: some requirements for implementation. Int J Oper Prod Manage 4(4):3–15

Lovell MC (2003) Optimal lot size, inventories, prices and JIT under monopolistic competition. Int J Prod Econ 81–82:59–66. doi:10.1016/S0925-5273(02)00349-3

Macbeth DK, Baxter L, Farguson N, Neil G (1988) Buyer-vendor relationship with just-in-time: lessons from US multinationals. Ind Eng 20(9):38–41

Macheka L, Manditsera FA, Ngadze RT, Mubaiwa J, Nyanga LK (2013) Barriers, benefits and motivation factors for the implementation of food safety management system in the food sector in Harare Province. Zimbabwe Food Control 34(1):126–131. doi:10.1016/j.foodcont.2013.04.019

Maiga AS, Jacobs FA (2009) JIT performance effects: a research note. Adv Acc 25(2):183–189. doi:10.1016/j.adiac.2009.06.003

Martel M (1993) The role of just-in-time purchasing in Dynapert`s transition to world class manufacturing. Prod Inventory Manage J 34:71–76

McLachlin R (1997) Management initiatives and just-in-time manufacturing. J Oper Manage 15(4):271–292. doi:10.1016/S0272-6963(97)00010-7

Meybodi MZ (2003) Using principles of just-in-time to improve new product development process. Adv Compet Res 11(1):116–138

Miltenbirg G (1990) Changing MRP's costing procedure to suit JIT. Prod Inventory Manage J 31(2):77–83

Miltenburg J (1993) A theoretical framework for understanding why JIT reduces cost and cycle time and improves quality. Int J Prod Econ 30–31:195–204. doi:10.1016/0925-5273(93)90091-X

Min W, Sui Pheng L (2006) EOQ, JIT and fixed costs in the ready-mixed concrete industry. Int J Prod Econ 102(1):167–180. doi:10.1016/j.ijpe.2005.03.002

Moattar Husseini SM, O'Brien C, Hosseini ST (2006) A method to enhance volume flexibility in JIT production control. Int J Prod Econ 104(2):653–665. doi:10.1016/j.ijpe.2005.05.025

Monden Y (2002) The relationship between mini profit-center and JIT system. Int J Prod Econ 80(2):145–154. doi:10.1016/S0925-5273(02)00314-6

Muralidharan C, Anantharaman N, Deshmukh SG (2001a) Vendor rating in purchasing scenario: a confidence interval approach. Int J Oper Prod Manage 21(10):1305–1326. doi:10.1108/01443570110404736

Muralidharan C, Anantharaman N, Deshmukh SG (2001b) Vendor rating in purchasing scenario: a confidence interval approach. Int J Oper Prod Manage 21(9/10):1305–1326

Nassimbeni G (1996) Factors underlying operational JIT purchasing practices: results of an empirical research. Int J Prod Econ 42(3):275–288. doi:10.1016/0925-5273(95)00190-5

O'Neal CR (1989) JIT procurement and relationship marketing. Ind Mark Manage 18(1):55–63. doi:10.1016/0019-8501(89)90021-7

Oral EL, Mıstıkoglu G, Erdis E (2003) JIT in developing countries—a case study of the Turkish prefabrication sector. Build Environ 38(6):853–860. doi:10.1016/S0360-1323(03)00021-0

Padukone H, Subba R (1993) Global status of JIT—implication for developing countries. Prod J 34(3):419–429

Parnaby J (1988) A systems approach to the implementation of JIT methodologies in lucas industries. Int J Prod Res 26(3):483–492. doi:10.1080/00207548808947878

Pong CKM, Mitchell F (2012) Inventory investment & control: how have UK companies been doing? Br Acc Rev 44(3):173–188. doi:10.1016/j.bar.2012.07.008

Priestman S (1985) SQC and JIT: partnership in quality, does culture make a difference? Qual Prog 18(5):31–35

Primrose PL (1992) Evaluating the introduction of JIT. Int J Prod Econ 27(1):9–22. doi:10.1016/0925-5273(92)90122-N

Prodipto R (1999) Data warehousing for textiles, texinfotech-99. Paper presented at the Techno park (I) Ltd., New Delhi

Rentroia-Bonito MA, Tribolet J, Jorge JA, Ghaoui C (2005) Just-in-time approach to learning: arguing the case for cost-effective knowledge dissemination. In: Ghaoui C, Jain M, Bannore V, Jain CL (eds) Knowledge-based virtual education, vol 178. Studies in fuzziness and soft computing. Springer, Berlin, pp 1–22. doi:10.1007/11399520_1

Roy R, KK G (1996) JIT: World scenario and possibility of its applicability in Indian industries. Paper presented at the National conference on operation research in modern technology, REC Kurukshetra, India, 8–9 Mar

Sakuri K (1986) Japanese worker atitudes: a key factor to productivity. Int J Oper Prod Manage 6(1):42–53

Sandanayake YG, Oduoza CF, Proverbs DG (2008) A systematic modelling and simulation approach for JIT performance optimisation. Robot Comput Integr Manuf 24(6):735–743. doi:10.1016/j.rcim.2008.03.013

Sewell G (1990) Management information systems for JIT production. Omega 18(5):491–503. doi:10.1016/0305-0483(90)90005-T

Singh A (1989) Just-in-time system: an integrated system. Prod J 30(3):309–314

Singh P, Bhandarkar A (1996) Paradigm shift in Indian industries: the need for tolerance of ambiguity. MDI Manage J 9:106–112

Singhvi S (1992) Employee involvement in JIT success: Eicher experience. Prod J 33:366–369

Vora JA (1992) Applying a theory of organization change to adopting JIT. Omega 20(2):193–199. doi:10.1016/0305-0483(92)90073-G

Voss CA (1986) Just-in-time manufacture. IFS Limited. Springer, London

Voss C, Robinson S (1987) Application of justin-time manufacturing techniques in United Kingdom. Int J Oper Prod Manage 7(4):46–52

Vrat P, Mittal S, Tyagi K (1993) Implementation of JIT in Indian environment: a Delhi study. Prod J 34:251–256

Vuppalapati K, Ahire SL, Gupta T (1995) JIT and TQM: a case for joint implementation. Int J Oper Prod Manage 15(5):84–94. doi:10.1108/01443579510083686

White RE, Pearson JN (2001) JIT, system integration and customer service. Int J Phys Distrib Logistics Manage 31(5):313

White RE, Prybutok V (2001) The relationship between JIT practices and type of production system. Omega 29(2):113–124. doi:10.1016/S0305-0483(00)00033-5

Wisner JD, Stanley LL (1994) Forecasting practices of JIT and non-JIT purchasers. Eur J Purchasing Supply Manage 1(4):219–225. doi:10.1016/0969-7012(95)00011-9

Withers BE, Ebrahimpour M, Hikmet N (1997) An exploration of the impact of TQM and JIT on ISO 9000 registered companies. Int J Prod Econ 53(2):209–216. doi:10.1016/S0925-5273(97)00118-7

Xiong G, Nyberg TR (2000) Push/pull production plan and schedule used in modern refinery CIMS. Robot Comput Integr Manuf 16(6):397–410. doi:10.1016/S0736-5845(00)00023-5

Yasin MM, Small M, Wafa MA (1997) An empirical investigation of JIT effectiveness: an organizational perspective. Omega 25(4):461–471. doi:10.1016/S0305-0483(97)00005-4

Yasin MM, Small MH, Wafa MA (2003) Organizational modifications to support JIT implementation in manufacturing and service operations. Omega 31(3):213–226. doi:10.1016/S0305-0483(03)00024-0

Zámečník R (2014) The measurement of employee motivation by using multi-factor statistical analysis. Proc Soc Behav Sci 109:851–857. doi:10.1016/j.sbspro.2013.12.553

Chapter 4
Causes of Slow Implementation of JIT

Implementing JIT in a production system may not always be straightforward, and disappointing situations for managers and stakeholders can frequently arise, such as an early abandonment of the technique, its slow implementation, or the absence of convincing results that do not match with those reported in the literature. However, information regarding JIT benefits could often be misinterpreted, since the literature addresses them in different contexts. On the other hand, the literature approaching problems and errors associated with JIT implementation process is not equally disseminated.

Some factors that cause slow JIT implementation are related to the inexistence of some elements mentioned in Chap. 2, such as the commitment from the senior management to achieve the established goals or trained and well capable employees. However, the level of maturity that JIT can reach in companies can also be low due to deficiencies in one or more of these elements.

This chapter studies the reasons that lead companies to either slow JIT implementation or its abandonment. Table 4.1 illustrates the main obstacles reported in the literature reviewed for this book, although many others can also be found.

1. *High cost of implementation*: As any other technique to implement in a production system, JIT entails investing in fees at the beginning of its operation, which is the major reason why companies end up canceling it. Besides, when companies lack experts who can properly carry out the implementation process, they are usually forced to hire the services of affluent staff of external specialists.

   Issues during the process of JIT implementation are generally due to a lack of necessary elements, which are often difficult to obtain or achieve, and most of them are related to economic costs. For instance, companies require an efficient distribution plant or layout to implement JIT in their production systems; however, moving the actually installed equipment and machines generally represents high expenses, not to mention that distances in materials transportation can also be large (Steele 2001). Similarly, proper JIT implementation demands strategies and organizational changes that are often expensive and not easy to accept by administration departments, which may be used for working in a *status quo*, while JIT always implies administrative

J.L. García-Alcaraz and A.A. Maldonado-Macías, *Just-in-Time Elements and Benefits*,
Management and Industrial Engineering, DOI 10.1007/978-3-319-25919-2_4

Table 4.1 Causes of slow JIT implementation

Causes of slow JIT implementation	Authors	Total
High cost of the implementation	Garg et al. (1996b), Roy and Guin (1996, 1999), Kumar and Garg (2000), Kumar et al. (2001)	5
Shortage of multifunctional workers	Kumar et al. (2001), Kumar and Garg (2000), Balakrishnan and Linsmeier (1996), Kumar (2010)	4
Lack of teamwork	Garg (1997a), Singh and Garg (2011b), Kumar (2010), Singhvi (1992)	4
Lack of management support in quality programs	Nandi (1988), Singh (1989), Kumar et al. (2001)	3
Lack of training	Garg (1997b), Garg et al. (1994), Vrat et al. (1993)	3
Use of traditional methods for quality control	Singh (1989), Singh and Garg (2011b), Kumar and Garg (2000)	3
Lack of support from R&D departments	Kumar (2010), Kumar et al. (2001, 2004)	3
Negative attitude of workers	Kumar (2010), Kumar et al. (2001, 2004)	3
Casual and informal quality audits	Garg et al. (1996a), Kumar and Garg (2000), Singh and Garg (2011a), Kumar et al. (2004)	4
Lack of communication among levels	Nandi (1988), Vrat et al. (1993)	2
Lack of consumer awareness of quality policies	Nandi (1988), Kumar and Garg (2000)	2
Lack of understanding of JIT techniques	Garg (1997b), Roy and Guin (1999)	2
Poor and inadequate maintenance of machinery and equipment	Garg (1997b), Kumar and Garg (2000)	2

changes (Yasin et al. 2003). Furthermore, JIT requires the integration of suppliers and any other entity providing a service to the manufacturing company. This can increase the number of responsibilities to delegate for its implementation as well as the cost of the implementation itself (Lee and Paek 1995). Further information on those activities related to the implementation of JIT can be found in Sohal et al. (1993). Similarly, specific cases of JIT implemented in different countries may be obtained from Oral et al. (2003) in the case of Turkey, or (Alcaraz et al. 2014) as far as Mexico is concerned.

2. *Absence of multifunctional workers*: For a production philosophy or system to be of quality, it must be accepted by workers. However, human nature tends to remain in a comfort zone and might evade changes, which may be an obstacle for JIT implementation. Thus, operators who are reluctant to embrace an organizational or operational change can hardly accept new educational systems that would allow them to become multifunctional employees and be able

to develop diverse activities and work in several lines of production, which could favor the progress of companies.

In a case study among Mexican maquiladoras, García-Alcaraz et al. (2015) analyzed by means of a Structural Equation Model (SEM) the impact of human resources on their benefits. Furthermore, the work of Johnson and Manoochehri (1990) pioneered the identification of the roles of human resources when JIT was implemented in companies. Likewise, Im et al. (1994) thoroughly analyzed the changes that companies needed to pursue in the management processes of human resources to implement JIT in production systems. The work emphasized on the fact that education and the creation of multifunctional workers should always be goals for companies, although they may often be challenging to achieve.

Studies of human resources management in JIT implementation can be found in Damien and Amrik (2000) for the specific case of Australia, and (Salaheldin 2005) in the case of manufacturing companies of Egypt. Similarly, authors Deshpande and Golhar (1996) addressed the particular case of Canadian industries. On the other hand, further information on the training needs for JIT environments can be found in Jones (2001).

3. *Lack of teamwork*: Since JIT is considered a philosophy, its success is based on factors associated with human resources. The causes for a slow implementation discussed above derive from a lack of initiatives from the human resources of a company to develop multifunctional skills, which is also associated with a lack of teamwork.

   One of the basic principles of quality control is to acknowledge every process or subsequent operation in a company as a client. Therefore, all departments in companies should be considered as clients among themselves. This principle is well embraced when members recognize that they all work for the same company and their own success impacts on the success and status of that company. Nevertheless, achieving these levels of integration is not an easy task and requires extensive structural changes.

   In their work Blaga and Jozsef (2014b) and Boer and Blaga (2012) explained the crucial role of human resources in the success of any program or technique that may be applied in a production system. They emphasized on the integration of work teams, especially among operators who tend to be more familiar with the production processes and the variables involved in them. Similarly, Blaga and Jozsef (2014a) stated that the best form of integrating operators into multidisciplinary teams was through the creation of Quality Circles (QC). It is fully demonstrated that most innovations in production processes originate in QC. Finally, Power and Sohal (2000) also reported the benefits of team integration in JIT environments and proposed a number of strategies to achieve the formation of such work teams.

4. *Lack of manager support in quality programs*: Several models that researched on the implementation of production techniques—including mean manufacturing and JIT—found that Management Commitment was essential for the

success of such implementations (Karatepe and Karadas 2012), since managers and supervisors are usually a reflection of the attitudes and commitment of senior managers or directors. Therefore, a lack of commitment and involvement in the activities of a company from top managers can imply no moral authority to demand employees and subordinates to follow certain quality policies and rules. Thus, senior management must play the important role of being an example of what the company wishes to achieve. If companies look for employee commitment, top managers should seek to express theirs (Mokhtar and Yusof 2010).

In a literature review performed by Nasim et al. (2014) on the background required by companies to succeed in quality programs, authors found crucial the commitment from managers in leading and participating in these quality projects. Similarly, authors Wen-Hai and Yu-An (2009) also stated the importance of commitment from top managers in their study of a high technology industry in Taiwan. They found that reliability and commitment of the top management department must be very high to guarantee the success of the quality program.

Additional information on management and senior management commitment can be found in Jamaluddin et al. (2015) and Bou and Beltrán (2005). Authors relied on Structural Equation Models (SEM) to find causal relationships between commitment and performance indices of a company. The major contributions of this work were the identification of Management Commitment as a latent variable and the proposal of variables to be measured or evaluated to estimate management commitment.

5. *Lack of training*: This issue is not recent and refers to the preparation that companies provide to human resources, as well as a lack of multifunctional employees. Since the decade of 1970, Morsch and Griest (1969) addressed the need for proper training programs for employees in the industry. They also mentioned that providing this training should not be merely a commitment from companies, but a legal matter where policies and normative frameworks from governments demanded these organizations to provide their employees with a minimum number of hours of training to enhance their qualifications. Thus, as Meriküll et al. (2012) recently stated, training human resources should be a completely planned activity between the government system and the industrial needs of companies. It would allow the workforce to represent the organizational strength of the businesses to which they belong.

   Finally, a company with untrained and unpowered employees may be doomed to failure, and sometimes might hire employees from competition companies, which is often considered unethical. Hence, authors such as Jones (1980) have also recommended that government and industrial sectors jointly invest in education of the workforce by establishing relations and agreements to generate continuous education and training programs. Similarly, JIT as a philosophy lies in the way human beings act, and when these are not properly trained and guided toward a goal, education efforts would probably be useless.

6. *Use of traditional methods for quality control*: As it was mentioned in the first chapter, JIT is one of the pillars for quality in lean manufacturing environments. A lack of methods to continually assess quality could prevent operators and administrative staff from knowing the complexity of the product or whether it would meet the customer specifications. It seems that lean manufacturing with its techniques, methodologies, and activities really focuses on generating products of the highest quality and JIT is merely a tool for quality assurance. The support that JIT provides for Total Quality (TQ) programs companies is a widely studied topic. For instance, Withers et al. (1997) proposed quality certifications according to ISO standards as quality measurements for companies. Likewise, Cua et al. (2001) introduced an analysis that related total quality (TQ) programs, JIT, and total productive maintenance (TPM) with the performance indices of companies. Authors managed to integrate the three techniques into a single analysis. Similarly, Rahman and Bullock (2005) carried out an analysis of what they named "hard quality" and "soft quality" techniques. They emphasized on differentiating the methodologies and the capabilities of individuals, as well as the training and the information technologies used in order to monitor quality during the production system. Both hard and soft quality techniques were directly related to the performance levels of companies from an economic point of view.
   Finally, Valmohammadi and Roshanzamir (2015) recently established a clear difference between the traditional quality assurance techniques and those modern techniques that made extensive use of information and communication technologies along the production process. They included organizational culture as a moderator variable to analyze the performance of a company, and results inferred that even the best information technologies and modern methods for quality assurance could fail if companies did not rely on an organizational culture to support them. Thus, culture and information technology assimilation are issues on which companies must extensively work to assure quality in their products.
7. *Lack of support from research development (R&D) departments*: The previous paragraphs discussed how commitment from the senior management department is crucial to ensure the success of JIT. Nevertheless, companies must also draw upon the support of their research and development (R&D) departments, since they provide suggestions for new products that could eventually culminate in new designs. Besides, since JIT focuses on the elimination of waste, it may be suitable to seek its elimination from the design and construction phase of the product. Thus, the (R&D) department can seek the creation of new goods with a minimum amount of waste (Schniederjans and Schniederjans 2015).
   Furthermore, some authors believe that an R&D department should be responsible for innovations that could differentiate the company from its competitors, which is why all companies should count on an R&D program to propose new products on the market (Xu et al. 2007). This requires a great amount of manpower trained in various aspects, such as the prediction of future

customer needs. Thus, companies that are unable to predict the needs of customers and the status of their markets could perish in a short time, since their competitors would likely provide innovative or advanced products (Rammer et al. 2009). However, if companies rely on consolidated R&D departments, they must work to empower and train their human resources, since they are the spirit of the enterprise and the R&D department (Noruzy et al. 2013).

8. *Negative attitude of workers*: The support from workers in the production lines is also essential for the successful implementation of JIT. Negative attitudes from employees in companies have been extensively researched from different aspects, such as motivation and personal and professional satisfaction. A specific case that studied the attitude of employees in a JIT environment was proposed by Groebner and Merz (1994), while Bouville and Alis (2014) studied the same situation in the health sector. Furthermore, Lipińska-Grobelny and Papieska (2012) carried out an analysis on the attitudes of workers, measured their satisfaction, and mentioned how it affected their attitudes toward the projects undertaken by the company and their integration in them. Similarly, Dawson (2006) introduced the case study of a company where the positive attitude of employees had helped the rapid implementation of lean manufacturing programs.
   Thus, companies can implement several activities to reduce or avoid negative attitudes from employees. For instance, they can initiate motivational programs for staff, although sharing information about the situation of the company and making them feel part of it may equally help develop a sense of appropriateness and commitment. Similarly, it is helpful to promote a healthy social life for workers and diminish the levels of the hierarchical structure within the organization (Tabassi and Bakar 2009). To do this, companies must encourage a harmonious environment based on trust among all employees (Rusu and Avasilcai 2014).
9. *Casual and informal quality audits*: It is necessary to mention that audits do not generate quality by themselves. However, they do provide information to the company in order to be familiar with and understand the state of quality programs and the level of compliance (Power and Terziovski 2007). For these reasons, several authors believe that companies must know the role of the auditor, and for further information regarding the responsibilities of this position, readers may consult (Kim and Yi 2009).
   On the other hand, some authors Velthuis and van Asseldonk (2011) have proposed that companies perform audits along the production system instead of at the end. This might be one of the most suitable options, although it requires a high level of education from employees, which may not always be possible. It is worth mentioning that despite all the quality audits and inspections that could be run in a production system, certain errors may not always be avoidable as a result of common human–system interactions (Powell et al. 2013). Also, companies should design systematic audits to be conducted by external personnel to avoid complicity among departments. However, hiring external

auditors represents an economic investment for the company and some managers might actually consider it more a cost than an investment. The lack of official audits may be the cause of a slow implementation of JIT or any other technique or lean manufacturing tool.

10. *Lack of communication among levels*: A huge amount of information and communication technologies can nowadays be installed in production systems in order to monitor and track the entire flow of materials to locate the exact place of a part or product. This information should be accessible among all employees and people in the hierarchical structure to create awareness of the real situation of the production system. Otherwise, a lack of communication will prevent operators from being familiar with the goals to achieve, the late deliveries, or the commitments from on-time deliveries. For instance, if a delivery is delayed and employees are unaware of the situation, they may be unable to prioritize activities and tasks of the particular work order. For this reason, visual management has currently gained wide acceptance in the field of industry, since it allows operators to access real-time information on the production system by using any signal or electronic device or written media (Sevier 1992).
    Several researchers have studied the crucial role of communication during JIT implementation in production systems. For instance, since the decade of the 1990s it was mentioned that communication was the key to JIT success (Helms 1990). Also, other studies stated that JIT should not be considered non-exclusive for production systems, but rather a methodology to be applied in the whole supply chain to achieve effective integration and communication among all levels of the hierarchical structure in companies (Schneider and Leatherman 1992). Finally, as for the communication with operators in the production line, authors Banerjee and Golhar (1993) carried out a study comparing the levels of motivation from workers and working environment between organizations who diffused and did not diffuse their results.
11. *Lack of customer awareness of quality policies*: The major value for a customer regarding a product is usually its cost or price, while information on the production process and quality policies needed to offer that product into the market is often neglected. Therefore, companies should encourage customer awareness of their quality policies that ensure a useful product (Shi et al. 2014). Similarly, since certain quality specifications and other terminologies include technical concepts, communication, and social relations departments to carry out a proper dissemination of internal policies using a simple language that can be understood by all customers (Rennung et al. 2014). They can successfully disseminate this information through the points of sale or include it in advertising brochures, guarantees, and other documents distributed with the sale of the final product (Purnasari and Yuliando 2015).
12. *Lack of understanding of JIT techniques*: When JIT was applied in Western companies, its results were not similar to those obtained by most Eastern organizations where it was conceived, especially because of those cultural

differences that lead to different interpretations of the technique. For instance, while product quality is mostly viewed as a philosophy and a way of life in the East, Western culture tends to approach it only as a technique for lean manufacturing. Thus, herein lies the main difference when understanding JIT; while for some companies it is a philosophy and a way of life, others view it solely as a tool (Ahmad et al. 2012). This phenomenon has been partially solved over the years by empowering employees to open up to new interpretations of philosophies implemented in their companies; however, managers are to the greatest extent accountable for the training of their workers, since they are responsible for decision making.

13. *Poor and inadequate maintenance of equipment and machinery*: Companies must rely on available and reliable equipment to ensure a proper flow of materials along the production system and perform valuable operations. Otherwise, an interruption in the flow of materials due to the unavailability of equipment can prevent organizations from achieving their goals of timely deliveries.
    Several researchers and scholars have focused their attention on the relationship between preventive and corrective maintenance and the benefits companies obtain from JIT implementation (McCarthy and Rich 2015). For instance, Cua et al. (2001) studied the relationship among total quality (TQ) programs, total productive maintenance (TPM), JIT philosophy, and certain economic performance indices of companies. Authors found preventive maintenance as a vital element for the survival of JIT philosophy. Finally, the concept of maintenance has greatly evolved to the extent that authors McCarthy and Rich (2004) have named it lean maintenance. It seems that the path to success for companies can be lean philosophy, which brings together a great number of techniques directed toward the same objective, to generate products of the highest quality.

References

Ahmad MF, Zakuan N, Jusoh A, Takala J (2012) Relationship of TQM and business performance with mediators of SPC, lean production and TPM. Proc Soc Behav Sci 65:186–191. doi:10.1016/j.sbspro.2012.11.109

Alcaraz JLG, Maldonado AA, Iniesta AA, Robles GC, Hernández GA (2014) A systematic review/survey for JIT implementation: Mexican maquiladoras as case study. Comput Ind 65 (4):761–773. doi:10.1016/j.compind.2014.02.013

Balakrishnan R, Linsmeier TJ (1996) Financial benefits from JIT adoption: effects of customer concentration and cost structure. Acc Rev 71(2):183–205

Banerjee S, Golhar DY (1993) EDI implementation in JIT and non-JIT manufacturing firms: a comparative study. Int J Oper Prod Manage 13(3):25–37

Blaga P, Jozsef B (2014a) Human resources, quality circles and innovation. Proc Econ Financ 15:1458–1462. doi:10.1016/S2212-5671(14)00611-X

Blaga P, Jozsef B (2014b) Increasing human resource efficiency in the production process. Proc Technol 12:469–475. doi:10.1016/j.protcy.2013.12.516

Boer J, Blaga P (2012) A more efficient production using quality tools and human resources management. Proc Econ Financ 3:681–689. doi:10.1016/S2212-5671(12)00214-6
Bou JC, Beltrán I (2005) Total quality management, high-commitment human resource strategy and firm performance: an empirical study. Total Qual Manage Bus Excellence 16(1):71–86. doi:10.1080/1478336042000309875
Bouville G, Alis D (2014) The effects of lean organizational practices on employees' attitudes and workers' health: evidence from France. Int J Human Res Manage 25(21):3016–3037. doi:10.1080/09585192.2014.951950
Cua KO, McKone KE, Schroeder RG (2001) Relationships between implementation of TQM, JIT, and TPM and manufacturing performance. J Oper Manage 19(6):675–694. doi:10.1016/S0272-6963(01)00066-3
Damien P, Amrik SS (2000) An empirical study of human resource management strategies and practices in Australian just-in-time environments. Int J Oper Prod Manage 20(8):932–958
Dawson B (2006) Lauren employees' efforts help lean succeed. Rubber Plast News 35(15):22–22
Deshpande SP, Golhar DY (1996) HRM practices of JIT firms in Canada. Prod Plann Control 7 (1):79
García-Alcaraz J, Prieto-Luevano D, Maldonado-Macías A, Blanco-Fernández J, Jiménez-Macías E, Moreno-Jiménez J (2015) Structural equation modeling to identify the human resource value in the JIT implementation: case maquiladora sector. Int J Adv Manuf Technol 77(5–8): 1483–1497. doi:10.1007/s00170-014-6561-5
Garg D (1997a) Relevance of JIT purchasing in Indian industries. Kurukshetra University, Kurukshetra, India
Garg D (1997b) Relevance of JIT purchasing in Indian industries. Kurukshetra University, India
Garg S, Vart P, Kanda A (1994) Work culture in JIT environment. Prod J 35(3):463–466
Garg D, Deshmukh S, Kaul O (1996a) Critical analysis in JIT purchasing in Indian context. Prod J 37:271–279
Garg D, Deshmukh SG, Kaul ON (1996b) Parametric analysis of inventory models under JIT purchasing analysis of inventory models under JIT purchasing agreements. Paper presented at the National Conference Operational Research in Modern Technology, REC Kurukshetra, March 8–9, 1996
Groebner DF, Merz CM (1994) The impact of implementing JIT on employees' job attitudes. Int J Oper Prod Manage 14(1):26–37
Helms MM (1990) Communication: the key to JIT success. Prod Inventory Manage J 31(2):18–21
Im JH, Hartman SJ, Bondi PJ (1994) How do JIT systems affect human resource management? Prod Inventory Manage J 35(1):1–4
Jamaluddin Z, Razali AM, Mustafa Z (2015) The relationship between quality management practices and organisational performance: a structural equation modelling approach. In: Proceedings of AIP conference, vol 1643(1), pp 494–504. doi:10.1063/1.4907485
Johnson TW, Manoochehri GH (1990) Adopting JIT: implications for worker roles and human resource management. Ind Manage 32(3):2
Jones HG (1980) Manpower implications of capital investment in Sweden. Omega 8(4):451–457. doi:10.1016/0305-0483(80)90075-4
Jones MJ (2001) Just-in-time training. Adv Dev Human Resour 3(4):480
Karatepe OM, Karadas G (2012) The effect of management commitment to service quality on job embeddedness and performance outcomes. J Bus Econ Manage 13(4):614–636. doi:10.3846/16111699.2011.620159
Kim J-B, Yi CH (2009) Does auditor designation by the regulatory authority improve audit quality? Evidence from Korea. J. Account. Public Policy 28(3):207–230. doi:10.1016/j.jaccpubpol.2009.04.006
Kumar V (2010) JIT based quality management: concepts and implications in indian context. Int J Eng Sci Technol 2(1):40–50
Kumar V, Garg D (2000) JIT elements in Indian context: an analysis. Prod J 41(2):217–222

Kumar V, Garg D, Mehta N (2001) JIT based quality management in Indian industries: prospectus and future directions. Paper presented at the national conference supply chain management, Institute of Public Enterprise, Hyderabad, India, 19–20 Jan

Kumar V, Garg D, Mehta N (2004) JIT elements in Indian context: An analysis. 63 (August): 665–662

Lee SM, Paek JH (1995) An enlarged JIT programme: its impact on JIT implementation and performance of the production system. Prod Plann Control 6(2):185

Lipińska-Grobelny A, Papieska E (2012) Readiness for change and job satisfaction in a case of lean management application—a comparative study. Int J Occup Med Environ Health 25 (4):418–425. doi:10.2478/S13382-012-0044-4

McCarthy D, Rich DN (2004) 2—lean TPM. In: McCarthy D, Rich DN (eds) Lean TPM. Butterworth-Heinemann, Oxford, pp 24–48. doi:10.1016/B978-075065857-7/50005-6

McCarthy D, Rich N (2015) Chapter Two - The Lean TPM Master Plan. In: McCarthy D, Rich N (eds) Lean TPM (Second Edition). Butterworth-Heinemann, Oxford, pp 27–54. doi:10.1016/B978-0-08-100090-8.00002-1

Meriküll J, Eamets R, Humal K, Espenberg K (2012) Power without manpower: Forecasting labour demand for Estonian energy sector. Energy Policy 49:740–750. doi:10.1016/j.enpol.2012.07.018

Mokhtar SSM, Yusof RZ (2010) The influence of top management commitment, process quality management and quality design on new product performance: a case of Malaysian manufacturers. Total Qual Manage Bus Excellence 21(3):291–300. doi:10.1080/14783360903553198

Morsch WC, Griest J (1969) A model for manpower, employment, training and education. Socio-Econ Plann Sci 2(2–4):225–250. doi:10.1016/0038-0121(69)90020-2

Nandi SN (1988) Implementing quality control circles in India. Prod J 28(4):419–424

Nasim K, Iqbal MZ, Khan IA (2014) Antecedents of TQM implementation capability: a review with a conceptual model. Total Qual Manage Bus Excellence 25(11/12):1395–1409. doi:10.1080/14783363.2013.807682

Noruzy A, Dalfard V, Azhdari B, Nazari-Shirkouhi S, Rezazadeh A (2013) Relations between transformational leadership, organizational learning, knowledge management, organizational innovation, and organizational performance: an empirical investigation of manufacturing firms. Int J Adv Manuf Technol 64(5–8):1073–1085. doi:10.1007/s00170-012-4038-y

Oral EL, Mıstıkoglu G, Erdis E (2003) JIT in developing countries—a case study of the Turkish prefabrication sector. Build Environ 38(6):853–860. doi:10.1016/S0360-1323(03)00021-0

Powell DA, Erdozain S, Dodd C, Costa R, Morley K, Chapman BJ (2013) Audits and inspections are never enough: a critique to enhance food safety. Food Control 30(2):686–691. doi:10.1016/j.foodcont.2012.07.044

Power D, Sohal AS (2000) Human resource management strategies and practices in just-in-time environments: Australian case study evidence. Technovation 20(7):373–387. doi:10.1016/S0166-4972(99)00151-0

Power D, Terziovski M (2007) Quality audit roles and skills: perceptions of non-financial auditors and their clients. J Oper Manage 25(1):126–147. doi:10.1016/j.jom.2006.02.005

Purnasari H, Yuliando H (2015) How relationship quality on customer commitment influences positive e-WOM. Agric Agric Sci Proc 3:149–153. doi:10.1016/j.aaspro.2015.01.029

Rammer C, Czarnitzki D, Spielkamp A (2009) Innovation success of non-R&D-performers: substituting technology by management in SMEs. Small Bus Econ 33(1):35–58. doi:10.1007/s11187-009-9185-7

Rennung F, Luminosu C, Draghici A (2014) Evaluation of methods for customer integration to the quality of IT Services. Proc Technol 16:101–109. doi:10.1016/j.protcy.2014.10.073

Roy R, Guin KK (1996) JIT: world scenario and possibility of its applicability in Indian industries. Paper presented at the National conference on operation research in modern technology, REC Kurukshetra, India, 8–9 Mar

Roy RN, Guin KK (1999) A proposed model of JIT purchasing in an integrated steel plant. Int J Prod Econ 59(1–3):179–187. doi:10.1016/S0925-5273(98)00099-1

Rusu G, Avasilcai S (2014) Linking human resources motivation to organizational climate. Proc Soc Behav Sci 124:51–58. doi:10.1016/j.sbspro.2014.02.459

Salaheldin SI (2005) JIT implementation in Egyptian manufacturing firms: some empirical evidence. Int J Oper Prod Manage 25(4):354–370. doi:10.1108/01443570510585543

Schneider JD, Leatherman MA (1992) Integrated just-in-time: a total business approach. Prod Inventory Manage J 33(1):78–82

Schniederjans D, Schniederjans M (2015) Quality management and innovation: new insights on a structural contingency framework. Int J Qual Innov 1(1):1–20. doi:10.1186/s40887-015-0004-8

Sevier AJ (1992) Managing employee resistance to jit: creating an atmosphere that facilitates implementation. Prod Inventory Manage J 33(1):83–87

Shi Y, Prentice C, He W (2014) Linking service quality, customer satisfaction and loyalty in casinos, does membership matter? Int J Hospitality Manage 40:81–91. doi:10.1016/j.ijhm.2014.03.013

Singh A (1989) Just-in-time system: an integrated system. Prod J 30(3):309–314

Singh S, Garg D (2011a) JIT system: concepts, benefits and motivation in Indian industries. Int J Manage Bus Stud 1(1):26–30

Singh S, Garg D (2011b) JIT system: concepts, benefits and motivations in Indian industries. Int J Manage Bus Stud 1(1):26–30

Singhvi S (1992) Employee involvement in JIT success: Eicher experience. Prod J 33:366–369

Sohal AS, Ramsay L, Samson D (1993) JIT manufacturing: industry analysis and a methodology for implementation. Int J Oper Prod Manage 13(7):22–56

Steele AL (2001) Cost drivers and other management issues in the jit supply chain environment. Prod Inventory Manage J 42(2):61–67

S-u Rahman, Bullock P (2005) Soft TQM, hard TQM, and organisational performance relationships: an empirical investigation. Omega 33(1):73–83. doi:10.1016/j.omega.2004.03.008

Tabassi AA, Bakar AHA (2009) Training, motivation, and performance: the case of human resource management in construction projects in Mashhad, Iran. Int J Project Manage 27(5):471–480. doi:10.1016/j.ijproman.2008.08.002

Valmohammadi C, Roshanzamir S (2015) The guidelines of improvement: relations among organizational culture, TQM and performance. Int J Prod Econ 164:167–178. doi:10.1016/j.ijpe.2014.12.028

Velthuis AGJ, van Asseldonk MAPM (2011) Process audits versus product quality monitoring of bulk milk. J Dairy Sci 94(1):235–249. doi:10.3168/jds.2010-3528

Vrat P, Mittal S, Tyagi K (1993) Implementation of JIT in Indian environment: a Delhi study. Prod J 34:251–256

Wen-Hai C, Yu-An L (2009) The study of the antecedent factors of organisational commitment for high-tech industries in Taiwan. Total Qual Manage Bus Excellence 20(8):799–815. doi:10.1080/14783360903128082

Withers BE, Ebrahimpour M, Hikmet N (1997) An exploration of the impact of TQM and JIT on ISO 9000 registered companies. Int J Prod Econ 53(2):209–216. doi:10.1016/S0925-5273(97)00118-7

Xu Q, Chen J, Xie Z, Liu J, Zheng G, Wang Y (2007) Total innovation management: a novel paradigm of innovation management in the 21st century. J Technol Transfer 32(1–2):9–25. doi:10.1007/s10961-006-9007-x

Yasin MM, Small MH, Wafa MA (2003) Organizational modifications to support JIT implementation in manufacturing and service operations. Omega 31(3):213. doi:10.1016/S0305-0483(03)00024-0

Part III
Methodology

Chapter 5
Research Methodology

This chapter proposes a methodology for the models that will relate the identified JIT elements and causes of its slow implementation with the benefits obtained by the companies surveyed. This relation could allow managers and administrators to identify the elements that are keys to the success of JIT, as well as those that cause its slow implementation. The methodology here proposed was divided into several stages and combines several techniques widely used in multivariate analyzes.

5.1 Stage 1: Questionnaire Development

The first step in order to associate JIT elements and causes of its slow implementation with the possible benefits consisted in developing a questionnaire that included these three objects of study previously identified and described in preceding chapters from the literature available. The questionnaire was developed under the following structure.

5.1.1 Section 1: Demographics

Demographic data served to categorize and classify the questionnaires obtained. The section demanded information regarding the email of respondents, their positions in the companies they worked, as well as the name of the company; the latter was an optional question since certain firms ought to remain anonymous. Additional demographic data concerned the gender of respondents and the industrial sector of companies. The latter had to be answered by selecting among the following: automotive, packaging, electronics, communications, plastic, metal, and medical. However, if companies did not belong to one of these sectors, respondents were required to specify the sector of their companies after marking the option

J.L. García-Alcaraz and A.A. Maldonado-Macías, *Just-in-Time Elements and Benefits*, Management and Industrial Engineering, DOI 10.1007/978-3-319-25919-2_5

"others." As for the size of companies, every respondent was solicited to state the average number of employees working in the company they represented in the research. Then, companies were classified by their size according to the classification stated by the Ministry of Economy of Mexico.

5.1.2 Section 2: JIT Elements

The second section of the questionnaire addressed the 43 JIT elements required for a successful JIT implementation that were previously identified.

- The subsection *Elements Associated with Human resources* contained 12 elements.
- The subsection *Elements Associated with the Processes* included 18 elements.
- The subsection *Elements Associated with Products* contained 13 elements.

5.1.3 Section 3: JIT Benefits

The third section of the questionnaire addressed the benefits previously identified from literature and offered by a successful JIT implementation in production lines. This section was composed of six subsections that corresponded to the six categories of JIT benefits previously identified. Thus, the 31 benefits reported in literature were all included in the questionnaire.

- The subsection *Benefits Associated with Human Resources* contained five benefits having an impact on human resources, whether suppliers, operators, or administrative personnel.
- The subsection *Benefits Associated with Production Processes* contained seven benefits to report, mainly concerning improved efficiency and productivity indexes along the production line.
- The subsection *Benefits Associated with Engineering* included four benefits to report and related to rapid product engineering changes as well as a better use of available space.
- The subsection *Benefits Associated with Quality* included three benefits to report and concerned improvements in quality characteristics attributed to or reflected on the products manufactured by companies.
- The subsection *Benefits Associated with Materials Handling* contained six benefits to report and mostly related to a better use of space and movement of materials along the company and an increased flexibility in their management.
- The subsection *Economic Benefits* included six benefits to report and related to costs reduction and increased profit margins or financial profits for the company.

5.1.4 Section 4: Causes of a Slow JIT Implementation

This fourth section addressed the causes of a slow JIT implementation previously identified in Chap. 4 and described in Table 4.1. However, this section did not include a subsection as such, since this study did not categorize them in any, although it should not be forgotten that, as Chap. 4 stated, a slow JIT implementation is related to the absence or deficiency of one or several JIT elements.

5.2 Measurement Scale

A Likert scale (Tastle and Wierman 2007; Al-Tahat and Bataineh 2012) was used to assess the second, third, and fourth sections of the questionnaire in order to indicate presence or absence of any JIT element or benefit and the cause of a slow JIT progress. Likert measurement has been widely used in different research works associated with supply chain, such as Swafford et al. (2006), Moon et al. (2012), Jakhar (2015), Gligor et al. (2015), Alfalla-Luque et al. (2015). The scale used for respond the questionnaire appears in Table 5.1.

5.3 Administration of the Questionnaire

This research relied on an electronic database of maquiladora companies from the INDEX (Maquiladoras Association AC) established in northern Mexico. Every materials manager, supply chain manager, materials and supplies planners, and managers related to the supply chain and material flow in the system production was contacted by questionnaire administrators. Next, the questionnaire was placed on a special website for questionnaires and surveys administration. Afterward, potential respondents were invited via email to participate in the research and

Table 5.1 Rating scale used for questionnaire administration

Value	1	2	3	4	5
Meaning	The element is not present	The element is slightly present	The element is regularly present	The element is almost always present	The element is always present
	The benefit was not obtained	The benefit is scantily obtained	The benefit is regularly obtained	The benefit is usually obtained	The benefit is always obtained
	This is not a cause of slow JIT implementation	This causes little slow of JIT implementation	This causes a regular slow JIT implementation	This almost always causes a slow JIT implementation	This always causes a slow JIT implementation

provided with the web link to answer the survey. The questionnaire was available for four months from January to April, 2014. Reminders were sent to respondents 15 days after the invitation, and if companies were still unable to respond to the questionnaire, a second reminder was sent 15 days later. If no response was obtained for the following 15 days, the invitation was discarded for a period of two months and the email of those particular potential respondents were corroborated. Following these three months, if emails were found valid, a third and final reminder was again sent to invite these managers to participate in the research. However, in cases were emails were found incorrect, the invitation was sent to another manager of the same company still related to supply chain management. Finally, after the four months, participants were thanked for their participation and acknowledged for the importance of their contribution.

5.4 Information Capture and Database Debugging

This research relied on the support of specialized software to analyze the large amount of information obtained from questionnaires. Therefore, statistical software SPSS 21® was employed, given the expertise of authors of this research with its management and applications as well as the availability of its license within the research laboratory. However, Excel spreadsheets were also employed. Thus, a database or template on software was designed for information capture, where every line represented one answered case or questionnaire and every column indicated every item or variables integrated in the questionnaire. This type of analysis and database has been used in other pieces of research related to supply chain, such as Akintoye et al. (2000), Ambulkar et al. (2015), Chin et al. (2014), Jackson and Singh (2015), Ketikidis et al. (2008), Kumar et al. (2015).

5.4.1 Missing Values

Several routine procedures were run before any data analysis in order to refine the database; they were especially emphasized on the elimination of missing values and identification of outliers that could cause a skewed analysis. As for missing values, since data was obtained through a Likert scale, missing values were replaced by the median, since authors make this recommendation. Similarly, they consider suitably a maximum of 10 % of missing values in variables and in questionnaires (Hair et al. 1987, 2009). Nevertheless, other techniques can be used to replace missing values as long as they do not exceed the maximum limit indicated above.

5.4.2 Extreme Values or Outliers

As for extreme values, variables were standardized. First, the standard deviation for every item was extracted. Then, every extreme value was divided by the standard deviation of the variable in every case where this variable was included. If data had an approximation to normal distribution, then standardized values ranging from −3 to 3 could be considered appropriate values for analysis. However, if absolute values were equal to or higher than 3, they could be considered outliers and should be directed to a more thorough analysis, although some authors recommend integrating into the analysis standardized values equal to or lower than 4 (Giaquinta 2009; Hair et al. 2009; Ala-Harja and Helo 2014; Rosenthal and Rosnow 1991a; Wold et al. 2001).

These missing values are often due to typographical errors. For instance, the scale used in this research included values between 1 and 5; thus, any value outside this range could be considered as an outlier. If a value 22 was observed in a cell, it could be a typographical mistake referring to value 2. Therefore, it is important to rely on techniques to identify every questionnaire obtained. In this case, every survey was labeled and assigned a value starting with 001.

Furthermore, boxplots were constructed with SPSS software described above as a means to graphically identify extreme values (Carter et al. 2009; Bruffaerts et al. 2014). Box-and-whisker- plots, as the name implies, are represented with a box that integrates from 25th percentile to 75th percentile values, which is traditionally known as the interquartile range. The bottom part of the boxplot includes values from 0 to 25th percentile, while the upper part includes values from 75th percentile to 100th percentile (Li et al. 2014; Simpson Jr et al. 1988). In the case of this study, software SPSS indicated the case number or questionnaire with values outside that range, which were therefore considered as outliers.

The identification of extreme values was crucial for this research, since data analysis relied on regression techniques and these outliers could skew the analysis when the coefficients were identified (Lem et al. 2013; Hansson et al. 1993; Chang et al. 2015). Even though analyzes previously presented were univariate, identifying outliers from multivariable analyzes is equally important. For this research, the square Mahalanobis distance was used to carry out a multivariable assessment, since it integrates the correlation among variables and generates a more appropriate index (Todeschini et al. 2013; Patil et al. 2015; Giménez et al. 2012).

5.4.3 Zero Variance

Identifying cases or questionnaires in which respondents—perhaps uninterested—measured all questions with the same scale value or little variation is another procedure frequently used to debug databases. For this research, the standard deviation for every questionnaire was first obtained (Wang et al. 2015; Manenti and

Buzzi-Ferraris 2009; Lourenço and Pires 2014). When respondents assigned the same value to every question, the standard deviation of that questionnaire was 0, the survey was discarded since this value suggested little interest or commitment. Although there is not a minimum acceptable variance or standard deviation determined, this research discarded questionnaires with standard deviation values equal to or lower than 0.5.

5.4.4 Normality Test

Since the analysis was relied on regression techniques, it was sought to meet all their possible requirements. Normal distribution of data is perhaps one of the most important conditions for regression methods whose aim is that the distribution of every variable resembles to a normal distribution without skewness and with adequate asymmetry and ideal kurtosis.

On the one hand, acceptable skewness for every variable analyzed was an absolute value ranging between −1 and 1, since values outside that range could indicate the presence of outliers. Off values also indicated they were skewed values, which is why they were analyzed by elaborating a graph of standardized values and adjusting a normal distribution curve (Rimoldini 2014; Loperfido 2013). However, confidence intervals for skewness of every variable were constructed. The standard deviation of every one of these variables was added and subtracted three times to the average skew value. If this interval included values −1 or 1, the variable was skewed and the possible values causing it had to be identified (Xiaojun and Morris 1991; Withers 1987; Godfrey and Orme 1991).

Finally, the kurtosis of a variable refers to platykurtic and leptokurtic distribution of data. An ideal distribution should be mesokurtic or similar to a normal distribution (Kim and White 2004). For the kurtosis analysis, values close to 1 were also sought, and confidence intervals were generated by multiplying the standard deviation of the parameter by 3 and adding and subtracting it to the mean value (Galvao et al. 2013; Kerman and McDonald 2013).

5.4.5 Homoscedasticity Test

Homoscedasticity is another test to data, since variables must maintain the same variance throughout their range of values. For this test, graph visual analyzes were carried out with variables and standardized residuals generated by simple linear regression. In this case, it was sought no patterns in residuals and that all could be homogeneous (Jarque and Bera 1980; Ohtani and Toyoda 1980). The variable analyzed was plotted on the axis of ordinates and the residual obtained on the axis of abscissas. It was always sought to obtain the same number of points on the left as on the right of the 0 value, since residuals were standardized (Bera and Jarque 1981;

Giles and Giles 1996). Analyzes with this type of graphics also allowed for the identification of some outliers to which they were assigned with a high percentage of the total error of residuals.

5.4.6 Multicollinearity Test

Multicollinearity occurs when several independent variables are analyzed in a multiple regression analysis. This problem can be the result of items or questions in the questionnaire that measured the same aspect and whose answer and value will be thus very similar. Consequently, some of the variables tested could be eliminated since they were explained by another variable. Moreover, if the correlation coefficients of these similar variables were analyzed, high values could be perceived (Mason and Brown 1975; Wang et al. 1990).

This study used inflation variance indexes to determine the level of multicollinearity among variables. It has been established that values above 3.3 indicate high levels of collinearity. This could be the result of low values in the analysis of the eigenvalues (Cortina 1993; Sarkar 1996; Ueki and Kawasaki 2013; Jadhav et al. 2014).

Condition index can also be used to determine collinearity between variables. It consists in dividing the maximum eigenvalue of the independent variables by the values obtained for every variable. Values in condition number above 1000 indicate a problem or high collinearity between variables analyzed (Troskie and Conradie 1986; Zimmermann 2015).

5.5 Descriptive Analysis of the Sample and Information

The descriptive analysis is divided into two parts. The first one addresses the description of the sample surveyed by indicating demographic information such as the position of respondents, seniority in their current positions, the industrial sector of the maquiladora company where they worked, and the average of employees working in the company according to the category established by the Secretariat of Economy of Mexico that defines the size of Mexican companies into small, medium, or large. Contingency tables were also elaborated to indicate the relationship between two or among more variables of the demographics section.

5.5.1 The Median as a Measure of Central Tendency

The median value was considered as a measure of central tendency for every JIT element, JIT benefit and cause of slow implementation, since data was obtained by means of an ordinal scale ranging from 1 to 5 and only represented the assessment of experts in the area (Iacobucci et al. 2015). The arithmetic mean could thus not be used, because values were not listed as intervals or ratio scales (Baxter et al. 2015). Missing values were replaced by the median due to the same reason.

Two cases were found after the interpretation of the median value (Tastle and Wierman 2007) of every item analyzed.

- High median values indicated that the JIT element, JIT benefit, or cause of slow JIT implementation was almost always present in companies surveyed.
- Low median values indicated that the JIT element, JIT benefit, or cause of slow JIT implementation was almost never present in companies surveyed.

Therefore, low median values were desired in the analysis of elements required for a JIT implementation, while high median values were expected for the JIT benefits. Similarly, low median values were desired in analysis of causes of slow JIT implementation. Thus, what was expected from the analysis was the requirement of few elements to implement JIT in order to obtain the maximum number of JIT benefits with minimal problems or causes of slow implementation.

Median as the measure of central tendency has been used in other pieces of research in the fields of supply chain and JIT, such as Avelar-Sosa et al. (2015), Villanueva-Ponce et al. (2015), García-Alcaraz et al. (2015), all have studied the maquiladora sector established in Mexico.

5.5.2 Interquartile Range as a Measure of Dispersion

This research used interquartile range as a measure of dispersion, which refers to the difference between the 75th percentile and the 25th percentile. Interquartile range values were used for the same reason as median values were considered as a measure of central tendency: data were obtained from a Likert scale with values ranging from 1 to 5. Two cases were found after the interpretation of interquartile range values of variables.

- A high interquartile range value in a variable indicated great dispersion of data and little consensus among respondents regarding the value of that item associated with a JIT element, JIT benefit, or cause of slow JIT implementation.
- A low interquartile range value in a variable indicated little data dispersion and thus great consensus among respondents on the value of that item associated with a JIT element, JIT benefit, or cause of slow JIT implementation. For instance, if all respondent assigned a value 2 to the same item, the variance of this value would be 0 and the interquartile range value would indicate that all respondents agreed with the median value assigned to the items. Additional research has also relied on interquartile range values as a measure of dispersion, such as Avelar-Sosa et al. (2015), Villanueva-Ponce et al. (2015), García-Alcaraz et al. (2015), Alcaraz et al. (2014), Withers et al. (1997).

5.6 Questionnaire Validation

Once data was cleared in the database and free from extreme values and missing values, then questionnaire and all latent variables included were validated. Several reliability indexes were used to determine whether data obtained was appropriate. These indices are described briefly below as well as the type of validity or reliability being measured.

For this research, reliability was referred as to obtaining same or compatible results in different experiments or statistical tests, and it is related to the reproducibility of the experiment. Validity/reliability was reported for latent variables. A latent variable is a single variable that includes related items reported in previous chapters concerning JIT elements, JIT benefits, and causes of slow JIT implementation in production systems.

5.6.1 Cronbach's Alpha Internal Consistency

This is one of the most widely reported indexes to measure consistency in a latent variable consisting of other variables that can be measured (Cronbach 1951). Cronbach's alpha index can be estimated based on either variance or the indexes of correlation among items that compose a latent variable (Adamson and Prion 2013). When analyzes rely on the variance method, the index is similar to the coefficient of determination in a simple linear regression, but is adjusted with the number of items that compose the latent variable. Moreover, it is an index ranging from 0 to 1 for all values. Lines below depict a description of the possible interpretation of values of this Cronbach's index (Rindskopf 2015):

Alpha coefficient >0.9 is excellent
Alpha coefficient >0.8 is good
Alpha coefficient >0.7 is acceptable
Alpha coefficient >0.6 is questionable
Alpha coefficient >0.5 is poor
Alpha coefficient <0.4 is unacceptable

Therefore, values close to 1 indicate that a latent variable has a good reliability, while lower values indicate that a latent variable has little reliability and the items within it are not adequately measured. When Cronbach's alpha indexes are low, they may be tested to determine whether these values could increase. For instance, it is common to analyze the possible index obtained with the elimination of an item of the latent variable (Kopalle and Lehmann 1997; Nunnally and Bernstein 1994). In fact, since Cronbach's alpha index is a modification of the coefficient of determination of a simple linear regression, it is possible that one of the items have little correlation with the others. It could thus be removed so that the remaining items support a better explanation of the latent variable. Therefore, the analysis carried out in this study considered the possibility of eliminating items from latent variables if they were not suitable in order to increase the internal reliability of the latent variable (Fornell and Larcker 1981).

Numerous methods have been proposed in order to achieve a robust Cronbach's alpha index (Christmann and Van Aelst 2006). However, this research relied on the method already integrated into SPSS 21 software® and established 0.7 as the minimum acceptable index value, although it has been largely debated on the minimum value that should be accepted (Kottner and Streiner 2010). Nevertheless, authors have also relied on simulations to determine an approximate acceptable value (Leontitsis and Pagge 2007). Additionally, available research has emphasized on the difference between exploratory index and confirmatory index. The former usually considers lower values than the latter, since unlike confirmatory studies in exploratory studies there is no specific idea of the measurable variables that could be integrated into a same latent variable (Pinto et al. 2014).

5.6.2 Average Variance Extracted (AVE), Discriminant Validity

Discriminant validity tests whether the constructs first believed to be unrelated are actuality related. Convergent validity, on the other hand, tests whether constructs believed to be related one another are actually not related. They are perhaps two similar concepts and widely used to define reliability of variables.

As for AVE, authors recommended that values be always higher than 0.5 (Fornell and Larcker 1981; Kock 2013b), even though it has also been stated that a latent variable is reliable; if in a correlation matrix of latent variables analyzed, the square root of AVE is higher than any of the indexes correlated in that matrix for both rows and columns. Therefore, when this is not possible, it indicates that some items or measurable variables in the latent variable have high factor loadings over other latent variables. Hence, these measurable variables are also associated with another latent variable, and factor loadings of all latent variables should be analyzed to determine where these measurable variables should be also located. Since such analysis may be complicated, nowadays they also can be reported by most statistical softwares.

5.6.3 Correlation Coefficient, Predictive Validity

Models are generated to be highly predictive and help predict the behavior of variables. Correlation coefficient is widely used in all techniques based on regression analysis to measure the explained variance of an independent variable or set of variables over a dependent variable (Lecchi 2011) is made. This research relied on three indexes to measure predictive validity

- *R-squared*: it is calculated merely for dependent variables and reports the amount of variance explained in one dependent variable by a set of independent variables (Gonzalez et al. 2013). High R-square values are expected in data analysis since they would indicate a highly explanatory model to predict events or phenomena. On the other hand, low R-squared values would point out a little explanatory model and perhaps its usefulness would be reduced. It is recommended analyzing models in which R-squared values obtained are lower than 0.02, since they have little explanatory power (Kock 2013b).
- *Chi-square*: it is similar to R-square, but its estimate depends on the size of the sample. Thus, it seems to be a more comprehensive index, able to detect spurious or false relationships (Frémont et al. 2012). R-squared and adjusted R-squared are parametric estimates that depend on certain conditions and characteristics of the data collected (Wooldridge 1991).
- *Q-square*: it is a nonparametric estimate of R-squared. It is mainly reported since data are often not attached to a normal distribution, which is a condition for regression analysis (Aboalkhair et al. 2013). Q-values must always be higher than 0 and similar to R-squared values. These similar values would indicate that data comes from a normal distribution.

5.6.4 Dillon-Goldstein Rho Index, Composite Reliability

Cronbach's alpha index was first developed in 1951 and, since then, has been the basis for many studies. However, authors have proposed several modifications and improvements, although always seeking for values higher than 0.7 in variables analyzed (Tenenhaus et al. 2005). One such indexes is composite reliability, which aims at correcting a number of deficiencies of the Cronbach's alpha index.

5.7 Integrating the Latent Variables (Factor Analysis)

The models of structural equations use latent variables, which can be defined as a special type of variable non-observed directly, but is measured through others, which if they can be measured or with an assessment (Spirtes 2015; Kohler et al. 2015).

The integration of those latent variables can be realized in different ways, nevertheless, in this research has followed the method of factor analysis that has been used in similar studies by Alcaraz et al. (2014), Garcia-Alcaraz and Oropeza-Vento (2014), Avelar-Sosa et al. (2014). In order to know the feasibility to apply this factor analysis, three main indices are used (Vandekerckhove 2014; Chen and Gan 2014):

- Kaiser-Meyer-Olkin measure of sampling adequacy
- Bartlett's test of sphericity
- Determinant in correlation matrix

In case the obtained is adequate for applying the factor analysis technique, then for each one of the JIT elements, JIT benefits and JIT causes of slow implementation categories, the factor analysis is applied using the rotation promax method (Kock and Lynn 2012; Kock 2013a), since in this research is not desired to generate independent or orthogonal latent variables, since otherwise the relationships between these would make no sense (Vandekerckhove 2014). Nevertheless, using a promax rotation certain level of dependency is conserved for applying the structural equation model.

In order to determine the number of latent variables or factors that are considered in the analysis of structural equations models that represent a category of JIT elements, JIT benefits, or causes of slow JIT implementation, it is considered whenever the eigenvalue of the matrix of correlations is greater than 1. Only in exceptional cases, where the eigenvalue is very near the unit and the variance explained by that factor is very high then its inclusion is considered (Schulze et al. 2015).

5.8 Hypothesis Formulation, Construction of Structural Equations Models

Several casual hypotheses were formulated between the latent variables or groups of items defined in Chaps. 2–4. In these casual hypotheses, it was assumed that a latent variable could illustrate another, which is illustrated in Fig. 5.1.

According to Fig. 5.1, latent variable 1 has a direct and positive effect on latent variable 2, which is one of the proposed hypotheses. However, structural equation models can be composed of more than two variables, which implies greater number of relations between them. A hypothesis to be proved was thus generated for every relation between variables (Temme et al. 2006). Figure 5.2 shows a structural equation model composed of three latent variables with thus more hypotheses or relations between them to be tested.

Figure 5.2 can help explain the different types of effects that will be measured in the structural equation models. These effects will be described below.

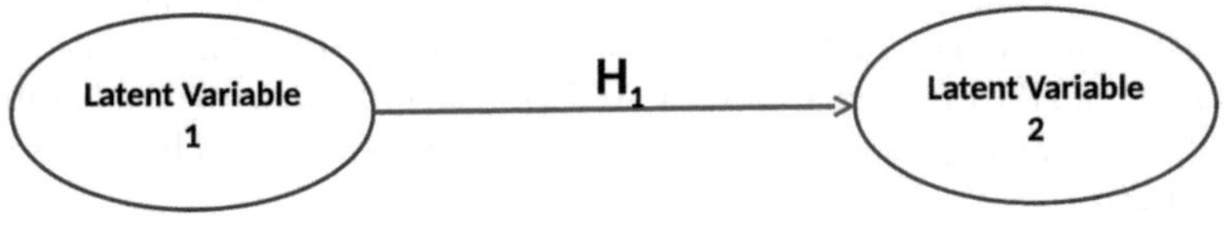

Fig. 5.1 Simple hypothesis approach

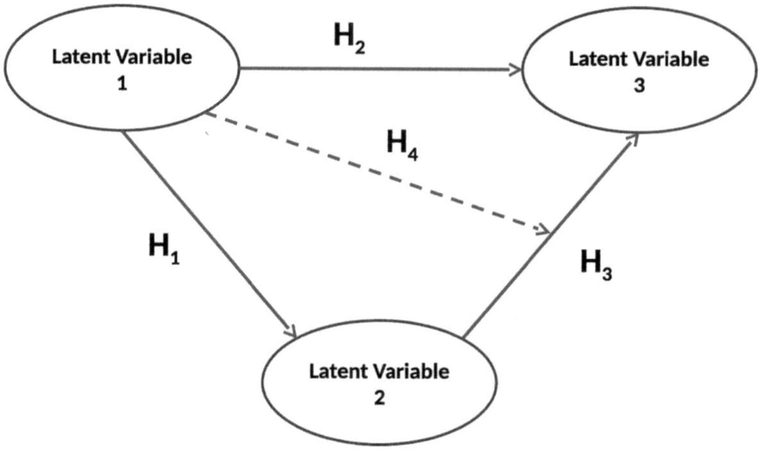

Fig. 5.2 Multiple hypotheses approach

5.8.1 Effects Between Variables

Different effects can occur between the latent variables that integrate a structural equation model:

- Direct effects
- Indirect effects (mediating effect)
- Sum of total effects
- Moderating effect

5.8.1.1 Direct Effects

Direct effects between two latent variables are indicated by an arrow that unites them. The latent variable from which the arrow emerges is known as independent or exogenous variable, while the latent variable receiving the point of the arrow is called dependent or endogenous variable (Wold et al. 2001). Thus, Fig. 5.2 shows three direct effects proposed by H1, H2, and H3.

Temporality (Chatelin et al. 2002) was considered for the generation of these hypotheses, since it helps determine the direction of the arrows or effects. For example, the process of raw materials supply may have an effect on the delivery time of the finished product, since raw materials are first delivered and then finished product is delivered afterward.

A beta (β) value will be estimated for every direct effect. This value is referred as to the intensity of change that exists in a latent variable in relation with the other one to which it is related (Wetzels et al. 2009). For instance, if a relationship between two variables has a beta value equal to 0.53, whenever the first latent variable increases its standard deviation one unit, the standard deviation of the second unit will increase 0.53 units. However, a beta value equal to −0.53 indicates that whenever the standard deviation of the first latent variable increases by one unit, the standard deviation of the second will decrease by 0.53 units. Notice that these two examples show that beta values can be either positive or negative. When the value is positive, the standard deviation of the dependent variable increases, but it decreases when the beta value is negative.

Similarly, models proposed in the next section show *P*-values added to beta values, which are used to determine statistical significance of the beta value or parameter (Wold et al. 2001). In other words, every beta value will be tested by means of a hypothesis test to identify whether the confidence interval includes 0. In this case, it is concluded that this parameter is not significant from a statistical point of view and this relation can be eliminated. However, when zero is not included in the confidence interval, the parameter is considered significant and should remain in the model. All statistical tests on models proposed here were performed with a 95 % confidence level, which indicates a 5 % significance level. Therefore, a parameter is significant when its *P*-value is always lower than or equal to 0.05.

5.8.1.2 Indirect Effects

Direct effects are related to the direct relationship between two variables. However, it is possible that two variables have indirect relationships between them, which thus include not only these two variables but also a third, fourth, or perhaps also a fifth variable (Willaby et al. 2015). Figure 5.2 shows that latent variable 1 has a direct effect on latent variable 3, which is why hypothesis 2 (H2) is established. Nevertheless, there is an indirect effect between both variables given through latent variable 2. That is, latent variable 3 could be reached through the segment of direct effect between latent variable 1 and 2, which is set by hypothesis 1 (H1); but latent variable 2 can reach latent variable 3 through the segment established by hypothesis 3 (H3) (Intakhan 2014). Therefore, latent variable 2 is called mediator variable, since it is the connection between latent variable 1 and 3 (Kaynak et al. 2015).

Likewise for direct effects, the model estimates values of indirect effects whose significance is also tested by means of a hypothesis test reflected through a *P*-value. Therefore, in the analysis carried out for the present research, a statistically significant indirect effect must have an estimate value lower than 0.05, since it was the value previously set (Preacher and Hayes 2004).

Finally, indirect effects may have several mediator variables and thus many segments to reach the dependent from the independent variable. For instance, while the previous example showed one mediator variable—latent variable 2—merely two segments could reach the dependent variable from the dependent variable. Nevertheless, if five segments could determine an indirect effect that would indicate the existence of four mediator variables that connect from the independent variable and the dependent variable (Hayes and Preacher 2010).

5.8.1.3 Sum of Total Effects

As previously stated, Fig. 5.2 shows that latent variable 1 has a direct effect on latent variable 3 through the hypothesis 2, but it also has an indirect effect given through the mediator latent variable 2 (Intakhan 2014). Therefore, the sum of total effects that latent variable 1 has on latent variable 3 is the addition of these two effects (direct and indirect). Total effects also have *P*-values to estimate their statistical significance.

5.8.1.4 Moderating Effects

In Fig. 5.2, hypothesis 4 (H4) is represented with a dotted or segmented arrow, since it has certain special features. It is called a moderating effect. Note that this dotted arrow or moderating effect starts in a latent variable but ends on a segment that is a direct effect. It does not reach a latent variable as the other arrows.

To explain this effect one may think of enzymes, which can speed up or slow down a chemical reaction. In this case, hypothesis 4 (H4) is actually testing the impact of latent variable 1 on the relationship between latent variable 2 and latent variable 3. Note that a moderating effect exists on the relationship between two latent variables and not directly on a latent variable.

Moderating effects can also have positive and negative values for the estimated beta value. Similarly, a hypothesis test is also carried out to determine whether these values are statistically significant. As mentioned above, moderating values were statistically significant when *P*-values were lower than or equal to 0.05, which demonstrated that a certain latent variable actually had an effect on the relationship being tested. On the other hand, cases where *P*-values were greater than 0.05 indicated that the latent variable analyzed did not have any moderating effect on the relationships tested, since confidence interval for the estimated beta value surely included 0.

Moderating effects have been widely studied in different areas of science due to their impact on relationships established by two latent variables. For instance, they have been used in the field of medicine, especially to analyze the effect of alcoholism on the synthesis of certain enzymes (Cho et al. 2004) and the social impact of disability as a condition (Rouquette et al. 2015). Similarly, hospital research has relied on moderating effects to measure the quality of patient care (Ro 2012), while the automotive sector has employed them to analyze the effect of quality certification on the performance indices of companies (Zakuan et al. 2012). Finally, moderating effects have also aimed to analyze the effect of socialization over work satisfaction (Song et al. 2015). Therefore, these effects have become keys in JIT research, since they can help understand how two variables are related to each other, and how the presence of a third variable can accelerate or decelerate the desired effect between the first two variables.

5.8.1.5 Effect Sizes

Multiple indexes are estimated to measure predictive validity in latent variables that are endogenous or dependent. For instance, Fig. 5.2 shows that latent variable 2 is explained by latent variable 1, as it is explained by the direction of the arrow. Therefore, it is expected that latent variable 2 have an acceptance percentage of variance due to latent variable 1. This index is called R-square (Wold et al. 2001; Kock 2013b). On the other hand, even though latent variable 3 depends on latent variables 1 and 2, it will have a single R-squared value obtained from the sum of the two R-squared values: one from latent variable 1 and the other from latent variable 2. Therefore, it must be estimated what percentage of the final R-squared is due to variable 1 and what percentage is due to variable 2. The sum of the two effect sizes due to two independent variables explaining a dependent variable must equal the R-square value of the dependent variable (Chatelin et al. 2002; Tenenhaus et al. 2005).

For instance, if it is assumed that a latent dependent variable is explained 83 % by two independent latent variables, latent variable 1 may explain 40 % and latent variable 2 would thus explain the remaining 43 %—or latent variable 1 may explain 22 % and the second variable would therefore explain 61 %. Note that in both cases, the sum of the percentages explained always equaled the R-square, which in this case was 83 %. Effect sizes have been widely reported in studies where dependent latent variables are explained by several independent latent variables. Examples can be consulted in Rouquette et al. (2015), Boon Sin et al. (2015), Ay et al. (2015).

5.8.2 Model Efficiency Ratios

As in models generated by regression analysis techniques, several efficiency indexes must be used in models of structural equations to estimate it before to determine its suitability. This book relied on several indexes to determine the fit of JIT structural equation model. References such as Kock and Lynn (2012), Kock (2013b) can be consulted for more details on methods of model estimation.

Major indexes for model estimation:

Average path coefficient (APC)
Average R-squared (ARS)
Average adjusted R-squared (AARS)
Average block variance inflation factor (AVIF)
Average full collinearity VIF (AFVIF)
Tenenhaus GoF (GoF)
Simpson's paradox ratio (SPR)
R-squared contribution ratio (RSCR)
Statistical suppression ratio (SSR)
Nonlinear bivariate causality direction ratio (NLBCDR)

This research tested all indexes with a 95 % confidence level. However, indexes APC, ARS, and AARS can be statistically tested by estimating their *P*-value. These *P*-values are estimated through a resampling process (Rosenthal and Rosnow 1991b). Some aspects considered to determine the model fit according to the parameters described above are:

- As for the indexes that could be tested with a *P*-value, the model fit was carried out with a 95 % confidence level, indicating a significance level of merely 5 %.
- The AARS value is generally lower than the ARS value, since the former integrates sample size in its estimation (Theil 1958; Wooldridge 1991).
- As for the AVIF and AFVIF indexes, values recommend are lower than 3.3, although values close to 5 have also been accepted (Kock and Lynn 2012; Kock 2011; García-Alcaraz et al. 2015).
- GoF, also known as Tenenhaus index or Tenenhaus GoF, is based on the communality of items found within a latent variable (Tenenhaus et al. 2005). Although the index has suffered from various modifications, its formula has remained similar. Perhaps one of the most relevant formulas to obtain GoF was reported by Wetzels et al. (2009), who stated that an index was small when its value is lower than or equal to 0.01 and an index was medium when the value obtained was equal to or greater than 0.25. Finally, indexes are large or appropriate if their values are higher than 0.36.
- The SPR index helps determine whether the proposed structural equation model is free from the Simpson paradox (Pearl 2009; Wagner 1982). Ideally, this index is expected to equal 1, which would indicate that the proposed model is fully exempt from the problem. However, it is accepted that at least 70 % of all values in the relations among latent variables be free from the Simpson paradox. Therefore, an acceptable value for the SPR index could be one higher than or equal to 0.7. Lower values would indicate that the model should be analyzed one more time.
- The RSCR index measures the extent to which the proposed model is free from negative R-squared values. R-squared can only take positive values since this is a squared index, and, conceptually, all square values and numbers are positive. Negative R-squared values occur in conjunction with the Simpson's paradox (Pearl 2009; Wagner 1982). Moreover, the RSCR index value recommended is 1, which would indicate that 100 % of latent dependent variables have positive R-square values. However, a value of 0.9 could be accepted and would indicate that at least 90 % latent variables do not have negative R-squared values.
- The SSR index expresses whether the proposed model is free from statistical suppressions, such as Simpson's paradox (MacKinnon et al. 2000). In this case, one may consider possible causality in the model—not just a relation—or an incorrect sense of the relation where the hypothesis should be proposed in the reverse direction (Spirtes et al. 1993). However, SSR index is still experimental; for the moment, it is proposed that acceptable index values must be higher than or equal to 0.7, which would indicate that 70 % of the relationships in the latent variables presented in the model are free from statistics suppressions.

- The NLBCDR index seeks to identify whether the direction established between latent variables is correct, for which the index makes an estimate of the errors obtained when a variable is independent and when it is dependent. Similar to the previous measure, NLBCR index is still at its experimental stage, although it is recommended to consider that acceptable values must be higher than or equal to 0.7, which would indicate that directionality between latent variables is weak.

References

Aboalkhair AM, Coolen FPA, MacPhee IM (2013) Nonparametric predictive reliability of series of voting systems. Eur J Oper Res 226(1):77–84. doi:10.1016/j.ejor.2012.11.001

Adamson KA, Prion S (2013) Reliability: measuring internal consistency using cronbach's α. Clin Simul Nurs 9(5):e179–e180. doi:10.1016/j.ecns.2012.12.001

Akintoye A, McIntosh G, Fitzgerald E (2000) A survey of supply chain collaboration and management in the UK construction industry. Eur J Purchasing Supply Manage 6(3–4):159–168. doi:10.1016/S0969-7012(00)00012-5

Ala-Harja H, Helo P (2014) Green supply chain decisions—case-based performance analysis from the food industry. Transp Res Part E: Logistics Transp Rev 69:97–107. doi:10.1016/j.tre.2014.05.015

Alcaraz JLG, Maldonado AA, Iniesta AA, Robles GC, Hernández GA (2014) A systematic review/survey for JIT implementation: Mexican maquiladoras as case study. Comput Ind 65 (4):761–773. doi:10.1016/j.compind.2014.02.013

Alfalla-Luque R, Marin-Garcia JA, Medina-Lopez C (2015) An analysis of the direct and mediated effects of employee commitment and supply chain integration on organisational performance. Int J Prod Econ 162:242–257. doi:10.1016/j.ijpe.2014.07.004

Al-Tahat MD, Bataineh KM (2012) Statistical analyses and modeling of the implementation of agile manufacturing tactics in industrial firms. Mathe Probl Eng. doi:10.1155/2012/731092

Ambulkar S, Blackhurst J, Grawe S (2015) Firm's resilience to supply chain disruptions: scale development and empirical examination. J Oper Manage 33–34:111–122. doi:10.1016/j.jom.2014.11.002

Avelar-Sosa L, García-Alcaraz J, Castrellon J (2014) The effect of some risk factors in the supply chain performance: a case of study. J Appl Res Technol 12(4):958–968

Avelar-Sosa L, García-Alcaraz J, Vergara-Villegas O, Maldonado-Macías A, Alor-Hernández G (2015) Impact of traditional and international logistic policies in supply chain performance. Int J Adv Manuf Technol 76(5–8):913–925. doi:10.1007/s00170-014-6308-3

Ay Y, Karadağ E, Acat MB (2015) The technological pedagogical content knowledge-practical (TPACK-practical) model: examination of its validity in the Turkish culture via structural equation modeling. Comput Educ 88:97–108. doi:10.1016/j.compedu.2015.04.017

Baxter K, Courage C, Caine K (2015) Chapter 10—Surveys. In: Caine KBC (ed) Understanding your users, 2nd edn. Morgan Kaufmann, Boston, pp 264–301. doi:http://dx.doi.org/10.1016/B978-0-12-800232-2.00010-9

Bera AK, Jarque CM (1981) Efficient tests for normality, homoscedasticity and serial independence of regression residuals: Monte Carlo evidence. Econ Lett 7(4):313–318. doi:10.1016/0165-1765(81)90035-5

Boon Sin A, Zailani S, Iranmanesh M, Ramayah T (2015) Structural equation modelling on knowledge creation in Six Sigma DMAIC project and its impact on organizational performance. Int J Prod Econ 168:105–117. doi:10.1016/j.ijpe.2015.06.007

Bruffaerts C, Verardi V, Vermandele C (2014) A generalized boxplot for skewed and heavy-tailed distributions. Stat Probab Lett 95:110–117. doi:10.1016/j.spl.2014.08.016

Carter NJ, Schwertman NC, Kiser TL (2009) A comparison of two boxplot methods for detecting univariate outliers which adjust for sample size and asymmetry. Stat Methodol 6(6):604–621. doi:10.1016/j.stamet.2009.07.001

Chang C-J, Li D-C, Huang Y-H, Chen C-C (2015) A novel gray forecasting model based on the box plot for small manufacturing data sets. Appl Mathe Comput 265:400–408. doi:10.1016/j.amc.2015.05.006

Chatelin YM, Vinzi VE, Tenenhaus M (2002) State-of-art on PLS path modeling through the available software. University of Connecticut, Connecticut

Chen J, Gan M (2014) A review of latent variable on urban travel behavior. J Transp Syst Eng Inf Technol 14(6):21–29. doi:10.1016/S1570-6672(13)60144-0

Chin TA, Hamid ABA, Raslic A, Heng LH (2014) The impact of supply chain integration on operational capability in Malaysian manufacturers. Proc Soci Behav Sci 130:257–265. doi:10.1016/j.sbspro.2014.04.030

Cho W-J, Chin W-K, Kuo C-T (2004) Effects of alcoholic moderators on anisotropic etching of silicon in aqueous potassium hydroxide solutions. Sens Actuators A Phys 116(2):357–368. doi:10.1016/j.sna.2004.04.033

Christmann A, Van Aelst S (2006) Robust estimation of Cronbach's alpha. J Multivar Anal 97 (7):1660–1674. doi:10.1016/j.jmva.2005.05.012

Cortina JM (1993) Interaction, nonlinearity, and multicollinearity: Implications for multiple regression. J Manage 19(4):915–922. doi:10.1016/0149-2063(93)90035-L

Cronbach L (1951) Coefficient alpha and the internal structure of tests. Psychometrika 16(3):297–334. doi:10.1007/BF02310555

Fornell C, Larcker D (1981) Evaluating structural equation models with unobservable variables and measurement error. J Mark Res 18(1):39–50

Frémont H, Duchamp G, Gracia A, Verdier F (2012) A methodological approach for predictive reliability: practical case studies. Microelectron Reliab 52(12):3035–3042. doi:10.1016/j.microrel.2012.07.016

Galvao AF, Montes-Rojas G, Sosa-Escudero W, Wang L (2013) Tests for skewness and kurtosis in the one-way error component model. J Multivar Anal 122:35–52. doi:10.1016/j.jmva.2013.07.002

Garcia-Alcaraz J, Oropeza-Vento M (2014) Impacto del Kaizen en el Desempeño Industrial. Paper presented at the Congreso Internacional de Academia Journals, Ciudad Juarez, Chihuahua, 3 Apr 2014

García-Alcaraz J, Prieto-Luevano D, Maldonado-Macías A, Blanco-Fernández J, Jiménez-Macías E, Moreno-Jiménez J (2015) Structural equation modeling to identify the human resource value in the JIT implementation: case maquiladora sector. Int J Adv Manuf Technol 77(5–8):1483–1497. doi:10.1007/s00170-014-6561-5

Giaquinta M (2009) Mathematical analysis: an introduction to functions of several variables. Springer, New York

Giles JA, Giles DEA (1996) Risk of a homoscedasticity pre-test estimator of the regression scale under LINEX loss. J Stat Plann Infer 50(1):21–35. doi:10.1016/0378-3758(95)00043-7

Giménez E, Crespi M, Garrido MS, Gil AJ (2012) Multivariate outlier detection based on robust computation of Mahalanobis distances. Application to positioning assisted by RTK GNSS networks. Int J Appl Earth Obs Geoinf 16:94–100. doi:10.1016/j.jag.2011.11.011

Gligor DM, Esmark CL, Holcomb MC (2015) Performance outcomes of supply chain agility: when should you be agile? J Oper Manage 33–34:71–82. doi:10.1016/j.jom.2014.10.008

Godfrey LG, Orme CD (1991) Testing for skewness of regression disturbances. Econ Lett 37 (1):31–34. doi:10.1016/0165-1765(91)90238-G

Gonzalez AA, Girotti M, Wakefield TW, Dimick JB (2013) Predictive reliability of hospital readmission rates in vascular surgery. J Vasc Surg 57(5, Supplement):52S. doi:http://dx.doi.org/10.1016/j.jvs.2013.02.186

Hair J, Anderson R, Tatham R (1987) Multivariate data analysis. Macmillan, New York

Hair J, Black W, Babin B, Anderson R (2009) Multivariate data analysis. Prentice Hall, Upper Saddle River

Hansson G, Stewart GAB, Sharp D, Lee A, Cotton DWK, Rogers S, Wilson APR (1993) Data presentation: box-plots for microbiologists? Lancet 341(8840):282. doi:10.1016/0140-6736(93)92626-5

Hayes AF, Preacher KJ (2010) Quantifying and testing indirect effects in simple mediation models when the constituent paths are nonlinear. Multivar Behav Res 45(4):627–660

Iacobucci D, Posavac SS, Kardes FR, Schneider MJ, Popovich DL (2015) Toward a more nuanced understanding of the statistical properties of a median split. J Consum Psychol 25(4):652–665. doi:http://dx.doi.org/10.1016/j.jcps.2014.12.002

Intakhan P (2014) Direct & indirect effects of top management support on abc implementation success: evidence from ISO 9000 certified companies in Thailand. Proc Soc Behav Sci 164:458–470. doi:10.1016/j.sbspro.2014.11.103

Jackson LA, Singh D (2015) Environmental rankings and financial performance: an analysis of firms in the US food and beverage supply chain. Tourism Manage Perspect 14:25–33. doi:10.1016/j.tmp.2015.02.002

Jadhav NH, Kashid DN, Kulkarni SR (2014) Subset selection in multiple linear regression in the presence of outlier and multicollinearity. Stat Methodol 19:44–59. doi:10.1016/j.stamet.2014.02.002

Jakhar SK (2015) Performance evaluation and a flow allocation decision model for a sustainable supply chain of an apparel industry. J Cleaner Prod 87:391–413. doi:10.1016/j.jclepro.2014.09.089

Jarque CM, Bera AK (1980) Efficient tests for normality, homoscedasticity and serial independence of regression residuals. Econ Lett 6(3):255–259. doi:10.1016/0165-1765(80)90024-5

Kaynak R, Sert T, Sert G, Akyuz B (2015) Supply chain unethical behaviors and continuity of relationship: using the PLS approach for testing moderation effects of inter-organizational justice. Int J Prod Econ 162:83–91. doi:10.1016/j.ijpe.2015.01.010

Kerman SC, McDonald JB (2013) Skewness–kurtosis bounds for the skewed generalized and related distributions. Stat Probab Lett 83(9):2129–2134. doi:10.1016/j.spl.2013.05.028

Ketikidis PH, Koh SCL, Dimitriadis N, Gunasekaran A, Kehajova M (2008) The use of information systems for logistics and supply chain management in South East Europe: current status and future direction. Omega 36(4):592–599. doi:10.1016/j.omega.2006.11.010

Kim T-H, White H (2004) On more robust estimation of skewness and kurtosis. Financ Res Lett 1(1):56–73. doi:10.1016/S1544-6123(03)00003-5

Kock N (2011) A mathematical analysis of the evolution of human mate choice traits: Implications for evolutionary psychologists. J Evol Psychol 9(3):219–247

Kock N (2013a) Advanced mediating effects tests, multi-group analyses, and measurement model assessments in PLS-based SEM. ScriptWarp Systems, Laredo

Kock N (2013b) Using WarpPLS in e-collaboration studies: what if I have only one group and one condition? Int J e-Collab 9(3):1–12

Kock N, Lynn GS (2012) Lateral collinearity and misleading results in variance-based SEM: an illustration and recommendations. J Assoc Inf Syst 13(7):546–580

Kohler M, Müller F, Walk H (2015) Estimation of a regression function corresponding to latent variables. J Stat Plann Infer 162:88–109. doi:10.1016/j.jspi.2014.12.006

Kopalle PK, Lehmann DR (1997) Alpha inflation? The impact of eliminating scale items on Cronbach's alpha. Organ Behav Human Decis Process 70(3):189–197. doi:10.1006/obhd.1997.2702

Kottner J, Streiner DL (2010) Internal consistency and Cronbach's α: a comment on Beeckman et al. (2010). Int J Nurs Stud 47(7):926–928. doi:http://dx.doi.org/10.1016/j.ijnurstu.2009.12.018

Kumar R, Singh RK, Shankar R (2015) Critical success factors for implementation of supply chain management in Indian small and medium enterprises and their impact on performance. IIMB Manage Rev 27(2):92–104. doi:10.1016/j.iimb.2015.03.001

Lecchi M (2011) Evaluation of predictive assessment reliability on corroded transmission pipelines. J Nat Gas Sci Eng 3(5):633–641. doi:10.1016/j.jngse.2011.07.005

Lem S, Onghena P, Verschaffel L, Van Dooren W (2013) The heuristic interpretation of box plots. Learn Instr 26:22–35. doi:10.1016/j.learninstruc.2013.01.001

Leontitsis A, Pagge J (2007) A simulation approach on Cronbach's alpha statistical significance. Mathe Comput Simul 73(5):336–340. doi:10.1016/j.matcom.2006.08.001

Li D-C, Huang W-T, Chen C-C, Chang C-J (2014) Employing box plots to build high-dimensional manufacturing models for new products in TFT-LCD plants. Neurocomputing 142:73–85. doi:10.1016/j.neucom.2014.03.043

Loperfido N (2013) Skewness and the linear discriminant function. Stat Probab Lett 83(1):93–99. doi:10.1016/j.spl.2012.08.032

Lourenço VM, Pires AM (2014) M-regression, false discovery rates and outlier detection with application to genetic association studies. Comput Stat Data Anal 78:33–42. doi:10.1016/j.csda.2014.03.019

MacKinnon DP, Krull JL, Lockwood CM (2000) Equivalence of the mediation, confounding and suppression effect. Prev Sci 1(4):173–181

Manenti F, Buzzi-Ferraris G (2009) Criteria for outliers detection in nonlinear regression problems. In: Jacek J, Jan T (eds) Computer aided chemical engineering, vol 26. Elsevier, Amsterdam, pp 913–917. doi:http://dx.doi.org/10.1016/S1570-7946(09)70152-X

Mason R, Brown WG (1975) Multicollinearity problems and ridge regression in sociological models. Soc Sci Res 4(2):135–149. doi:10.1016/0049-089X(75)90008-3

Moon KK-L, Yi CY, Ngai EWT (2012) An instrument for measuring supply chain flexibility for the textile and clothing companies. Eur J Oper Res 222(2):191–203. doi:10.1016/j.ejor.2012.04.027

Nunnally J, Bernstein I (1994) Psychometric theory. McGraw-Hill, New York

Ohtani K, Toyoda T (1980) Estimation of regression coefficients after a preliminary test for homoscedasticity. J Econ 12(2):151–159. doi:10.1016/0304-4076(80)90003-2

Patil N, Das D, Pecht M (2015) Anomaly detection for IGBTs using Mahalanobis distance. Microelectron Reliab 55(7):1054–1059. doi:10.1016/j.microrel.2015.04.001

Pearl J (2009) Causality: models, reasoning, and inference. Cambridge University Press, Cambridge

Pinto FST, Fogliatto FS, Qannari EM (2014) A method for panelists' consistency assessment in sensory evaluations based on the Cronbach's alpha coefficient. Food Qual Prefer 32(Part A 0):41–47. doi:http://dx.doi.org/10.1016/j.foodqual.2013.06.006

Preacher KJ, Hayes AF (2004) SPSS and SAS procedures for estimating indirect effects in simple mediation models. Behav Res Methods Instr Comput 36(4):617–731

Rimoldini L (2014) Weighted skewness and kurtosis unbiased by sample size and Gaussian uncertainties. Astron Comput 5:1–8. doi:10.1016/j.ascom.2014.02.001

Rindskopf D (2015) Reliability: measurement. In: Wright JD (ed) International encyclopedia of the social & behavioral sciences, 2nd edn. Elsevier, Oxford, pp 248–252. doi:http://dx.doi.org/10.1016/B978-0-08-097086-8.44050-X

Ro H (2012) Moderator and mediator effects in hospitality research. Int J Hospitality Manage 31 (3):952–961. doi:10.1016/j.ijhm.2011.11.003

Rosenthal R, Rosnow R (1991a) Essentials of behavioral research: methods and data analysis. Mc Graw Hill, Boston

Rosenthal R, Rosnow RL (1991b) Essentials of behavioral research: methods and data analysis. McGraw Hill, Boston

Rouquette A, Badley EM, Falissard B, Dub T, Leplege A, Coste J (2015) Moderators, mediators, and bidirectional relationships in the International Classification of Functioning, Disability and Health (ICF) framework: an empirical investigation using a longitudinal design and structural equation modeling (SEM). Soc Sci Med 135:133–142. doi:10.1016/j.socscimed.2015.05.007

Sarkar N (1996) Mean square error matrix comparison of some estimators in linear regressions with multicollinearity. Stat Probab Lett 30(2):133–138. doi:10.1016/0167-7152(95)00211-1

Schulze R, Hilger A-L, Engelberg PM (2015) Factor analysis and latent variable models in personality psychology. In: Wright JD (ed) International encyclopedia of the social &

behavioral sciences, 2nd edn. Elsevier, Oxford, pp 703–708. doi:http://dx.doi.org/10.1016/B978-0-08-097086-8.25010-1

Simpson Jr RJ, Johnson TA, Amara IA (1988) The box-plot: an exploratory analysis graph for biomedical publications. Am Heart J 116(6, Part 1):1663–1665. doi:http://dx.doi.org/10.1016/0002-8703(88)90791-0

Song Z, Chon K, Ding G, Gu C (2015) Impact of organizational socialization tactics on newcomer job satisfaction and engagement: core self-evaluations as moderators. Int J Hospitality Manage 46:180–189. doi:10.1016/j.ijhm.2015.02.006

Spirtes P (2015) Latent structure and causal variables. In: Wright JD (ed) International encyclopedia of the social & behavioral sciences, 2nd edn. Elsevier, Oxford, pp 394–397. doi:http://dx.doi.org/10.1016/B978-0-08-097086-8.42136-7

Spirtes P, Glymour C, Scheines R (1993) Causality, prediction and search. Springer, Berlin

Swafford PM, Ghosh S, Murthy N (2006) The antecedents of supply chain agility of a firm: Scale development and model testing. J Oper Manage 24(2):170–188. doi:10.1016/j.jom.2005.05.002

Tastle WJ, Wierman MJ (2007) Consensus and dissention: A measure of ordinal dispersion. Int J Approximate Reasoning 45(3):531–545. doi:10.1016/j.ijar.2006.06.024

Temme D, Kreis H, Hildebrandt L (2006) PLS path modeling—a software review. Humboldt University Berlin, Institute of Marketing, Berlin

Tenenhaus M, Vinzi VE, Chatelin YM, Lauro C (2005) PLS path modeling. Comput Stat Data Anal 48(1):159–205

Theil H (1958) Economic forecasts and policy. North-Holland, Amsterdam

Todeschini R, Ballabio D, Consonni V, Sahigara F, Filzmoser P (2013) Locally centred Mahalanobis distance: a new distance measure with salient features towards outlier detection. Analyt Chim Acta 787:1–9. doi:10.1016/j.aca.2013.04.034

Troskie CG, Conradie WJ (1986) The distribution of the ratios of characteristics roots (condition numbers) and their applications in principal component or ridge regression. Linear Algebra Appl 82:255–279. doi:10.1016/0024-3795(86)90156-4

Ueki M, Kawasaki Y (2013) Multiple choice from competing regression models under multicollinearity based on standardized update. Comput Stat Data Anal 63:31–41. doi:10.1016/j.csda.2013.01.019

Vandekerckhove J (2014) A cognitive latent variable model for the simultaneous analysis of behavioral and personality data. J Mathe Psychol 60:58–71. doi:10.1016/j.jmp.2014.06.004

Villanueva-Ponce R, Garcia-Alcaraz J, Cortes-Robles G, Romero-Gonzalez J, Jiménez-Macías E, Blanco-Fernández J (2015) Impact of suppliers' green attributes in corporate image and financial profit: case maquiladora industry. Int J Adv Manuf Technol 1–20. doi:10.1007/s00170-015-7082-6

Wagner CH (1982) Simpson's paradox in real life. Am Stat 36(1):46–48

Wang S-G, Tse S-K, Chow S-C (1990) On the measures of multicollinearity in least squares regression. Stat Probab Lett 9(4):347–355. doi:10.1016/0167-7152(90)90145-W

Wang C, Li J, Guo P (2015) The normalized interval regression model with outlier detection and its real-world application to house pricing problems. Fuzzy Sets Syst. 274:109–123. doi:10.1016/j.fss.2014.06.009

Wetzels M, Odekerken-Schroder G, van Oppen C (2009) Using PLS path modeling for assessing hierarchical construct models: Guidelines and empirical illustration. MIS Q. 33(1):177–196

Willaby HW, Costa DSJ, Burns BD, MacCann C, Roberts RD (2015) Testing complex models with small sample sizes: a historical overview and empirical demonstration of what Partial Least Squares (PLS) can offer differential psychology. Pers Individ Differ 84:73–78. doi:10.1016/j.paid.2014.09.008

Withers CS (1987) The bias and skewness of L1-estimates in regression. Comput Stat Data Anal 5(4):301–303. doi:10.1016/0167-9473(87)90053-3

Withers BE, Ebrahimpour M, Hikmet N (1997) An exploration of the impact of TQM and JIT on ISO 9000 registered companies. Int J Prod Econ 53(2):209–216. doi:10.1016/S0925-5273(97)00118-7

Wold S, Trygg J, Berglund A, Antti H (2001) Some recent developments in PLS modeling. Chemometr Intell Lab Syst 58(2):131–150

Wooldridge JM (1991) A note on computing r-squared and adjusted r-squared for trending and seasonal data. Econ Lett 36(1):49–54

Xiaojun L, Morris JM (1991) On measuring asymmetry and the reliability of the skewness measure. Stat Probab Lett 12(3):267–271. doi:10.1016/0167-7152(91)90089-A

Zakuan N, SrM Yusof, Saman MZM, Shaharoun AM, Laosirihongthong T (2012) The moderator effects of ISO/TS16949 certification in Thailand automotive industry. Proc Soc Behav Sci 40:141–145. doi:10.1016/j.sbspro.2012.03.173

Zimmermann R (2015) On the condition number anomaly of Gaussian correlation matrices. Linear Algebra Appl 466:512–526. doi:10.1016/j.laa.2014.10.038

Part IV
Descriptive Analysis

Chapter 6
Descriptive Analysis of the Sample

This chapter discusses a descriptive analysis related to the three objects of study: the elements of JIT, the benefits obtained after its successful implementation, and the causes of its slow implementation. A questionnaire for data collection was administered and validated in order to become familiar with the status of this lean manufacturing tool in the Mexican maquiladora sector. While this process of data collection and validation was explained in the methodology section of the previous chapter, the objective of this chapter is to characterize the sample following descriptive methods of research.

6.1 The Industrial Sectors Analyzed

The questionnaire was administered in several companies of the maquiladora industry in Juarez, Chihuahua, and it was conducted from September to November 2014. From a total of 1241 questionnaires administered in 326 companies, 144 were collected.

Table 6.1 illustrates the sectors surveyed considering the 144 valid questionnaires. Sectors were listed in descending order according to their frequency. For instance, a total of 56 questionnaires were collected in the medical sector, which represents 38.9 % of the sample. Similarly, the automotive sector provided 30 questionnaires, representing 20.8 % of the sample. Both sectors represent together 59.7 % of the sample. Additionally, only two questionnaires were collected from the packaging sector, which specializes in the design, manufacture, and distribution of media packaging and represents 1.4 % of the total sample.

J.L. García-Alcaraz and A.A. Maldonado-Macías, *Just-in-Time Elements and Benefits*,
Management and Industrial Engineering, DOI 10.1007/978-3-319-25919-2_6

Table 6.1 Industrial sectors surveyed

Sector	Frequency	Percentage	Valid percentage	Cumulative percentage
Medical	56	38.9	38.9	38.9
Automotive	30	20.8	20.8	59.7
Electronics-electrical	22	15.3	15.3	75.0
Manufacturing services	21	14.6	14.6	89.6
Plastics	6	4.2	4.2	93.8
Metals	4	2.8	2.8	96.5
Communications	3	2.1	2.1	98.6
Packaging and logistics	2	1.4	1.4	100.0
Total	144	100.0	100.0	

6.2 The Size of Companies

The Mexican city of Ciudad Juarez, Chihuahua, is known for its major manufacturing industrial center. Since it is ranked second in Mexico, several companies located in this are considerably large. Table 6.2 reports the size of the companies that were surveyed for this research. According to this information, the results about the sample are:

- From the total valid questionnaires received, 136 corresponded to companies with more than 51 employees, including operators and administrative personnel. These 136 companies were considered medium-sized and they represented 94.4 % of the total sample.
- From the total valid questionnaires obtained, 118 concerned companies with more than 501 employees. These 118 companies represented 81.9 % of the total sample.
- A total of 26 questionnaires obtained concerned companies with less than 500 employees. These companies represented 18.1 % of the sample.
- Merely 8 from the 144 questionnaires obtained corresponded to small companies who have less than 50 employees.

Table 6.2 Size of companies surveyed

Number of employees	Frequency	Percentage	Valid percentage
More than 501	118	81.9	81.9
201–500	9	6.2	6.2
Between 51 and 200	9	6.3	6.3
Less than 50	8	5.6	5.6
Total	144	100.0	100.0

Therefore, it can be stated that the majority of the surveyed companies were of large size, with enough mature logistics systems to implement JIT among their production systems.

6.3 Analysis of Size and Industrial Sector

Contingency tables, or commonly called crosstabs, are another important analyses approached for this research. This section provides cross tables between two variables: the size of companies (depending on the number of employees) and the industrial sector to which they belong. Table 6.3 depicts the intersection of these two variables from which the following conclusions can be drawn.

- Five out of the eight companies who employ less than 50 workers belong to the service sector.
- Seven out of the nine companies hiring between 51 and 200 belong to the services sector.
- As for companies employing between 201 and 500 employees, three of them belong to the plastics sector, two concern the automotive sector, two are related to the medical sector, and one company belongs to the metal sector.
- In companies with 501 employees, the largest number of questionnaires were obtained from the medical sector (54 out of 118 surveys). The automotive sector ranked second with 26 out of 118 questionnaires reported, and the electronic-electrical industry held the third place, with 19 out of 118 questionnaires obtained from it.

Table 6.3 Contingency table: Industrial sectors and number of employees

Sector	Less than 50	Between 51 and 200	201–500	More from 501	Total
Automotive	1	1	2	26	30
Communications	0	0	0	3	3
Electronics-electrical	2	0	1	19	22
Packaging and logistics	0	0	0	2	2
Medical	0	0	2	54	56
Metals	0	1	1	2	4
Plastics	0	0	3	3	7
Manufacturing services	5	7	0	9	21
Total	8	9	9	118	144

Table 6.4 Seniority of respondents

	Frequency	Percentage	Cumulative percentage
Between 2 and 5 years	44	30.6	71.5
Between 1 and 2 years	30	20.8	40.9
Less than 1 year	29	20.1	20.3
Between 5 and 10 years	26	18.1	89.6
More than 10 years	14	9.7	99.3
Subtotal	143	99.3	99.3
Not declared	1	0.7	100
Total	144	100.0	

6.4 Seniority and Position of Respondents

Information regarding the seniority and current position of respondents was crucial, since the context in which this research was conducted could be described as highly dynamic, due to frequent changes in job positions. People holding administrative positions are often promoted along the organizational structure in the company. Table 6.4 shows the seniority of respondents and allows for the following interpretations.

- There is a homogeneous distribution of seniority among respondents, since values do not show significant disparity from one category to another.
- Seniority from 2 to 5 years was reported by the majority of respondents, with a total of 44 individuals who represented 30.6 % of the entire sample.
- Seniority from 1 to 2 years was reported by 30 respondents that represented 20.8 % of the sample.
- One person did not report his/her seniority in his/her position and it represented 0.7 % of the total sample. Hence, the total sum of respondents was declared as 143.

As for the positions held, all respondents were reported to be managers in departments such as supply chain management, supplier management, and materials management, among others. Therefore, all individuals surveyed were involved in the implementation of JIT systems within their company and activities associated with the flow of materials along the supply chains.

6.5 Gender of Respondents

Gender parity in Mexico has nowadays gained a special relevance in political, governmental, and administrative positions, among many others. Since this research sought to identify the gender of all respondents, Table 6.5 introduces the contingency table showing the relation between the gender and the industrial sector to

Table 6.5 Contingency table: Industrial sector and gender

Industrial sector	Gender		Total
	Male	Female	
Medical	31	25	56
Automotive	15	15	30
Electronics-electrical	15	7	22
Manufacturing services	17	4	21
Plastics	5	1	6
Metals	3	1	4
Communications	3	0	3
Packaging and logistics	0	2	2
Total	89	55	144

which the respondents belonged. The following conclusions can be drawn according to the information provided:

- On the one hand, 89 from the 144 respondents were reported to be male managers and 55 as female managers. This does not represent gender parity, at least in positions related to supply chains and productions systems.
- On the other hand, two women occupied managerial positions in the packaging sector.
- As for the automotive sector—one of the most important sectors of the region, Table 6.5 depicts a more equilibrated distribution of positions by gender, since 15 males and 15 females were reported to hold managerial positions.
- The electrical and electronics sector shows disparity of gender among respondents. Table 6.5 reported a total of 15 male and 7 female directors, since the ratio is almost 2 male to 1 female manager.
- The medical sector also depicts a homogenous distribution of positions according to gender. Table 6.5 reported 31 male respondents and 25 female respondents holding managerial positions from a total of 56 people.

6.6 Chapter Summary

This may be one of the briefest chapters, since it merely describes the sample from which data was collected. However, relevant information is obtained from this section:

- The most representative sectors of the maquiladora industry in the region of Ciudad Juarez, Chihuahua concern the medical, automotive, and electrical-electronic sectors.
- There are a considerable number of maquiladora companies in the region of Ciudad Juarez offering a variety of services.
- The packaging industry has become an emerging sector in the region.

- Most of the surveyed companies were large organizations. They relied on well-defined supply chain models with high levels of maturity in the implementation and application of the JIT methodology. These companies represented 80 % of the total sample.
- Seniority of respondents was a homogeneous variable, since little disparity appeared in the different ranges of this variable.
- Despite efforts from the government of Mexico to promote and maintain gender parity in all contexts, this sample depicted an unequal distribution of positions based on the gender of respondents. For the maquiladora industry, this study demonstrated a possible bias toward the masculine gender in management positions associated with supply chains.

Chapter 7
Descriptive Analysis of the Elements of JIT

This third part of the book provides a descriptive analysis for every of the three objects of study: JIT elements, the benefits obtained after its successful implementation, and causes of its slow implementation. More specifically, this chapter introduces the descriptive analysis for the elements that several authors point out as essential or indispensable for the success of JIT.

As Chap. 2 mentioned, the elements required for a successful JIT implementation in production lines were divided into three categories for their study, which are reminded below:

- Elements associated with human resources
  - Flexible workforce
  - Zero defects programs
  - Multifunctional workers
  - Workers motivation
  - Short delivery times
  - Error prevention
  - Long-term contracts
  - Self-correction of errors
  - Employee empowerment in QC
  - Supplier quality certification
  - Evaluation and selection of suppliers
  - Effective communication
- Elements associated with the production process
  - Kanban system
  - Preparation time reduction (SMED)
  - Small batch sizes
  - Cellular manufacturing
  - Safety stock
  - Improved layout
  - Just-in-time purchasing
  - Process control
  - Standardized containers

J.L. García-Alcaraz and A.A. Maldonado-Macías, *Just-in-Time Elements and Benefits*, Management and Industrial Engineering, DOI 10.1007/978-3-319-25919-2_7

  - Technology groups
  - Process flexibility
  - Reduction of work in process
  - Specialized factories
  - Kaizen
  - Programming below capacity
  - Use of robots in production process
  - Pull system
- Elements associated with the product
  - Quality circles
  - Total quality control
  - Statistical quality control
  - Quality development programs
  - Continuous quality improvement
  - Zero defects
  - Quality-oriented training
  - High visibility of QC
  - QC long-term commitment
  - Regular quality and reliability of audits
  - Quality culture
  - 100 % quality inspection
  - Simplifying the process of total quality

7.1 JIT Elements Associated with Human Resources

Previous chapters found 12 elements associated with human resources that are required for a successful JIT implementation. Since in this study data were obtained by means of subjective assessments, the average or standard deviation for every element could not be reported. Therefore, the median of every element was considered as the central tendency measure, while the interquartile range represents the measure of dispersion. Table 7.1 illustrates the median and the interquartile range values obtained for these elements, which were listed in descending order according to their median values.

According data contained in Table 7.1, the following conclusions offered.

- Eleven elements registered median values higher than 4. This indicates that respondents considered them important for JIT implementation. Merely one element possessed a median value below 1 but close to 4.
- *Short delivery times*: This element registered the highest median value—4.57—and the lowest interquartile range—0.91—from the group of elements. Moreover, it is the only element with all quartile values superior to 4 and an interquartile range value lower than 1. All this demonstrates that short delivery

Table 7.1 Descriptive analysis of JIT elements associated with human resources

Order	JIT element	Percentiles			IQR
		25	50	75	
1	Short delivery times	4.02	4.57	4.93	0.91
2	Supplier quality certification	3.82	4.48	4.86	1.04
3	Error prevention	3.78	4.46	4.88	1.10
4	Evaluating and selecting suppliers properly	3.73	4.44	4.91	1.18
5	Multifunctional workers	3.67	4.39	4.98	1.31
6	Effective communication	3.61	4.34	4.93	1.32
7	Workers motivation	3.45	4.28	4.91	1.46
8	Zero deviation from production programs	3.45	4.26	4.88	1.43
9	Self-correction	3.51	4.25	4.85	1.34
10	Employee empowerment in QC	3.43	4.14	4.75	1.32
11	Strengthen training programs	3.41	4.14	4.76	1.35
12	Long-term contracts	3.01	3.87	4.64	1.63

times are considered by respondents as the most important element from human resources to ensure a successful JIT program. This could be to some extent explained since, in order to achieve short delivery times, companies need proper synchronization between the production planning system and the manufacturing system of manufacturing. However, this synchronization requires the implementation of many other elements, such just-in-time purchases, relying on certified suppliers, and above all, counting on highly motivated workers. Further information on this element within a JIT environment can be found in Gupta and Heragu (1991), Inman et al. (2011), and Miltenburg (1993).

- *Supplier quality certification*: The median value of this element is 4.48, merely nine hundredths lower than the previous element. Moreover, its interquartile range value is 1.04 and the second lowest among the group of elements. This indicates high consensus among respondents on the importance of relying on certified suppliers. In fact, when companies do not face problems with the quality of the materials purchased, they can expect a continuous flow of materials through their production systems. This also favors short-time deliveries of the final product. Additional information regarding the role of certified suppliers in JIT systems can be found in Beikkhakhian et al. (2015), Du et al. (2015), Gurel et al. (2015), Pazhani et al. (2015), and Rajesh and Malliga (2013).
- *Error prevention*: The median value of this element is 4.46—13 hundredths lower than the first element and two hundredths in comparison to the second element. Moreover, the interquartile range value is 1.10, which indicates consensus among the respondents on the importance of this element. The impact of error prevention on JIT systems is reported in Amasaka (2007), Bendul and Skorna (2015), Gupta and Heragu (1991), and Huq and Huq (1994).

- *Evaluation and selection of suppliers*: This element refers to the evaluation and selection of suppliers from the procurement departments of companies to proceed to a purchase order. The median value of this element is 14.44—13 hundredths lower than the first element—and the interquartile range value is 1.18. This demonstrates that all respondents agree on the importance of appropriate evaluation and selection of suppliers in JIT systems. Further information addressing this element is reported in Aksoy and Öztürk (2011), Araz and Ozkarahan (2007), Iwase and Ohno (2011), Kojima et al. (2008), and Prahinski and Benton (2004).
- *Multifunctional workers*: This is closely related to the seventh element that concerns workers motivation, since disposition from operators to become multifunctional is closely related to their motivation. Motivated workers may be eager to learn how operations are performed along the whole line of production; this way they become multifunctional. However, if operators are not fully motivated, they may not be interested in learning about the different positions along the production line of production and would likely focus on performing their corresponding tasks. Different discussions addressing multifunctional workers and their training required can be found in Alcaraz et al. (2014), Amasaka (2007), Monden (2002), and Villa and Taurino (2013).
- *Effective communication*: this may be considered crucial yet challenging to implement and achieve throughout the organizational structure of the company. Information should be as direct as possible in all levels to avoid distortion of the original message and misunderstandings between superiors and operators or executors of the order. It is unfortunate that well given information or instruction is frequently distorted throughout the different levels of the hierarchical structure. Further information addressing communication in the process of JIT implementation can be obtained from the following references: Rentroia-Bonito et al. (2005), Alcaraz et al. (2014), Amasaka (2014), Chaumette and Vignéras (2004), and Pragman (1996).
- *Zero deviation from production schedule*: Supervisors are responsible for executing production orders and making modifications and alterations to the programs when needed. However, these deviations can become a serious issue when the order executed turns out to be different from the order actually placed. Thus, it is crucial that companies adhere to the initial production orders to avoid later deviations. Similarly, there must always be a full understanding of the requirements for every production order, since a lack of the understanding often generates these deviations. References, such as Das and Patnaik (2015), Houghton and Portougal (1997), and Wang et al. (1999), address the subject of planning production and the consequent deviations that may arise from it.
- Elements 9, 10, and 11 are all associated with the performance of workers and relate one another to some extent. They must be able to autocorrect defects that they may find in production process to facilitate the flow of materials throughout the production system. However, to achieve this they must be provided with suitable training based on the needs of the sector where they labor. As a result, their acquired education will ensure their reliability, and supervisors may be

willing to grant greater responsibilities to them. Literature addressing the role of human resources in the process of JIT implementation can be found in de Menezes et al. (2010), Hiltrop (1992), Huber and Brown (1991), Jayaram et al. (1999), Lengnick-Hall et al. (2013), Power and Sohal (2000), and Santiago and Alcorta (2012).

- *Long-term contracts with workers*: This element was ranked by respondents of this survey as the least significant element for a successful JIT system and is the only items with a median value lower than 4. However, long-term contracts are usually effective in Japan, and Eastern firms in general, due to the high integration of employees in their companies. Hence, workers regularly count on lifetime employment. Additional information addressing the role of long-term contracts in the relationship between a company and its human resources can be consulted in Power and Sohal (2000) and Jabbour et al. (2013).

7.2 Elements of JIT Associated with the Production Process

This category integrated 18 elements linked with production process. Table 7.2 depicts the descriptive analysis of these elements which are ordered in descending order according to their median or 50th percentile. The conclusions drawn from this univariate is described below.

- Eleven elements of this category showed median values higher than 4, while seven have median values that were below 4.
- Three items show 25th percentile values inferior to 3, and 15 have 25th percentile values superior to 3 but inferior to 4.
- None of the elements analyzed in Table 7.2 shows interquartile range values lower than 1, yet 17 of them show interquartile range values higher than 1 and lower than 2. This indicates that less consensus was registered in this category regarding the median values of the elements. Moreover, element 18 shows an interquartile range value equal to 2, which represents the lowest value for this category.
- *Just-in-time purchasing*: It has been previously stated that a JIT system begins with the supply of raw materials. Thus, it begins with purchases requested at the right time, in the required amount, and with the expected quality. This element holds the first place in the list according to its median value of 4.3, which points out the importance of JIT purchases for respondents. Similar results were reported and can be found in Hong and Hayya (1992), Handfield (1993), Gilbert et al. (1994), Gélinas et al. (1996), Germain and Dröge (1997), Gunasekaran (1999), De Toni and Nassimbeni (2000), Kaynak and Hartley (2006), and Min and Sui Pheng (2007).

Table 7.2 Descriptive analysis of JIT elements associated with production process

Order	JIT Element	Percentiles			IQR
		25	50	75	
1	Just-in-time purchasing	3.51	4.30	4.91	1.40
2	Total preventive maintenance	3.46	4.28	4.90	1.44
3	Preparation time reduction (SMED)	3.48	4.25	4.86	1.38
4	Kaizen system	3.42	4.24	4.87	1.45
5	Kanban system	3.45	4.21	4.83	1.38
6	Pull system	3.34	4.19	4.85	1.51
7	Improved plant layout	3.46	4.18	4.79	1.33
8	Process control	3.47	4.18	4.78	1.31
9	Process flexibility	3.35	4.17	4.83	1.48
10	Reduction of work in process	3.37	4.12	4.75	1.38
11	Group technology	3.31	4.04	4.70	1.39
12	Safety stock	3.2	3.98	4.67	1.47
13	Small batch sizes	3.17	3.93	4.66	1.49
14	Cellular manufacturing	3.22	3.93	4.64	1.42
15	Standardized containers	3.08	3.78	4.52	1.44
16	Specialized factories	2.7	3.59	4.44	1.74
17	Scheduling below installed capacity	2.52	3.46	4.31	1.79
18	Use of robots	2.39	3.46	4.39	2.00

- *Continuous improvement*: Total Preventive Maintenance (TPM) must be applied to the machinery and equipment used for the manufacturing operations throughout the production system. Moreover, TPM can be crucial for a successful JIT implementation since when equipment and machines do not work or become damaged, the flow of materials can slow or the production system can stop, which could lead to very little inventory in process. Therefore, to ensure the continuous flow of materials throughout the production system and guarantee just-in-time deliveries, companies must rely on available machinery and equipment at any time. Besides, it shall not be forgotten that machines are useful and profitable only if they are in operation. This element has a median value of 4.28, which is two hundredths lower than the element holding the first place in Table 7.2. Literature regarding total preventive maintenance within a Just In Time and Lean Manufacturing environment can be found in the following references, which provide similar results to those found here: Cua et al. (2001), Eti et al. (2004), McCarthy and Rich (2004), Smith and Hawkins (2004), Rodrigues and Hatakeyama (2006), Thomas et al. (2006), Ahmad et al. (2012), Bakri et al. (2012), and Singh et al. (2013).
- *Preparation time reduction (SMED)*: Since machines are used to provide a service or produce certain quantities of production orders, they must often be calibrated and adjusted according to the needs of each product. The third element of this list refers to the reduction of preparation time of this machinery. It

shows a median value of 4.25, indicating that not only is essential to maintain machinery and equipment in a proper state with a suitable maintenance program, but also to make rapid changes to this equipment to ensure the flow of materials. The median value of this element is five hundredths lower than the first element, which demonstrates close values, and similar importance, among the elements of this category. For readers willing to consult additional literature on the application of SMED in Just In Time, lean manufacturing environment, and total productive maintenance programs, the following references can be recommended. They have reported similar results to those of this research Eti et al. (2004), Almomani et al. (2013), Ferradás and Salonitis (2013), Kemal Karasu et al. (2014), Kumar et al. (2014), and Sundar et al. (2014).

- *Kaizen*: Also referred to as continuous improvement, this element shows a median value of 4.24, with a difference of one hundredth in comparison with the third element and six hundredths if compared with the first element. These values indicate that responders considered continuous improvement as an essential element in just-in-time systems. In fact, the opinions of employees regarding the areas of opportunity within the production system could undoubtedly facilitate the flow of materials. Further information concerning the role of Kaizen in a production environment of lean manufacturing and just in time can be consulted in Lyu (1996), Radharamanan et al. (1996), Recht and Wilderom (1998), Taghizadegan (2006), Moore (2007), Kumiega and Van Vliet (2008), Farris et al. (2009), Ortiz (2010), and Knechtges and Decker (2014).
- *Kanban system*: This element was ranked as the fifth most important in Table 7.2 which demonstrates that one of the tools of lean manufacturing is a pillar for JIT systems. Also referred to as a cards system, Kanban ensures the flow of materials throughout the production system, since every workstation knows how the component or subassembly will be assembled and when to refill containers that reach the minimum level of inventory. This element shows a median value of 4.21, with a difference of three hundredths if compared to the previous element and nine hundredths in comparison with the first elements. The median value also highlights the importance of Kanban system for responders to ensure a successful JIT implementation. Additional literature concerning the impact of Kanban system on JIT environments can be found in Fiscus (1987), Reda (1987), Fukukawa and Hong (1993), Markham et al. (1998), Chan (2001), Liberopoulos and Koukoumialos (2005), and Wang and Sarker (2005, 2006).
- *Pull system*: This element shows a median value of 4.19 that is 11 hundredths lower than the median value of the first element and indicates the importance of pull systems in the process of JIT implementation for respondents. Pull system is closely related to Kanban systems and additional literature that addresses its role in a lean manufacturing and just in time environment can be consulted in Kim and Tang (1997), van der Laan et al. (1999), Ohno (2011), and Selçuk (2013).
- *Improved plant layout*: A properly organized distribution of production cells or work areas allows for little movement of materials that would not add them any

value. However, inappropriate distribution of production areas could cause large displacements of materials and assemblies. This element has a median value of 4.18 that is one hundredth lower than the median value from the previous element concerning the implementation of Pull system. The median value also indicates how significant respondents considered the improvement of plant layouts as a means to meet the objectives planned for the company when implementing JIT. Researches in Inman et al. (2011), Villa and Taurino (2013), Alcaraz et al. (2014) and Kia et al. (2014) highlight the importance of plant distribution systems in an environment where JIT was implemented as a strategy of lean manufacturing,

- *Process control*: Companies should properly document all materials flowing in their production lines; otherwise, uncontrolled production processes could fail to provide important information on the production order, such as its status, the personnel responsible for its execution, the machines that would operate, and, above all, the delivery time of the finished product. This element has the same median value as the previous one, and it is 12 hundredths lower than the median value of the first item. Thus, process control and knowing the status of production orders was considered of vital importance for managers surveyed in this research. References such as Banerjee and Armouti (1992), Sengupta et al. (1993), Benton and Shin (1998), Trentesaux et al. (1998), Huq (1999), Al-Tahat and Mukattash (2006), and Moattar Husseini et al. (2006) can provide additional information regarding the role of process control in production systems to compare and analyze the results with those provided in this book.
- *Process flexibility*: Production processes must be flexible enough to cope with quick changes and face uncertainty of demand; that is, meet unexpected orders from the customer. It is closely related to the third element concerning the reduction of preparation time of equipment. A system cannot be flexible if production lines are unable to handle quick changes from one product design to another. This element has a mean value of 4.17 that indicates its level of significance for the respondents surveyed in this research. The median value of this item is also one hundredth lower—almost negligible—if compared to the previous element and 13 hundredths in comparison with the median value of the first element. Additional literature addressing flexibility in a JIT environment can be found in Moattar Husseini et al. (2006), Narasimhan et al. (2004), Maiga and Jacobs (2009), Inman et al. (2011), Weng et al. (2012), and Green et al. (2014).
- *Reduced work in process*: The reduction of inventory in process along the production line facilitates the flow of materials. This type of inventory is usually generated when components are missing and the production cannot continue. However, inventory can also be generated when human resources struggle to handle deviation of production programs, since when companies receive an urgent order, orders executed at that time are usually suspended and can generate inventory in process. It is important to note that several elements required in a successful JIT implementation are closely linked. References such as

Houghton and Portougal (1997), Baykoç and Erol (1998), Hou and Hu (2011), Alcaraz et al. (2014), and Deif and ElMaraghy (2014) provide additional information regarding the impact of inventory in process on the performance indexes of a JIT system.

- *Group technology*: A continuous flow of materials can be achieved not only with a proper organization of production, but also by effectively organizing the equipment used to transform the raw materials. Group technology concerns groups of machines with similar operations. Products manufactured within in these groups do not need to move along the whole industrial plant, since they are manufactured in one small and single place. Thus, avoiding unnecessary movement of materials can be redeemable to companies. Moreover, personnel these groups become experts in the activities of the group and specialize in the use of certain machinery and equipment. This high level of expertise in a number of activities ensures that workers become multifunctional. Additional literature addressing group technology can be consulted in Jensen et al. (1996), Yasin et al. (1997, 2003), Amasaka (2002), Sandanayake et al. (2008), Pourbabai (1988), and Spencer (1998).
- *Safety stocks*: This is the first element with a median value lower than 4, and it was ranked as the twelfth most important element of production process to ensure a successful JIT implementation. Even though JIT systems seek to minimize inventories, there are sometimes necessary despite the cost that they represent, which could be higher if companies did not rely on safety stock to deal with uncertainty in demand. For instance, if an error occurred in one of the operations of a certain production order, certain components of the assembly could become useless and would have to be replaced. Thus, obtaining the new piece from the safety inventory implies that companies cover the maintenance cost of stock. However, if the piece is not available in inventory, the production line could stop, which would generate inventories in process and late deliveries to customers. Safety inventories can become a challenge in JIT environments, since they are opposed to some of the principles of the philosophy. For a further comprehension of the role of safety inventories in environments of Just In Time and lean manufacturing, readers may consult the following references: Vellani (2007), Louly and Dolgui (2013), and Arıkan et al. (2014).
- *Small batch sizes*: This element was ranked as the thirteenth most important element from production process to ensure a successful JIT implementation. Small lots of orders of supplies can save costs of inventory maintenance, since it may be more suitable to request several orders of small quantities than few orders of large quantities. However, in order to work with small bath sizes, companies must be able to either forecast future demand, or rely on one that is sufficiently stable. Additional research regarding the impact of small lot sizes on the efficiency indexes of a JIT program can be found in Cao and Schniederjans (2004) and Mackelprang and Nair (2010).
- *Cellular manufacturing*: While group technology manufactures products with certain similarities, cell manufacturing groups equipment according the

activities that will be performed in every cell. This allows highly trained personal in the execution of such activities, and the consequences of human errors become fewer (McLaughlin and Durazo-Cardenas 2013). Cellular manufacturing is related to several JIT elements from human resources. In addition, movement of materials diminishes since nowadays products within a cellular manufacturing structure can be completely manufactured in a single place. Information related to cellular manufacturing can be consulted in the following literature: Pattanaik and Sharma (2009), Chiarini (2013), and Marodin et al. (2015).

- *Standardized containers*: Standardization can become one of the most challenging stages in the implementation of a 5S system, since it seeks that all tasks be performed similarly in a production system. It is crucial for manufacturing systems, and especially maquiladoras, to rely on standardized containers for the processing of material. This could allow operators to understand the amount of material required for each product as well as its manufacturing process. Standardized containers must be used from the moment raw material is received until the distribution of the finished product, although they can be different for every activity in the production process. References such as Yasin et al. (1997, 2003), Benton and Shin (1998), and Villa and Taurino (2013) can provide additional information regarding modifications required in companies during the JIT implementation process.
- *Specialized factories and use of robots in the production process*: Although the latter element was ranked last among elements from Table 7.2, it seems appropriate to address it at this moment since they both concern the involvement of managers. Companies should rely on specialized factories and the use of robots, since they guarantee that machinery, equipment, and human resources in the production line are specialized, which consequently improve the flow of materials throughout the system. However, when equipment and human resources become extremely specialized in certain tasks, flexibility can be compromised and the new technology may become challenging to adapt to new lines of production when a product is no longer manufactured. It must be mentioned that elements 16, 17, and 18 show first quartile values lower than 3, which indicates that respondents considered them regularly significant. Further literature regarding modifications required during the process of JIT implementation can be consulted in Gupta et al. (1992), Oral et al. (2003), Yasin et al. (2003), Amasaka (2007), and Villa and Taurino (2013).
- *Scheduling below installed capacity*: This may bring conflicts along the production system, since it could be believed that companies may be unable to supply all the requested orders on time. However, these companies could draw upon an outsourcing system, always ensuring the quality of their product from contractors. Note that the median value of this element is 0.84 hundredths lower than the median value of the first element, which is a considerable difference. Moreover, this element has the maximum interquartile range value. Additional information on outsourcing systems in JIT environments is found in Collins et al. (1997), Gonzalez et al. (2006), and Baraldi et al. (2014).

7.3 Elements Associated with the Product

This category is composed of 13 elements required for a successful JIT implementation and associated with the product generated. They mainly concern quality and its characteristics. Table 7.3 lists these elements in descending order based on their median or 50th percentile values.

The analysis in relation to Table 7.3 allowed for the inference of the following conclusions:

- All interquartile ranges of this category are higher than 1 but lower than 1.5, which indicates consensus from respondents on the median values obtained for every elements and little dispersion in data.
- All elements have first quartile or 25th percentile values higher than 3 but lower than 4, and since none of them exceeds this value, it can be concluded that all elements associated with the product are regularly important for respondents.
- All elements show second quartile or 50th percentile higher than 4. It can thus be concluded that respondents consider all these elements as highly significant, since they are associated with product quality.
- Seven elements show third quartile or 75th percentile values that equal or are higher than 4.90. Thus, according to respondents, these elements are of utmost importance for the execution and implementation of a JIT program. Note that the second element concerning quality culture shows the highest third quartile value that is 5, which is also the maximum permissible value.
- *Continuous quality improvement*: This element shows a median value of 4.47, which demonstrates its high importance to respondents. Moreover, the element has the lowest interquartile range value reported in this list. This indicates the elevated consensus from respondents concerning its high value. A program

Table 7.3 Descriptive analysis of JIT elements associated with the product

Order	JIT Elements	Percentiles			IQR
		25	50	75	
1	Continuous quality improvement	3.79	4.47	4.92	1.13
2	Quality culture	3.67	4.40	5.00	1.33
3	Total quality control	3.56	4.34	4.95	1.39
4	Statistical quality control	3.6	4.33	4.91	1.31
5	Zero defects	3.57	4.33	4.93	1.36
6	Long-term QC commitment	3.56	4.31	4.92	1.36
7	Quality-oriented training	3.51	4.28	4.89	1.38
8	Quality development programs	3.45	4.27	4.90	1.45
9	High visibility of QC	3.5	4.25	4.86	1.36
10	Simplifying the process of total quality	3.47	4.25	4.87	1.40
11	Quality circles	3.42	4.22	4.85	1.43
12	Regulated quality and reliability of audits	3.36	4.15	4.80	1.44

seeking to continuously improve quality involves the execution of short-, medium-, and long-term plans. This means that quality does not simply emerge but it is the result of appropriate planning. Continuous quality improvement in JIT systems prevents quality-related problems throughout the production process and therefore favors a smooth flow of materials. In fact, numerous defects in parts or assemblies can compromise the complete lot of a product, and although many of these pieces could be placed in special containers along the production system for their reprocessing, they would generate inventory in process, which companies would later seek to minimize. Additional references such as Withers et al. (1997), Hipkin and De Cock (2000), Cua et al. (2001), Rahman and Bullock (2005), Agus and Hassan (2011), Inman et al. (2011), and Ahmad et al. (2012) emphasize on the role of quality planning in an industrial production environment and can be recommended for a particular interest in the topic.

- *Quality culture*: A quality plan by itself guarantees neither the quality of the product nor the continuous flow of materials throughout the production system. Quality plans are usually designed by senior management and executives, who must also be responsible for promoting quality culture among human resources, especially operators and middle managers, since they execute the quality plans implemented. This element shows a median value of 4.40, which demonstrates its level of significance, which was also reported by other authors, such as Gupta and Heragu (1991), Vora (1992), and Yasin et al. (2003). In these works, authors addressed the modifications and organizational adjustments required for a successful quality program in JIT systems, since the benefits obtained from the implementation of this philosophy are often perceived from a mere economic point of view (Brox and Fader 1997; Oral et al. 2003), and its implementation can be challenging due to aspects that are purely cultural (Oral et al. 2003; Machuca 2002). Therefore, companies with philosophies different from Eastern beliefs may endeavor to ensure the success of their JIT systems, since cultural reasons can be the cause of its early abandonment.
- *Total quality control*: Companies must monitor their quality plans as well as the appropriate dissemination of quality culture. This implies a total quality control along the production system which must be carried out from the reception of materials until the distribution of finished products. It is important to note that this control does not guarantee quality; it merely informs about its status in comparison with its plan. Thus, when deviations occur, companies must immediately take corrective actions, since these deviations can indicate quality errors in the production system that may generate additional work, inventory in process, poor quality products, and a discontinuous flow of materials throughout the production system. Additional information concerning the role of total quality control in a lean manufacturing and Just In Time environment can be found in Ahmad et al. (2012), Bou and Beltrán (2005), Flynn et al. (1995), Forza and Filippini (1998), Hipkin and De Cock (2000), Jun et al. (2006), Ooi (2014), Prajogo and Sohal (2001), Rahman and Bullock (2005), Vuppalapati et al. (1995), and Withers et al. (1997).

- *Statistical quality control*: This element shows a median value of 4.34 that is merely three hundredths below the median value of the first element. These point out its high level of significance for respondents. In order to implement a program of total quality control and provide the appropriate follow-up, companies must capture and analyze information obtained from the production process. This is traditionally referred as to statistical quality control, it provides with an overview of the status of the originally established quality plan and allows for taking corrective actions when needed. The absence of statistical quality control would represent an unexecuted plan for companies and would prevent them from being informed of the real situation of quality in a production system. Readers particularly interested in statistical quality control within an environment of lean manufacturing and Just in Time may wish to consult the following references: Ahmad et al. (2012), Fullerton and McWatters (2002), Gordon et al. (1994), and Lim et al. (2014).
- *Zero defects*: This element was ranked as the fourth most significant element of this category. Moreover, it shows the same median value as the previous element, which indicates its level of significance for managers of the manufacturing sector. Achieving zero defects in production lines is part of quality culture since it ensures the appropriate flow of materials. However, the implementation of a zero-defect philosophy requires a high training of human resources. They must be able to detect errors and solve them immediately—or propose solutions to solve them—in order to avoid that, defective products receive more added value in subsequent operations. Thus, this element is closely linked to elements of human resources who, as it was previously mentioned, become responsible for carrying out quality plans and generating quality in the product. Additional information regarding the impact of the zero defects philosophy on a production system can be found in: Benton and Shin (1998), Fullerton and McWatters (2002), Green et al. (2014), Gupta and Heragu (1991), Gupta et al. (1992), Inman et al. (2011), and Nassimbeni (1996).
- *Long-term commitment to QC*: Quality is generated through human resources, and companies must commit to long-term quality control. This corresponds mostly to senior management, which is responsible for the creation of quality plans. Long-term commitment to quality control has been widely reported and has also been referred as to managerial commitment. Similarly, several models nowadays explain its impact over quality indices and lean manufacturing. Many of these models were based on causal analyses (Alcaraz et al. 2014; Alfalla-Luque et al. 2015) either from an economic point of view or according to the organizational changes that must be performed (Lemak et al. 1997; Abdul-Nour et al. 1998; Huq 1999; Inman et al. 2011).
- *Quality-oriented training*: This element was ranked seventh in this category and concerns operating personnel, not administrative or executive staff, as it was in the case of previous elements. Moreover, it shows a median value of 4.28 that is three hundredths lower than the previous element and 19 hundredths if it is compared with the first element of table. Operators are the direct generators of

quality, who require statistical knowledge of the production process that could be gained through specialized training. Therefore, training provided to operators must be chiefly based on quality aspects of the product they manufacture, without neglecting other aspects that are also important. Causal models have associated education of employees with the performance indices of JIT and supply chains (Alcaraz et al. 2014), and they have reported that from an operational point of view, efficiency is improved with the amount of training provided to operators (Inman et al. 2011). Similarly, other approaches have emphasized on the role of training and education of employees in production process (Guidetti and Mazzanti 2007; Jacobs 2010; Trad and Kalpić 2014).

- *Quality development programs*: It is not recommended that companies rely on a single and static plan to guarantee product quality and the continuous flow of materials throughout the production system. It has been reported that regular meetings must be scheduled to continue developing and adjusting the quality plan; otherwise, the company could not undertake necessary modifications to remain in the market. Therefore, senior executives of companies should organize meetings periodically to ensure the dynamism of their production systems. This element has a median value of 4.27, which indicates its significance for respondents. This value is also one hundredth lower than the median value of the previous element but 20 hundredths lower if it is compared with the first element of table. Readers particularly interested in the development of quality programs from managers can refer to Ahmad et al. (2012), Barve et al. (2009), Cua et al. (2001), Hipkin and De Cock (2000), Jun et al. (2006), Ooi (2014), Rahman and Bullock (2005), and Salleh et al. (2012).
- *High visibility of HQ*: This element has a median value of 4.25 that is two hundredths lower than the previous element and 22 hundredths if it is compared with the median value of the first element in table. In order to ensure an adequate flow of materials throughout the production system and meet the expected efficiency indices of a JIT program, companies must endeavor to promote the flow of information throughout the system. This can be achieved when not only the senior management but also quality generators and operators become familiar with the quality indices of the company. It has been demonstrated that this visibility of quality indices encourages employees to endeavor to achieve the goals established in the quality plants, which also favors a greater amount of deliveries. Readers particularly interested in this topic can consult (Zairi 1991a, c, 1993; Ghobadian and Gallear 1996; Ahmed and Rafiq 2002) that address tools and techniques to implement in quality programs.
- *Simplifying the process of total quality*: This element shows a median value of 4.25 that is 22 hundredths lower if it is compared with the firs element of the list. The implementation of a total quality program may result tedious and cumbersome for administrators since a great amount of information must be registered and continuously controlled. Several authors have addressed this issue and ISO 9000 certification programs have also proposed a standardized form of presenting and controlling such information. Thus, total quality process must be

simplified in a way that is clear and understandable to all operators and human resources from all levels of the organizational structure. Quality should be a universal language within companies instead of privileged information merely available to the administrative department. A simplification of the process of total quality could bring on greater understanding of quality, improve communication, and ensure greater commitment. Additional literature studying the simplification of quality programs in an environment Just in Time and lean manufacturing can be consulted in Zairi (1991b), Fisher et al. (1995), Dooley and Flor (1998), and Forza and Filippini (1998).
- *Quality circles*: They have become an important part of successful quality programs. Employees in quality circles propose solutions to problems associated with quality and the flow of materials throughout the production system, since they are highly familiar with the production process. This element has a median value of 4.22 that is 3 hundredths lower than the previous element and 20 hundredths if it is compared with the median value of the first element in table. Readers particularly interested in the role of quality circles in the assurance of quality programs may consult the following references: Elmuti and Kathawala (1990), Rosenfeld et al. (1991), Dale et al. (2001), Kent (2009), Blaga and Jozsef (2014), and Huang et al. (2015).
- *Regulated quality and reliability of audits*: This study found that reliability and quality of audits play an important role in the efficiency indexes expected from a JIT implementation. Quality audits may be internal or external. Internal audits inform of the status of quality within the company; they are usually run to obtain a certain certification of accreditation. However, it is important to regularly run this type of audits by specialists that are external to the company, since they can analyze a problem that administrators and operators could not distinguish previously. Additional references that have studied quality and credibility of audits can be consulted in Rahmina and Agoes (2014), Sarwoko and Agoes (2014), Ball et al. (2015), Jiang et al. (2015), Laitinen and Laitinen (2015), Maroun (2015), and Miko and Kamardin (2015).

7.4 Chapter Conclusions

This chapter addressed the descriptive analyses of the JIT elements associated with human resources, production process, and products. The median value for every element was obtained as a measure of central tendency and the interquartile range was considered as the measure of dispersion. The analyses allowed for the inference of the following conclusions:

- Short deliveries to customers were considered the most important element associated with human resources to ensure a successful JIT implementation, since it demonstrates proper coordination between push and pull systems.

- Just-in-time purchasing was considered the most important element associated with the production process to ensure a successful JIT implementation. This demonstrates that JIT philosophy does begin with timely acquisition of raw materials.
- Quality programs were considered the most important element associated with the product to ensure a successful JIT implementation, since they can guarantee the flow of materials with zero defects within production systems.

References

Abdul-Nour G, Lambert S, Drolet J (1998) Adaptation of jit phylosophy and kanban technique to a small-sized manufacturing firm; a project management approach. Comput. Ind. Eng. 35(3–4): 419–422. doi:10.1016/S0360-8352(98)00123-5

Agus A, Zf Hassan (2011) Enhancing Production Performance and Customer Performance Through Total Quality Management (TQM): Strategies For Competitive Advantage. Procedia - Social and Behavioral Sciences 24:1650–1662. doi:10.1016/j.sbspro.2011.09.019

Ahmad MF, Zakuan N, Jusoh A, Takala J (2012) Relationship of TQM and Business Performance with Mediators of SPC, Lean Production and TPM. Procedia - Social and Behavioral Sciences 65:186–191. doi:10.1016/j.sbspro.2012.11.109

Ahmed PK, Rafiq M (2002) Chapter six - Total quality management (TQM) and internal marketing. In: Ahmed PK, Rafiq M (eds) Internal Marketing. Butterworth-Heinemann, Oxford, pp 91–124. doi:http://dx.doi.org/10.1016/B978-0-7506-4838-7.50011-9

Aksoy A, Öztürk N (2011) Supplier selection and performance evaluation in just-in-time production environments. Expert Syst. Appl. 38(5):6351–6359. doi:10.1016/j.eswa.2010.11.104

Al-Tahat MD, Mukattash AM (2006) Design and analysis of production control scheme for Kanban-based JIT environment. J. Franklin Inst. 343(4–5):521–531. doi:10.1016/j.jfranklin.2006.01.001

Alcaraz JLG, Maldonado AA, Iniesta AA, Robles GC, Hernández GA (2014) A systematic review/survey for JIT implementation: Mexican maquiladoras as case study. Comput. Ind. 65 (4):761–773. doi:10.1016/j.compind.2014.02.013

Alfalla-Luque R, Marin-Garcia JA, Medina-Lopez C (2015) An analysis of the direct and mediated effects of employee commitment and supply chain integration on organisational performance. Int. J. Prod. Econ. 162:242–257. doi:10.1016/j.ijpe.2014.07.004

Almomani MA, Aladeemy M, Abdelhadi A, Mumani A (2013) A proposed approach for setup time reduction through integrating conventional SMED method with multiple criteria decision-making techniques. Comput. Ind. Eng. 66(2):461–469. doi:10.1016/j.cie.2013.07.011

Amasaka K (2002) "New JIT": A new management technology principle at Toyota. Int. J. Prod. Econ. 80(2):135–144. doi:10.1016/S0925-5273(02)00313-4

Amasaka K (2007) Applying New JIT—Toyota's global production strategy: Epoch-making innovation of the work environment. Robotics and Computer-Integrated Manufacturing 23(3): 285–293. doi:10.1016/j.rcim.2006.02.001

Amasaka K (2014) New JIT, New Management Technology Principle: Surpassing JIT. Procedia Technology 16:1135–1145. doi:10.1016/j.protcy.2014.10.128

Araz C, Ozkarahan I (2007) Supplier evaluation and management system for strategic sourcing based on a new multicriteria sorting procedure. Int. J. Prod. Econ. 106(2):585–606. doi:10.1016/j.ijpe.2006.08.008

Arıkan E, Fichtinger J, Ries JM (2014) Impact of transportation lead-time variability on the economic and environmental performance of inventory systems. Int. J. Prod. Econ. 157:279–288. doi:10.1016/j.ijpe.2013.06.005

Bakri AH, Rahim ARA, Yusof NM, Ahmad R (2012) Boosting Lean Production via TPM. Procedia - Social and Behavioral Sciences 65:485–491. doi:10.1016/j.sbspro.2012.11.153

Ball F, Tyler J, Wells P (2015) Is audit quality impacted by auditor relationships? Journal of Contemporary Accounting & Economics 11(2):166–181. doi:10.1016/j.jcae.2015.05.002

Banerjee P, Armouti HA (1992) JIT approach to integrating production order scheduling and production activity control. Computer Integrated Manufacturing Systems 5(4):283–290. doi:10.1016/0951-5240(92)90045-E

Baraldi E, Proença JF, Proença T, de Castro LM (2014) The supplier's side of outsourcing: Taking over activities and blurring organizational boundaries. Ind. Mark. Manage. 43(4):553–563. doi:10.1016/j.indmarman.2014.02.012

Barve A, Kanda A, Shankar R (2009) The role of human factors in agile supply chains. European Journal of Industrial Engineering 3(1):2–20. doi:10.1504/EJIE.2009.021582

Baykoç ÖF, Erol S (1998) Simulation modelling and analysis of a JIT production system. Int. J. Prod. Econ. 55(2):203–212. doi:10.1016/S0925-5273(98)00061-9

Beikkhakhian Y, Javanmardi M, Karbasian M, Khayambashi B (2015) The application of ISM model in evaluating agile suppliers selection criteria and ranking suppliers using fuzzy TOPSIS-AHP methods. Expert Syst. Appl. 42(15–16):6224–6236. doi:10.1016/j.eswa.2015.02.035

Bendul JC, Skorna ACH (2015) Exploring impact factors of shippers' risk prevention activities: A European survey in transportation. Transportation Research Part E: Logistics and Transportation Review (0). doi:http://dx.doi.org/10.1016/j.tre.2015.05.008. Available online 29 June 2015 (in Press)

Benton WC, Shin H (1998) Manufacturing planning and control: The evolution of MRP and JIT integration. Eur. J. Oper. Res. 110(3):411–440. doi:10.1016/S0377-2217(98)00080-0

Blaga P, Jozsef B (2014) Human Resources, Quality Circles and Innovation. Procedia Economics and Finance 15:1458–1462. doi:10.1016/S2212-5671(14)00611-X

Bou JC, Beltrán I (2005) Total quality management, high-commitment human resource strategy and firm performance: an empirical study. Total Quality Management & Business Excellence 16(1):71–86. doi:10.1080/1478336042000309875

Brox JA, Fader C (1997) Assessing the impact of JIT using economic theory. Journal of Operations Management 15(4):371–388. doi:10.1016/S0272-6963(97)00005-3

Cao Q, Schniederjans MJ (2004) A revised EMQ/JIT production-run model: An examination of inventory and production costs. Int. J. Prod. Econ. 87(1):83–95. doi:10.1016/S0925-5273(03)00094-X

Chan FTS (2001) Effect of kanban size on just-in-time manufacturing systems. J. Mater. Process. Technol. 116(2–3):146–160. doi:10.1016/S0924-0136(01)01022-6

Chaumette S, Vignéras P (2004) Extensible and customizable just-in-time security (JITS) management of client-server communication in Java: Illustration on a simple electronic wallet. In: G.R. Joubert WENFJP, Walter WV (eds) Advances in Parallel Computing, vol Volume 13. North-Holland, pp 321–327. doi:http://dx.doi.org/10.1016/S0927-5452(04)80043-1

Chiarini A (2013) The Main Methods of Lean Organization: Kanban, Cellular Manufacturing, SMED and TPM. In: Lean Organization: from the Tools of the Toyota Production System to Lean Office, vol 3. Perspectives in Business Culture. Springer Milan, pp 81–116. doi:10.1007/978-88-470-2510-3_6

Collins R, Bechler K, Pires S (1997) Outsourcing in the automotive industry: From JIT to Modular Consortia. Eur. Manag. J. 15(5):498–508. doi:10.1016/S0263-2373(97)00030-3

Cua KO, McKone KE, Schroeder RG (2001) Relationships between implementation of TQM, JIT, and TPM and manufacturing performance. Journal of Operations Management 19(6):675–694. doi:10.1016/S0272-6963(01)00066-3

Dale BG, Elkjaer MBF, van der Wiele A, Williams ART (2001) Fad, fashion and fit: An examination of quality circles, business process re-engineering and statistical process control. Int. J. Prod. Econ. 73(2):137–152. doi:10.1016/S0925-5273(00)00165-1

Das S, Patnaik A (2015) 4 - Production planning in the apparel industry. In: Nayak R, Padhye R (eds) Garment Manufacturing Technology. Woodhead Publishing, pp 81–108. doi:http://dx.doi.org/10.1016/B978-1-78242-232-7.00004-7

de Menezes LM, Wood S, Gelade G (2010) The integration of human resource and operation management practices and its link with performance: A longitudinal latent class study. Journal of Operations Management 28(6):455–471. doi:10.1016/j.jom.2010.01.002

De Toni A, Nassimbeni G (2000) Just-in-time purchasing: an empirical study of operational practices, supplier development and performance. Omega 28(6):631–651. doi:10.1016/S0305-0483(00)00016-5

Deif AM, ElMaraghy H (2014) Impact of Dynamic Capacity Policies on WIP Level in Mix Leveling Lean Environment. Procedia CIRP 17:404–409. doi:10.1016/j.procir.2014.01.132

Dooley KJ, Flor RF (1998) Perceptions of success and failure in TQM initiatives. Journal of Quality Management 3(2):157–174. doi:10.1016/S1084-8568(99)80111-4

Du B, Guo S, Huang X, Li Y, Guo J (2015) A Pareto supplier selection algorithm for minimum the life cycle cost of complex product system. Expert Syst. Appl. 42(9):4253–4264. doi:10.1016/j.eswa.2015.01.056

Elmuti D, Kathawala Y (1990) Effects of computer aided quality circles on organizational productivity and satisfaction. Inf. Manag. 19(1):33–40. doi:10.1016/0378-7206(90)90012-7

Eti MC, Ogaji SOT, Probert SD (2004) Implementing total productive maintenance in Nigerian manufacturing industries. Appl. Energy 79(4):385–401. doi:10.1016/j.apenergy.2004.01.007

Farris JA, Van Aken EM, Doolen TL, Worley J (2009) Critical success factors for human resource outcomes in Kaizen events: An empirical study. Int. J. Prod. Econ. 117(1):42–65. doi:10.1016/j.ijpe.2008.08.051

Ferradás PG, Salonitis K (2013) Improving Changeover Time: A Tailored SMED Approach for Welding Cells. Procedia CIRP 7:598–603. doi:10.1016/j.procir.2013.06.039

Fiscus JE (1987) Kanban and Just-in-Time at Toyota: Edited by the Japanese management association, translated by David J. Lu. Productivity Press, Cambridge, MA, U.S.A., 1986. 192 pages, 40 figures, hardcover. ISBN 0-915299-08-0. List price: $29.95. Engineering Management International 4(2):157. doi:10.1016/0167-5419(87)90022-6

Fisher J, Kirk C, Taylor D (1995) The implications of TQM for R&D strategy in New Zealand firms. Technovation 15(1):1–9. doi:10.1016/0166-4972(95)96607-U

Flynn B, Salakibara S, Schroeder R (1995) Relationship between JIT and TQM Practices and Performance. Academy Management Journal 38(5):1325–1360

Forza C, Filippini R (1998) TQM impact on quality conformance and customer satisfaction: A causal model. Int. J. Prod. Econ. 55(1):1–20. doi:10.1016/S0925-5273(98)00007-3

Fukukawa T, Hong S-C (1993) The determination of the optimal number of kanbans in a Just-In-Time production system. Comput. Ind. Eng. 24(4):551–559. doi:10.1016/0360-8352(93)90197-6

Fullerton RR, McWatters CS (2002) The role of performance measures and incentive systems in relation to the degree of JIT implementation. Acc. Organ. Soc. 27(8):711–735. doi:10.1016/S0361-3682(02)00012-0

Gélinas R, Jacob R, Drolet J (1996) Just-in-time purchasing and the partnership strategy. European Journal of Purchasing & Supply Management 2(1):39–45. doi:10.1016/0969-7012(95)00015-1

Germain R, Dröge C (1997) Effect of just-in-time purchasing relationships on organizational design, purchasing department configuration, and firm performance. Ind. Mark. Manage. 26(2):115–125. doi:10.1016/S0019-8501(96)00113-7

Ghobadian A, Gallear DN (1996) Total quality management in SMEs. Omega 24(1):83–106. doi:10.1016/0305-0483(95)00055-0

Gilbert FW, Young JA, O'Neal CR (1994) Buyer-seller relationships in just-in-time purchasing environments. J. Bus. Res. 29(2):111–120. doi:10.1016/0148-2963(94)90015-9

Gonzalez R, Gasco J, Llopis J (2006) Information systems outsourcing: A literature analysis. Inf. Manag. 43(7):821–834. doi:10.1016/j.im.2006.07.002

Gordon ME, Philpot JW, Bounds GM, Long WS (1994) Factors associated with the success of the implementation of statistical process control. The Journal of High Technology Management Research 5(1):101–121. doi:10.1016/1047-8310(94)90016-7

Green Jr KW, Inman RA, Birou LM, Whitten D (2014) Total JIT (T-JIT) and its impact on supply chain competency and organizational performance. International Journal of Production Economics 147, Part A (0):125–135. doi:http://dx.doi.org/10.1016/j.ijpe.2013.08.026

Guidetti G, Mazzanti M (2007) Firm-level training in local economic systems: Complementarities in production and firm innovation strategies. The Journal of Socio-Economics 36(6):875–894. doi:10.1016/j.socec.2007.01.021

Gunasekaran A (1999) Just-in-time purchasing: An investigation for research and applications. Int. J. Prod. Econ. 59(1–3):77–84. doi:10.1016/S0925-5273(98)00025-5

Gupta YP, Heragu S (1991) Implications of implementing just-in-time systems. Technovation 11(3):143–162. doi:10.1016/0166-4972(91)90031-X

Gupta YP, Lonial SC, Glynn Mangold W (1992) Empirical assessment of the strategic orientation of JIT manufacturers versus non-JIT manufacturers. Computer Integrated Manufacturing Systems 5(3):181–189. doi:10.1016/0951-5240(92)90029-C

Gurel O, Acar AZ, Onden I, Gumus I (2015) Determinants of the Green Supplier Selection. Procedia - Social and Behavioral Sciences 181:131–139. doi:10.1016/j.sbspro.2015.04.874

Handfield RB (1993) A resource dependence perspective of Just-in-Time purchasing. Journal of Operations Management 11(3):289–311. doi:10.1016/0272-6963(93)90005-A

Hiltrop JM (1992) Just-in-time manufacturing: Implications for the management of human resources. Eur. Manag. J. 10(1):49–55. doi:10.1016/0263-2373(92)90043-4

Hipkin IB, De Cock C (2000) TQM and BPR: lessons for maintenance management. Omega 28(3):277–292. doi:10.1016/S0305-0483(99)00043-2

Hong J-D, Hayya JC (1992) Just-In-Time purchasing: Single or multiple sourcing? Int. J. Prod. Econ. 27(2):175–181. doi:10.1016/0925-5273(92)90011-U

Hou T-H, Hu W-C (2011) An integrated MOGA approach to determine the Pareto-optimal kanban number and size for a JIT system. Expert Syst. Appl. 38(5):5912–5918. doi:10.1016/j.eswa.2010.11.032

Houghton E, Portougal V (1997) Trade-offs in JIT production planning for multi-stage systems: balancing work-load variations and WIP inventories. International Transactions in Operational Research 4(5–6):315–326. doi:10.1016/S0969-6016(97)00032-4

Huang Y-W, Chang C-K, Lin W-C, Wang Y-F, Lu C-H, Lan C-H, Mah C, Huang MT, Shieh S-T, Liu Y-C, Wang H-w (2015) Quality Control Circle (QCC) strategy on reduction of the rate of needle stick injury and bloodborne exposure. Journal of Microbiology, Immunology and Infection 48 (2, Supplement 1):S99-S100. doi:http://dx.doi.org/10.1016/j.jmii.2015.02.352

Huber VL, Brown KA (1991) Human resource issues in cellular manufacturing: A sociotechnical analysis. Journal of Operations Management 10(1):138–159. doi:10.1016/0272-6963(91)90039-Z

Huq Z (1999) Conventional shop control procedures to approximate JIT inventory performance in a job shop. Journal of Manufacturing Systems 18(3):161–174. doi:10.1016/S0278-6125(99)80029-6

Huq Z, Huq F (1994) Embedding JIT in MRP: The case of job shops. Journal of Manufacturing Systems 13(3):153–164. doi:10.1016/0278-6125(94)90001-9

Inman RA, Sale RS, Green KW Jr, Whitten D (2011) Agile manufacturing: Relation to JIT, operational performance and firm performance. Journal of Operations Management 29(4):343–355. doi:10.1016/j.jom.2010.06.001

Iwase M, Ohno K (2011) The performance evaluation of a multi-stage JIT production system with stochastic demand and production capacities. Eur. J. Oper. Res. 214(2):216–222. doi:10.1016/j.ejor.2011.04.014

Jabbour CJC, Jabbour ABLdS, Govindan K, Teixeira AA, Freitas WRdS (2013) Environmental management and operational performance in automotive companies in Brazil: the role of human resource management and lean manufacturing. J. Clean. Prod. 47:129–140. doi:10.1016/j.jclepro.2012.07.010

Jacobs RL (2010) Training and Learning in the Workplace. In: McGaw PPB (ed) International Encyclopedia of Education (Third Edition). Elsevier, Oxford, pp 328–336. doi:http://dx.doi.org/10.1016/B978-0-08-044894-7.00773-9

Jayaram J, Droge C, Vickery SK (1999) The impact of human resource management practices on manufacturing performance. Journal of Operations Management 18(1):1–20. doi:10.1016/S0272-6963(99)00013-3

Jensen JB, Malhotra MK, Philipoom PR (1996) Machine dedication and process flexibility in a group technology environment. Journal of Operations Management 14(1):19–39. doi:10.1016/0272-6963(95)00030-5

Jiang H, Habib A, Zhou D (2015) Accounting restatements and audit quality in China. Advances in Accounting 31(1):125–135. doi:10.1016/j.adiac.2015.03.014

Jun M, Cai S, Shin H (2006) TQM practice in maquiladora: Antecedents of employee satisfaction and loyalty. Journal of Operations Management 24(6):791–812. doi:10.1016/j.jom.2005.09.006

Kaynak H, Hartley JL (2006) Using replication research for just-in-time purchasing construct development. Journal of Operations Management 24(6):868–892. doi:10.1016/j.jom.2005.11.006

Kemal Karasu M, Cakmakci M, Cakiroglu MB, Ayva E, Demirel-Ortabas N (2014) Improvement of changeover times via Taguchi empowered SMED/case study on injection molding production. Measurement 47:741–748. doi:10.1016/j.measurement.2013.09.035

Kent M (2009) Quality Circles: An Innovation to Optimize Quality Care in the Perioperative Setting. Journal of PeriAnesthesia Nursing 24(3):e3. doi:10.1016/j.jopan.2009.05.013

Kia R, Khaksar-Haghani F, Javadian N, Tavakkoli-Moghaddam R (2014) Solving a multi-floor layout design model of a dynamic cellular manufacturing system by an efficient genetic algorithm. Journal of Manufacturing Systems 33(1):218–232. doi:10.1016/j.jmsy.2013.12.005

Kim I, Tang CS (1997) Lead time and response time in a pull production control system. Eur. J. Oper. Res. 101(3):474–485. doi:10.1016/S0377-2217(96)00174-9

Knechtges P, Decker MC (2014) Application of Kaizen Methodology to Foster Departmental Engagement in Quality Improvement. Journal of the American College of Radiology 11 (12, Part A):1126–1130. doi:http://dx.doi.org/10.1016/j.jacr.2014.08.027

Kojima M, Nakashima K, Ohno K (2008) Performance evaluation of SCM in JIT environment. Int. J. Prod. Econ. 115(2):439–443. doi:10.1016/j.ijpe.2007.11.017

Kumar SV, Mani VGS, Devraj N (2014) Production Planning and Process Improvement in an Impeller Manufacturing Using Scheduling and OEE Techniques. Procedia Materials Science 5:1710–1715. doi:10.1016/j.mspro.2014.07.360

Kumiega A, Van Vliet B (2008) 30 - Kaizen: Continuous Improvement. In: Kumiega A, Vliet BV (eds) Quality Money Management. Academic Press, Burlington, pp 271–277. doi:http://dx.doi.org/10.1016/B978-0-12-372549-3.00030-6

Laitinen EK, Laitinen T (2015) A probability tree model of audit quality. Eur. J. Oper. Res. 243 (2):665–677. doi:10.1016/j.ejor.2014.12.021

Lemak DJ, Reed R, Satish PK (1997) Commitment to total quality management: Is there a relationship with firm performance? Journal of Quality Management 2(1):67–86. doi:10.1016/S1084-8568(97)90022-5

Lengnick-Hall ML, Lengnick-Hall CA, Rigsbee CM (2013) Strategic human resource management and supply chain orientation. Human Resource Management Review 23(4):366–377. doi:10.1016/j.hrmr.2012.07.002

Liberopoulos G, Koukoumialos S (2005) Tradeoffs between base stock levels, numbers of kanbans, and planned supply lead times in production/inventory systems with advance demand information. Int. J. Prod. Econ. 96(2):213–232. doi:10.1016/j.ijpe.2004.04.006

Lim SAH, Antony J, Albliwi S (2014) Statistical Process Control (SPC) in the food industry – A systematic review and future research agenda. Trends Food Sci. Technol. 37(2):137–151. doi:10.1016/j.tifs.2014.03.010

Louly M-A, Dolgui A (2013) Optimal MRP parameters for a single item inventory with random replenishment lead time, POQ policy and service level constraint. Int. J. Prod. Econ. 143(1): 35–40. doi:10.1016/j.ijpe.2011.02.009
Lyu J Jr (1996) Applying Kaizen and automation to process reengineering. Journal of Manufacturing Systems 15(2):125–132. doi:10.1016/0278-6125(96)82337-5
Mackelprang AW, Nair A (2010) Relationship between just-in-time manufacturing practices and performance: A meta-analytic investigation. Journal of Operations Management 28(4):283–302. doi:10.1016/j.jom.2009.10.002
Machuca JAD (2002) JIT facing the New Millennium. Int. J. Prod. Econ. 80(2):131–134. doi:10.1016/S0925-5273(02)00312-2
Maiga AS, Jacobs FA (2009) JIT performance effects: A research note. Advances in Accounting 25(2):183–189. doi:10.1016/j.adiac.2009.06.003
Markham IS, Mathieu RG, Wray BA (1998) A rule induction approach for determining the number of kanbans in A just-in-time production system. Comput. Ind. Eng. 34(4):717–727. doi:10.1016/S0360-8352(98)00099-0
Marodin G, Saurin T, Tortorella G, Denicol J (2015) How context factors influence lean production practices in manufacturing cells. Int. J. Adv. Manuf. Technol. 79(5–8):1389–1399. doi:10.1007/s00170-015-6944-2
Maroun W (2015) Reportable irregularities and audit quality: Insights from South Africa. Accounting Forum 39(1):19–33. doi:10.1016/j.accfor.2014.03.002
McCarthy D, Rich DN (2004) 2 - Lean TPM. In: McCarthy D, Rich DN (eds) Lean TPM. Butterworth-Heinemann, Oxford, pp 24–48. doi:http://dx.doi.org/10.1016/B978-075065857-7/50005-6
McLaughlin P, Durazo-Cardenas I (2013) Cellular Manufacturing Applications in MRO Operations. Procedia CIRP 11:254–259. doi:10.1016/j.procir.2013.07.077
Miko NU, Kamardin H (2015) Impact of Audit Committee and Audit Quality on Preventing Earnings Management in the Pre- and Post- Nigerian Corporate Governance Code 2011. Procedia - Social and Behavioral Sciences 172:651–657. doi:10.1016/j.sbspro.2015.01.415
Miltenburg J (1993) A theoretical framework for understanding why JIT reduces cost and cycle time and improves quality. Int. J. Prod. Econ. 30–31:195–204. doi:10.1016/0925-5273(93)90091-X
Min W, Sui Pheng L (2007) Modeling just-in-time purchasing in the ready mixed concrete industry. Int. J. Prod. Econ. 107(1):190–201. doi:10.1016/j.ijpe.2006.08.010
Moattar Husseini SM, O'Brien C, Hosseini ST (2006) A method to enhance volume flexibility in JIT production control. Int. J. Prod. Econ. 104(2):653–665. doi:10.1016/j.ijpe.2005.05.025
Monden Y (2002) The relationship between mini profit-center and JIT system. Int. J. Prod. Econ. 80(2):145–154. doi:10.1016/S0925-5273(02)00314-6
Moore R (2007) 8 - Kaizen. In: Moore R (ed) Selecting the Right Manufacturing Improvement Tools. Butterworth-Heinemann, Burlington, pp 159–172. doi:http://dx.doi.org/10.1016/B978-075067916-9/50009-6
Narasimhan R, Talluri S, Das A (2004) Exploring flexibility and execution competencies of manufacturing firms. Journal of Operations Management 22(1):91–106. doi:10.1016/j.jom.2003.12.003
Nassimbeni G (1996) Factors underlying operational JIT purchasing practices: Results of an empirical research. Int. J. Prod. Econ. 42(3):275–288. doi:10.1016/0925-5273(95)00190-5
Ohno K (2011) The optimal control of just-in-time-based production and distribution systems and performance comparisons with optimized pull systems. Eur. J. Oper. Res. 213(1):124–133. doi:10.1016/j.ejor.2011.03.005
Ooi K-B (2014) TQM: A facilitator to enhance knowledge management? A structural analysis. Expert Syst. Appl. 41(11):5167–5179. doi:10.1016/j.eswa.2014.03.013
Oral EL, Mıstıkoglu G, Erdis E (2003) JIT in developing countries—a case study of the Turkish prefabrication sector. Build. Environ. 38(6):853–860. doi:10.1016/S0360-1323(03)00021-0
Ortiz C (2010) Kaizen vs. Lean: Distinct but related. Met. Finish. 108(1):50–51. doi:10.1016/S0026-0576(10)80011-X

Pattanaik LN, Sharma BP (2009) Implementing lean manufacturing with cellular layout: a case study. Int. J. Adv. Manuf. Technol. 42(7–8):772–779. doi:10.1007/s00170-008-1629-8

Pazhani S, Ventura JA, Mendoza A (2015) A Serial Inventory System with Supplier Selection and Order Quantity Allocation considering Transportation Costs. Applied Mathematical Modelling (0). doi:http://dx.doi.org/10.1016/j.apm.2015.06.008. Available online 3 July 2015 (in Press)

Pourbabai B (1988) A Just-In-Time/Group Technology Based Scheduling Model. In: Radharamanan R (ed) Robotics and Factories of the Future '87. Springer Berlin Heidelberg, pp 105–110. doi:10.1007/978-3-642-73890-6_14

Power D, Sohal AS (2000) Human resource management strategies and practices in Just-In-Time environments: Australian case study evidence. Technovation 20(7):373–387. doi:10.1016/S0166-4972(99)00151-0

Pragman CH (1996) JIT II: A purchasing concept for reducing lead times in time-based competition. Bus. Horiz. 39(4):54–58. doi:10.1016/S0007-6813(96)90052-X

Prahinski C, Benton WC (2004) Supplier evaluations: communication strategies to improve supplier performance. Journal of Operations Management 22(1):39–62. doi:10.1016/j.jom.2003.12.005

Prajogo DI, Sohal AS (2001) TQM and innovation: a literature review and research framework. Technovation 21(9):539–558. doi:10.1016/S0166-4972(00)00070-5

Radharamanan R, Godoy LP, Watanabe KI (1996) Quality and productivity improvement in a custom-made furniture industry using Kaizen. Comput. Ind. Eng. 31(1–2):471–474. doi:10.1016/0360-8352(96)00177-5

Rahman S-u, Bullock P (2005) Soft TQM, hard TQM, and organisational performance relationships: an empirical investigation. Omega 33(1):73–83. doi:10.1016/j.omega.2004.03.008

Rahmina LY, Agoes S (2014) Influence of Auditor Independence, Audit Tenure, and Audit Fee on Audit Quality of Members of Capital Market Accountant Forum in Indonesia. Procedia - Social and Behavioral Sciences 164:324–331. doi:10.1016/j.sbspro.2014.11.083

Rajesh G, Malliga P (2013) Supplier Selection based on AHP QFD Methodology. Procedia Engineering 64:1283–1292. doi:10.1016/j.proeng.2013.09.209

Recht R, Wilderom C (1998) Kaizen and culture: on the transferability of Japanese suggestion systems. Int. Bus. Rev. 7(1):7–22. doi:10.1016/S0969-5931(97)00048-6

Reda HM (1987) A review of "kanban"—the Japanese "Just-In-Time" production system. Engineering Management International 4(2):143–150. doi:10.1016/0167-5419(87)90019-6

Rentroia-Bonito MA, Tribolet J, Jorge JA, Ghaoui C (2005) Just-in-Time Approach to Learning: Arguing the Case for Cost-Effective Knowledge Dissemination. In: Ghaoui C, Jain M, Bannore V, C. Jain L (eds) Knowledge-Based Virtual Education, vol 178. Studies in Fuzziness and Soft Computing. Springer Berlin Heidelberg, pp 1–22. doi:10.1007/11399520_1

Rodrigues M, Hatakeyama K (2006) Analysis of the fall of TPM in companies. J. Mater. Process. Technol. 179(1–3):276–279. doi:10.1016/j.jmatprotec.2006.03.102

Rosenfeld Y, Warszawski A, Laufer A (1991) Quality circles in temporary organizations: lessons from construction projects. Int. J. Project Manage. 9(1):21–27. doi:10.1016/0263-7863(91)90051-V

Salleh NAM, Kasolang S, Jaffar A (2012) Simulation of Integrated Total Quality Management (TQM) with Lean Manufacturing (LM) Practices in Forming Process Using Delmia Quest. Procedia Engineering 41:1702–1707. doi:10.1016/j.proeng.2012.07.371

Sandanayake YG, Oduoza CF, Proverbs DG (2008) A systematic modelling and simulation approach for JIT performance optimisation. Robotics and Computer-Integrated Manufacturing 24(6):735–743. doi:10.1016/j.rcim.2008.03.013

Santiago F, Alcorta L (2012) Human resource management for learning through knowledge exploitation and knowledge exploration: Pharmaceuticals in Mexico. Structural Change and Economic Dynamics 23(4):530–546. doi:10.1016/j.strueco.2011.11.002

Sarwoko I, Agoes S (2014) An Empirical Analysis of Auditor's Industry Specialization, Auditor's Independence and Audit Procedures on Audit Quality: Evidence from Indonesia. Procedia - Social and Behavioral Sciences 164:271–281. doi:10.1016/j.sbspro.2014.11.077

Selçuk B (2013) Adaptive lead time quotation in a pull production system with lead time responsive demand. Journal of Manufacturing Systems 32(1):138–146. doi:10.1016/j.jmsy.2012.07.017

Sengupta S, Davis RP, Ferrell WG (1993) Production planning and control in a JIT environment. Appl. Math. Model. 17(1):41–46. doi:10.1016/0307-904X(93)90126-2

Singh R, Gohil AM, Shah DB, Desai S (2013) Total Productive Maintenance (TPM) Implementation in a Machine Shop: A Case Study. Procedia Engineering 51:592–599. doi:10.1016/j.proeng.2013.01.084

Smith R, Hawkins B (2004) 3 - Total Productive Maintenance (TPM). In: Smith R, Hawkins B (eds) Lean Maintenance. Butterworth-Heinemann, Burlington, pp 55–104. doi:http://dx.doi.org/10.1016/B978-075067779-0/50003-0

Spencer MS (1998) Group Technology at John Deere: Production Planning and Control Issues. In: Suresh N, Kay J (eds) Group Technology and Cellular Manufacturing. Springer US, pp 501–516. doi:10.1007/978-1-4615-5467-7_29

Sundar R, Balaji AN, Kumar RMS (2014) A Review on Lean Manufacturing Implementation Techniques. Procedia Engineering 97:1875–1885. doi:10.1016/j.proeng.2014.12.341

Taghizadegan S (2006) Chapter 6 - Design for Lean/Kaizen Six Sigma. In: Taghizadegan S (ed) Essentials of Lean Six Sigma. Butterworth-Heinemann, Burlington, pp 59–101. doi:http://dx.doi.org/10.1016/B978-012370502-0/50008-4

Thomas AJ, Jones GR, Vidales P (2006) An Integrated Approach to TPM and Six Sigma Development in the Castings Industry. In: Pham DT, Eldukhri EE, Soroka AJ (eds) Intelligent Production Machines and Systems. Elsevier Science Ltd, Oxford, pp 620–625. doi:http://dx.doi.org/10.1016/B978-008045157-2/50108-5

Trad A, Kalpić D (2014) The Selection and Training Framework (STF) for Managers in (e-) Business Innovation Transformation Projects – Managerial Recommendations. Procedia Technology 16:768–774. doi:10.1016/j.protcy.2014.10.026

Trentesaux D, Tahon C, Ladet P (1998) Hybrid production control approach for JIT scheduling. Artif. Intell. Eng. 12(1–2):49–67. doi:10.1016/S0954-1810(96)00052-0

van der Laan E, Salomon M, Dekker R (1999) An investigation of lead-time effects in manufacturing/remanufacturing systems under simple PUSH and PULL control strategies. Eur. J. Oper. Res. 115(1):195–214. doi:10.1016/S0377-2217(98)00108-8

Vellani KH (2007) Chapter 2 - Asset identification and security inventory. In: Vellani KH (ed) Strategic Security Management. Butterworth-Heinemann, Burlington, pp 11–25. doi: http://dx.doi.org/10.1016/B978-012370897-7/50005-0

Villa A, Taurino T (2013) From JIT to Seru, for a Production as Lean as Possible. Procedia Engineering 63:956–965. doi:10.1016/j.proeng.2013.08.172

Vora JA (1992) Applying a theory of organization change to adopting JIT. Omega 20(2):193–199. doi:10.1016/0305-0483(92)90073-G

Vuppalapati K, Ahire SL, Gupta T (1995) JIT and TQM: a case for joint implementation. International Journal of Operations & Production Management 15(5):84–94. doi:10.1108/01443579510083686

Wang S, Sarker BR (2005) An assembly-type supply chain system controlled by kanbans under a just-in-time delivery policy. Eur. J. Oper. Res. 162(1):153–172. doi:10.1016/j.ejor.2003.10.038

Wang S, Sarker BR (2006) Optimal models for a multi-stage supply chain system controlled by kanban under just-in-time philosophy. Eur. J. Oper. Res. 172(1):179–200. doi:10.1016/j.ejor.2004.10.001

Wang W, Wang D, Ip WH (1999) JIT production planning approach with fuzzy due date for OKP manufacturing systems1. Int. J. Prod. Econ. 58(2):209–215. doi:10.1016/S0925-5273(98)00122-4

Weng W, Wei X, Fujimura S (2012) Dynamic routing strategies for JIT production in hybrid flow shops. Comput. Oper. Res. 39(12):3316–3324. doi:10.1016/j.cor.2012.04.022

Withers BE, Ebrahimpour M, Hikmet N (1997) An exploration of the impact of TQM and JIT on ISO 9000 registered companies. Int. J. Prod. Econ. 53(2):209–216. doi:10.1016/S0925-5273(97)00118-7

Yasin MM, Small M, Wafa MA (1997) An empirical investigation of JIT effectiveness: an organizational perspective. Omega 25(4):461–471. doi:10.1016/S0305-0483(97)00005-4

Yasin MM, Small MH, Wafa MA (2003) Organizational modifications to support JIT implementation in manufacturing and service operations. Omega 31(3):213–226. doi:10.1016/S0305-0483(03)00024-0

Zairi M (1991a) 5 - DEFINING TQM: TOOLS AND TECHNIQUES. In: Zairi M (ed) Total Quality Management for Engineers. Woodhead Publishing, pp 87–123. doi:http://dx.doi.org/10.1533/9781845698911.87

Zairi M (1991b) 7 - TOTAL SAFETY SYSTEMS FOR TQM. In: Zairi M (ed) Total Quality Management for Engineers. Woodhead Publishing, pp 145–165. doi:http://dx.doi.org/10.1533/9781845698911.145

Zairi M (1991c) 8 - ESSENTIALS OF TQM: LEADERSHIP. In: Zairi M (ed) Total Quality Management for Engineers. Woodhead Publishing, pp 166–189. doi:http://dx.doi.org/10.1533/9781845698911.166

Zairi M (1993) Competitive manufacturing: Combining total quality with advanced technology. Long Range Plan. 26(3):123–132. doi:10.1016/0024-6301(93)90014-7

Chapter 8
Descriptive Analysis of JIT Benefits

This chapter analyzes the benefits that can be obtained from a successful JIT implementation. Since data was obtained from subjective assessments, the median values were considered the measure of central tendency, while the measure of dispersion was represented by the interquartile range value and not by the traditional standard deviation used for interval values. According to Chap. 3, JIT benefits were classified in six categories:

- Benefits associated with human resources
  - Improved worker motivation
  - Increased teamwork
  - Reduced classification of positions
  - Increased resource utilization
  - Increased communication
- Benefits associated with the production process
  - Increased productivity
  - Reduced waste and rework
  - Increased efficiency
  - Increased process flexibility
  - Reduced production lead time
  - Reduced work in process
  - Integration of different manufacturing activities
- Benefits associated with engineering
  - Reduced space requirements
  - Reduced distance movements
  - Quick responses to engineering changes
  - Increased innovation
- Benefits associated with quality
  - Increased product quality
  - Paperwork reduction
  - Higher quality in production process

J.L. García-Alcaraz and A.A. Maldonado-Macías, *Just-in-Time Elements and Benefits*, Management and Industrial Engineering, DOI 10.1007/978-3-319-25919-2_8

- Benefits associated with material handling
  - Reduced part numbers
  - Reduced inventory
  - Reduced material handling
  - Close supplier–customer relationship
  - Reduced sizes of lots purchased
  - Increased inventory rotation
- Economic benefits
  - Reduced production costs
  - Reduced labor costs
  - Reduced general expenses
  - Increased competitiveness
  - Increased profit margin

8.1 Benefits Associated with Human Resources

Literature reviewed reported five benefits for human resources from a successful JIT implementation. Table 8.1 depicts the results after the interquartile analysis of the elements,which were described in the methodology section. The table also allows for the following interpretation of data.

- All elements listed show first quartile or 25th percentile values higher than 3 but lower than 3.5.
- Four elements listed in the table show median or second quartile values that are higher than 4, while one element has a median value lower than 4. This indicates that four out of these five benefits were obtained after the implementation of JIT in the manufacturing companies surveyed.
- All interquartile range values range from 1.4 to 1.6, which indicates dispersion in data reported by respondents and a moderate consensus on the median values obtained.
- *Increased resource utilization*: This element reports the highest median value, which indicates that it is usually obtained from a JIT implementation. The element refers to the appropriate use of resources available and needed to generate a product. These resources can be associated with time, human workforce, and raw materials, among others. One may not be surprised to notice this was the first benefit reported by respondents, since the main objective of a JIT program is the elimination of waste, and when the program is successfully implemented, companies can benefit from a better use of raw materials and increased efficiency in their production processes. Thus, when errors diminish, inventory in process or rework is reduced and human resources opportunities are enhanced. Moreover, when fewer errors occur, products remain in companies for shorter spaces of time from the moment they are designed until their

Table 8.1 Descriptive analysis of JIT benefits associated with human resources

Order	JIT benefit	Percentile			RI
		25	50	75	
1	Increased resource utilization	3.39	4.20	4.85	1.46
2	Increased teamwork	3.39	4.18	4.83	1.44
3	Increased communication	3.33	4.14	4.81	1.48
4	Improved worker motivation	3.28	4.09	4.77	1.49
5	Reduced classification of positions	3.1	3.89	4.66	1.56

commercialization. However, companies must make certain modifications to their structures in order to obtain these benefits from the appropriate use of resources (Yasin et al. 2003). For readers particularly interested in this topic can consult references such as Fullerton and McWatters (2001), Inman et al. (2011), Maiga and Jacobs (2009), White and Prybutok (2001) which thoroughly studied the benefits that an appropriate use of resources can offer to companies.

- *Increased teamwork*: This element shows a median value of 4.81, which indicates that this benefit is usually obtained from JIT systems in companies surveyed. One of the principal elements of a JIT program is communication inside and among all levels of the organization. Quality circles have become an example to improve teamwork in companies, since they enable workers to express their viewpoints and propose ideas for the improvement of production systems (Fullerton and McWatters 2001). Companies have even approached quality circles from a different perspective by including Kaizen tasks that must be performed through teamwork (Amasaka 2002). Nowadays, several works have associated collaborative work with the amount of motivation that quality circle leaders provide to workers. They have measured the effects of the former over the latter by means of casual analysis (Alcaraz et al. 2014; Green Jr et al. 2014). This has allowed for the emergence of new perspectives regarding production systems and the so-called birth of a new JIT philosophy (Amasaka 2007, 2014).

Companies can ensure increased communication through different approaches. First of all, increased communication can be achieved with high education, and it has been previously mentioned that training focused on quality and the needs of the production process is key to obtain the expected JIT benefits. Moreover, quality circles and Kaizen tasks promote an environment of collaboration where employees work together, although this can only be achieved through effective communication. Similarly, Kanban facilitates efficient communication between personnel responsible for the planning of an order and workers in production lines. Furthermore, companies nowadays rely on a number of information and communication technologies to maintain communication among all levels of the organization, be familiar with the status or orders, and avoid misunderstandings, since this could compromise the efficiency of the production process. For instance, if an order from managers were not transmitted to the suitable person in the appropriate way, distortions would surely arise and companies would not meet the objectives of JIT. Several studies have studied the use of information technologies in JIT systems, such as Fong et al. (2013), Bayo-Moriones and Lera-López (2007), Colombo et al. (2013). Also, Yu et al. (2008), Tong et al. (2012), Ali and Kumar (2011), Silvestre (2015), Mensah et al. (2015) have addressed the use of these technologies in supply chains.

- *Increased worker motivation*: This element shows a median value of 4.09, that is, higher than 4, five hundredths lower than the previous element, and 11 hundredths if it is compared with the first element. Motivation can be the product of high training that workers received and which is focused on quality aspects or personal needs for their successful performance (Alcaraz et al. 2014; Huq 1999). However, workers can also become motivated when they perceive high commitment from the top management to facilitate all the possible resources so that training does impact on the efficiency indexes of the company, but also on the integral development of human resources (Anderman and Gray 2015; Diefendorff and Seaton 2015; Dwivedula and Bredillet 2010). Nevertheless, top managers must demonstrate their commitment to middle managers or supervisors who possess a direct relation with the workers, since these should be able to recognize a leader or person who directs them to an objective (Rusu and Avasilcai 2014; Singh et al. 2015; Singh and Garg 2011; Tabassi and Bakar 2009).
- *Reduction of the classification of positions*: This is the only element with a median value lower than 4. This indicates that responders considered it a less frequent benefit. Its median value is also 20 hundredths lower than the median value of the previous element and 31 hundredths if it is compared with the first element. JIT environments require workers become multifunctional. This would allow them to occupy several positions at once and companies would not need many workers to perform many tasks, although this does not allow for high specialization of human resources. If one employee is able to hold any position in the production process, the flow of materials is guaranteed in case another operator is absent. This could prevent production lines from pausing or stopping, especially because they cause inventories in process. Similar results concerning the reduction of classification in positions have been reported in Alcaraz et al. (2014), Amasaka (2014), Shi et al. (2014). Furthermore, other authors have associated this element with a major flexibility of the productive system (Inman et al. 2011).

8.2 Benefits Associated with Production Process

This category includes seven JIT benefits associated with production process, which begins with the process of transformation of raw materials, includes all the operations that add value to the product, and finishes with product storage or distribution

Table 8.2 Descriptive analysis of JIT benefits associated with production process

Order	JIT benefit	Percentile			IQR
		25	50	75	
1	Increased efficiency	3.62	4.36	4.94	1.32
2	Reduced waste and rework	3.51	4.31	4.92	1.41
3	Increased of productivity	3.42	4.21	4.86	1.44
4	Increased process flexibility	3.38	4.2	4.85	1.47
5	Reduced production lead time	3.38	4.15	4.79	1.41
6	Reduced work in process	3.35	4.15	4.8	1.45
7	Integration of different manufacturing activities	3.2	4.01	4.72	1.52

to the final customer. Table 8.2 depicts the results of the descriptive analysis and lists the benefits in descending order according to their median or 50th percentile value.

Based on the results obtained, it is possible to state the following conclusions:

- All benefits listed in Table 8.2 show first quartile or 25th percentile values that are higher than 3 but lower than 4. Therefore, according to respondents, they were regularly obtained after a JIT implementation.
- All benefits listed in Table 8.2 show medians or 50th percentile values that are higher than 4 but lower than 4.5. This indicates that respondents regularly obtained this benefits on time.
- All elements listed in Table 8.2 show third quartile or 75th percentile values that equal or are higher than 4.94 but lower than 4.72. This indicates that all benefits were always obtained for respondents, since all values are close to 5, which is the maximum value in the scale used to obtain the information.
- All elements that show interquartile range values are lower than 1.52, but higher than 1.32, which indicates a moderate dispersion in data, a regular consensus on the median values of the elements.

As for the JIT benefits, the following conclusions are given:

- *Increased efficiency*: This element shows a median value of 4.36, which indicated that it is a benefit always obtained by the companies surveyed. If efficiency is defined as the aptitude to appropriately perform and fulfill a function, it is possible to conclude that companies surveyed reported an increased efficiency in their systems and deliveries. Similar results were obtained for the performance indexes of a supply chain in a JIT environment (Alcaraz et al. 2014). Moreover, Green Jr et al. (2014) proposed a model of structural equations to point out that JIT increases many performance indexes of companies, especially those related to supply chains. Additionally, Inman et al. (2011) associated agility achieved by means of a JIT implementation with the operational performance indexes reflected on the production process. Additional studies that report performance and efficiency improvements from JIT implementation can be consulted in Kojima et al. (2008), Maiga and Jacobs (2009), Fullerton et al. (2003), Sandanayake et al. (2008).

- *Reduced waste and rework*: This benefit shows a median value of 4.31, that is, five hundredths lower than the median of the first element listed in the table. One of the main objectives of JIT philosophy is the elimination of waste related to time, the inappropriate use of the resources, or the inappropriate performance of tasks. Elements that can provide this benefit can be related to the implementation of quality programs based on training for workers and supervisors. In fact, it is expected that properly trained employees generate less waste reprocessing. Similarly, specialized factories can ensure a reduction of waste and rework, since human resources in these factories can be specialized in specific production operations. Similar results were presented by Brox and Fader (1997), Benton and Shin (1998), Fullerton and McWatters (2001), Azadeh et al. (2005), Sandanayake et al. (2008), who can be consulted as additional information.
- *Increased productivity*: This benefit has a median value of 4.21, that is, 10 hundredths lower than the median of the previous element and 15 hundredths if it is compared with the first element in the list. In this book, productivity is understood as the relation between the amount of products obtained by a production system and the resources used to obtain these products. Productivity implies that companies make less use of resources to generate their products, which can be achieved by quality programs, productive maintenance programs, the implementation and use of cards system, and the use of continuous improvement programs, among others. For instance, more than 25 years ago, Paek (1990) emphasized on JIT philosophy as a means to improve productivity indexes, which would allow companies to better use their available resources. Similarly, Lawrence and Hottenstein (1995) pointed out that Mexican companies within JIT environments and affiliated to corporations in the United States were increasing their productivity indexes. Finally, (Fullerton and McWatters 2001, 2002; Kojima et al. 2008; Maiga and Jacobs 2009; Inman et al. 2011; Iwase and Ohno 2011) provided additional information concerning the increase of productivity and performance indices in JIT environments.
- *Increased process flexibility*: It has been stated that there are different types of flexibility, from which the most significant for companies may concern the production process. This element is ranked fourth in this category and shows a median value of 4.2, that is, one hundredth lower than the median value of the previous benefit but 16 hundredths lower if compared with the first element. Process flexibility can be the result of several factors. First of all, it is achieved; thanks to the ability of workers to become multifunctional in certain tasks. However, companies must also be able to make rapid changes from one product prototype to another and use information and communication technologies in the production system, such as robots. Finally, special distribution of plants may also be suitable to increase flexibility in the production process. Thus, companies must rely on cellular manufacture and group technology, which are also essential elements for JIT philosophy. Certain authors relied on structural equation modeling to point out the direct relation between flexibility and the economic indexes of companies (Alcaraz et al. 2014; Green Jr et al. 2014). However, it shall not be forgotten that flexibility is related to and reflect on other

factors, such as agility (Maiga and Jacobs 2009; Inman et al. 2011). Moreover, process flexibility is not a fortuitous event; it must be promoted from the moment raw materials are purchased (González-Benito et al. 2000; Moattar Husseini et al. 2006).

- *Reduced production lead time*: This element was ranked as fifth in the list of JIT benefits associated with the production process. It shows a median value of 4.15, that is, five hundredths lower than the median value of previous element and 21 hundredths if it is compared with the first element. Reduced production lead time can be the result of many factors. For instance, it can be achieved when companies emphasize on quality programs that must be established, since when they manage to produce with zero defects in one time, orders can quickly leave the production process, since they do not require rework or repairs (Ahmad et al. 2012; Ahmed and Rafiq 2002; Jun et al. 2006). Similarly, successful maintenance programs for machinery and equipment can reduce production lead time, since this equipment can operate for longer periods of time; otherwise, damaged machines could diminish the speed of production (Ahmad et al. 2012; Cua et al. 2001). Finally, producing below the real capacity of the process could reduce lead times in production, since it may prevent from having two production runs for the same product.
- *Reduced work in process*: This element shows the same median value as the previous element. Work in process reduction is closely related to the reduction of waste and rework and can be the result of several JIT elements. For instance, when workers are provided with effective training, they become more efficient and fewer errors may occur inside the process of production; and if errors do occur in the tasks these workers perform, they are empowered to self-correct after demonstrating enough knowledge and training to take appropriate decisions. Similarly, the implementation of zero defects can also reduce work in process and quality programs, since when companies appropriately follow quality standards, defects along the production disappear and inventory in process becomes inexistent. Further literature concerning the impact of JIT on the reduction of works or inventory in process can be consulted in Gupta and Heragu (1991), Gupta et al. (1992), Nassimbeni (1996), Kim and Ha (2003), Lovell (2003).
- *Integration of different manufacturing activities*: This element was ranked last in this category of JIT. It shows a median value of 4.01, that is, 14 hundredths lower than the median value of the previous element and 35 hundredths lower if compared with the first element of Table 8.2. The integration of different manufacturing activities can be chiefly the result of group technology and cellular manufacturing, since in this type of plant organization similar activities are performed in a single place and are often integrated into a single activity or groups of activities. Additional literature on the integration of manufacturing activities can be consulted in Brown (2014), Jensen et al. (1996), Pourbabai (1988), Spencer (1998).

8.3 Benefits Associated with Engineering

This category includes four JIT benefits that reflect on the production process in the strict sense of engineering. Table 8.3 depicts the results obtained from the descriptive analysis where the median values of elements appear as measurement of central tendency, while interquartile range values were considered as a deviation measure. Benefits listed in the table are arranged in the descending order, according to their median value.

Based on the results provided by Table 8.3, it is possible to provide with the following interpretations:

- The 25th percentile or the first quartile values of elements range from 3.17 to 3.32, this indicates that these benefits are regularly obtained by respondents of the survey.
- All elements show median, 50th percentile, or second quartile values that are equal or higher than 4, but are lower than 4.15. Thus, the range between the highest and the lowest median value is 15 hundredths, which indicates that, in average, these benefits are regularly obtained in companies surveyed.
- All elements show third quartile or 75th percentile values close to 5, higher than 4.73.
- All elements show interquartile range values that are higher than one but lower than 1.61. This suggests moderate consensus on the value of the median.
- *Reduced space requirements*: This was ranked first in the category and has a median value of 4.15. The reason of its significance may be due to the organization and distribution that companies set up for their plants (Monden 2002; Ahmad et al. 2003; White and Prybutok 2001), as well as the integration of machinery and equipment in manufacture cells and technological groups (McLaughlin and Durazo-Cardenas 2013; Pattanaik and Sharma 2009; Seifermann et al. 2014; Spencer 1998).
- *Reduced movement distances*: This benefit shows a median of 4.13, that is, two hundredths lower than the median of the first element. The reduction of movement distances can be the result not only of an effective distribution of machinery and equipment, but also the integration of this machinery and equipment in technology groups and manufacturing cells, since all operations can be executed in one place, and companies can avoid large flow of materials

Table 8.3 Descriptive analysis of JIT benefits associated with engineering in the production process

Order	JIT benefit	Percentile			RI
		25	50	75	
1	Reduced space requirements	3.32	4.15	4.83	1.51
2	Reduced movement distances	3.32	4.13	4.78	1.46
3	Quick responses to changes in engineering	3.18	4.10	4.79	1.61
4	Increased innovation	3.17	4.00	4.73	1.56

around the complete facilities. For instance, manufacturing cells can perform the whole production process from the reception of raw materials to the final product. For readers with a particular interest in the topic, Chen et al. (2011), Huang and Lin (2014), Ou-Yang and Utamima (2013) address the impact of the physical distribution of machinery and equipment on production system.
- *Quick responses to changes in engineering*: This benefit was ranked fourth in this category. It shows a median value of 4.10, that is, three hundredths lower than the median of the previous benefit, but five hundredths inferior if compared with the median value of the first element. This indicates that, for respondents of the survey, this benefit was regularly obtained after a JIT implementation. However, it must be noted that it shows the highest interquartile range value among the elements of this category. The ability of companies to quickly respond to engineering changes of a product can be the result of several JIT elements. For instance, companies make rapid adjustments on machinery and equipment in order to reduce preparation time between product designs. However, training also plays a key role, since the level of expertise of workers, as well as their knowledge of the capacities and abilities of machinery and equipment prepares them for sudden adjustments in this equipment. Additional literature studying this benefit can be found in Almomani et al. (2013), Ferradás and Salonitis (2013), Kemal Karasu et al. (2014).
- *Increased innovation*: This element has a median value of 4.00, that is, 10 hundredths lower than the median value of the previous element and 15 hundredths lower if compared with the median value of the first element. Innovation is referred as to the ability to improve existing products or propose others that are completely new. Training provided to workers is crucial to increase innovation in companies, especially if part of it focuses on the improvement of both, quality and product (Amasaka 2007; Berghman et al. 2012). Moreover, quality circles also promote innovation, since they aim to propose solutions to problems in the production process, which often result in interesting novel proposals (Blaga and Jozsef 2014; Cheng et al. 2013; Evanschitzky et al. 2012).

8.4 Benefits Associated with Quality

This category includes three JIT benefits that can be perceived in the quality of the product. Table 8.4 introduces the descriptive analysis of these elements that were listed in the descending order according to their median values.

Based on the results appearing in Table 8.4, it is possible to offer the following conclusions:

- All elements show first quartile or 25th percentile values that are higher than 3 but lower than 4, which indicates that at least 75 % of companies surveyed have regularly obtained these benefits.

Table 8.4 Descriptive analysis of JIT benefits associated with quality

Order	JIT benefit	Percentile			RI
		25	50	75	
1	Increased product quality	3.45	4.26	4.89	1.44
2	Higher quality in production process	3.32	4.10	4.76	1.44
3	Paperwork reduction	3.02	3.87	4.68	1.66

- Two JIT benefits associated with product quality show 50th percentile, median, or second quartile values that are higher than 4. On the other hand, one JIT benefit associated with quality has a median value lower than 4. This indicates that companies surveyed have at least regularly obtained the above-mentioned benefits.
- All elements show third quartile or 75th percentile values that are higher than four, which indicates that companies surveyed clearly have obtained these benefits with a JIT implementation.
- All elements show interquartile range values higher than 1 but lower than 2, which indicates a moderate distribution of the information and a regular consensus on the values of the medians.
- *Increased product quality*: This element shows a median value of 4.26, which is the highest value obtained in this category. A major quality in products can be the result of quality programs established as basic working elements (Ahmad et al. 2012; Amasaka 2002; Cua et al. 2001; Flynn et al. 1995), and education and training provided to workers (The Value of JIT-Structured On-the-Job Training 2011; Jacobs 2010; Jones 2001). However, zero defects programs that companies implement as part of their quality programs (Ferretti et al. 2013), and worker empowerment (Cheung et al. 2012; Chu 2003; de Macedo-Soares and Lucas 1995) are also essential to increase product quality.
- *Higher quality in the production process*: This benefit was ranked second in the category of JIT benefits associated with quality; moreover, it shows a median value of 4.10, that is, 16 hundredths lower than the median value of the first element. Improved quality in production processes can be the result of elements associated with quality plans and programs, especially statistical process control, which provides statistical information about the production process and hence informs of its status so that companies make decisions based on accurate information when deviations occur (Ahmad et al. 2012; Amasaka 2002; Cua et al. 2001). Production processes can be strengthened when machines and equipment are provided with suitable maintenance, since they are calibrated and properly fitted. Thus, robust and statistically stable production process will allow for the continuous flow of raw materials, since rejections or reprocesses will not occur. However, when production processes are not reliable, the cost of poor quality for companies could be high and cause the manufacturing of products in parallel production lines, which implies absorbing costs to correct errors (Ahmad et al. 2012; Lim et al. 2014).

- *Paperwork reduction*: This element was ranked as the third most obtained JIT benefit in terms of quality. It shows a median value of 3.87, that is, 23 hundredths lower than the median value of the previous element but 39 hundredths if it is compared with the median value of the first element. A reduction of paperwork can be the result of several elements, such as a trustful relationship between companies and their suppliers, since these relationships are usually source of great amount of administrative processes. Since reliable suppliers tend to be recognized by their raw material lots, companies could be comforted by the fact that returns or rejections due to quality may not occur from this relation (Aigbedo 2007; Aksoy and Öztürk 2011). This would simplify the relationship between the two parts and reduce costs for customer companies. However, internal functioning of companies can also support paperwork reduction. All levels of the organization should rely on the quality principle that addresses every subsequent activity as a client, since levels of trust similar to those established with customers may be required among departments to considerably reduce amount of administrative work. Similarly, information and communication technologies nowadays have allowed for a great reduction and simplification of paperwork, which previously performed physically (Bayo-Moriones and Lera-López 2007; Fong et al. 2013; Tams et al. 2014). Several of these reduction steps have become standardized by official norms such as ISO, and companies seek to become certified and accredited in these norms to monitor their quality processes (Dowlatshahi and Urias 2004).

8.5 Benefits Associated with Material Handling

The following category includes six JIT benefits that were previously reported and related to the handling of materials in the production system. Table 8.5 shows the results from the analysis of these elements whose median values were considered as the central tendency measure, while interquartile range values were considered as dispersion measure. Elements are also sorted in the descending order according to their median values.

Based on Table 8.5, it is possible to offer the following interpretation of data:

- The first five elements show first quartile or 25th percentile values lower than 4, while the last benefit shows a value that is inferior to 3. Thus, it can be stated that these benefits are regularly obtained in companies surveyed. However, it must be mentioned the sixth element is the first JIT benefit reporting such a low value.
- Four elements show median values higher than 4, while the remaining two elements show much lower values, which indicates that, in average, the four first elements are benefits regularly obtained in the companies surveyed.
- All elements show third quartile or 75th percentile values higher than 4 and close to 5. It can be thus stated that some companies surveyed always obtain these benefits.

Table 8.5 Descriptive analysis of JIT benefits associated with material handling

JIT benefit	Percentile			RI
	25	50	75	
Reduced inventory	3.38	4.22	4.88	1.50
Close supplier–customer relationships	3.33	4.15	4.81	1.48
Increased inventory turnover	3.3	4.09	4.77	1.47
Reduced material handling	3.27	4.06	4.74	1.47
Reduces lot sizes purchased	3.09	3.89	4.66	1.57
Reduced part numbers	2.96	3.80	4.60	1.64

- *Reduced inventory*: This element was ranked as the first JIT Benefit obtained in this category, which supports one of the main objectives of JIT, that is, the reduction of waste. Although reduction of inventory can be the result of several elements, it may be mainly related to just-in-time purchases, since material is purchased in the exact amount needed for the production process. Thus, companies avoid storage of raw materials that only increase the production cost (Aigbedo 2007; Billesbach 1991; Fazel et al. 1998; Hong et al. 1992; Kim and Ha 2003). Similarly, training and quality programs play key roles in the reduction of inventory, since operators in production processes become highly qualified and more efficient (Ahmad et al. 2012; Flynn et al. 1995; Forza and Filippini 1998). Thus, errors diminish and inventory in process is reduced and sometimes becomes inexistent (Jones 2001). Furthermore, companies can seek to reduce inventory resulting from the storage of products by relying on just-on-time deliveries and implementing a pull system produce the exact amount of product arranged (Iyogun 1991; Kojima et al. 2008).
- *Close supplier–customer relationship*: This is the second element of this category with the highest median value reported. It shows a median value of 4.15, that seven hundredths lower than the firs element. Inventories of raw material initiate with the amount of these materials received from suppliers. Therefore, it is important that manufacturing companies and their suppliers maintain a close relation based on trust and entire integration (David and Eben-Chaime 2003; Dey and Giri 2014; Dumond and Newman 1990). Moreover, this integration should rely on communication and information technologies available so that reliable suppliers guarantee full availability of raw materials required and urgent orders due to a lack of supplier–customer communication (Gélinas et al. 1996; Gilbert et al. 1994; Humphreys et al. 2003).
- *Increased inventory turnover*: This benefit was ranked third and shows a median value of 4.09, that is, six hundredths lower than the median value of the previous element but 13 hundredths if it is compared with the median of the first element. According to values and respondents, increased inventory turnover is generally obtained in JIT systems. Moreover, it can be the result of many factors, such as the production system implemented when it produces the exact amount requested by means of a production order. That is to say, companies do not

produce to store components or completed products (Chan 2001) which implies the purchase of the exact amount of raw material required (Krause and Ellram 1997; Wuttke et al. 2013). Similarly, training is significant to increase inventory turnover, since workforce becomes highly qualified and specialized in their tasks, and comprehension issues concerning transformation methodology or methods that will be applied for the transformation of raw materials, which reduces damages or breakdowns of machinery and equipment. Finally, the flow of materials can be also ensured with the implementation of appropriate programs of productive maintenance (Chiarini 2013; Cua et al. 2001; Rodrigues and Hatakeyama 2006).

- *Reduced material handling*: This benefit shows a median value of 4.06, that is, three hundredths lower than the median value of the previous element and 16 hundredths if compared with the median value of the first element. Less material handling reduces economic costs for companies, since this movement of materials does not add value to the product. Moreover, it also diminishes risks of industrial accidents, since it has been reported that a great percentage of accidents in companies occur at this stage (Bennett 1984; Niskanen and Lauttalammi 1989; Lortie 2012). Reduction of materials handling can be ensured by many factors, although two of the most significant concerns the physical distribution of machinery and equipment used to transform raw materials (Azadeh et al. 2014; Chen et al. 2011; Kia et al. 2014) as well as the plant organization for the production, either in technology groups or cellular manufacturing (Kia et al. 2014; Pattanaik and Sharma 2009), since these organizations allow for the entire product manufacturing within a small space.
- *Reduced lot sizes purchased*: This element was ranked as the fifth most obtained JIT benefit for material handling. It shows a median value of 3.9, that is, 17 hundredths lower if it is compared with the median value of the previous element and 33 hundredths in comparison with median of the first element. Companies can benefit from small lot sizes with a JIT program since one of its main objectives is the reduction of inventory. Therefore, with just-in-time purchases, companies are able to acquire the minimum amount of material needed and reduce inventory costs (Miocevic and Crnjak-Karanovic 2012). Similarly, it is important that manufacturing companies hold close relationships with their suppliers, who must be highly reliable and few to ensure a complete integration from them with the manufacturing companies. However, companies must be careful with the way the approach the reduction of lot sizes. Although costs of inventory maintenance could be avoided with the purchase of small lots, purchasing several orders of small amounts of material could highly increase administrative costs (Aigbedo 2007). Fortunately, information and communication technologies nowadays can provide companies with solutions to avoid these secondary administrative expenses. For instance, several models have been proposed to identify the ideal size of a lot within a JIT environment (Bayo-Moriones and Lera-López 2007; Fong et al. 2013).
- *Reduced par numbers*: This element holds the last place in the category of JIT benefits associated with material handling. It shows a median value of 3.8, that

is, only nine hundredths lower than the median value of the previous element and 42 hundredth if it is compared with the median value of the first element. Moreover, this is also the second JIT benefit reported with a median value lower than 4, a first quartile value lower than three, and the highest interquartile range value. Finally, it has been stated that standardized processes as JIT elements can be crucial for the reduction of parts in production companies (Lovell 2003).

8.6 Economic Benefits

This category is integrated by six JIT benefits previously reported and associated with economic and financial benefits. Some of them concerned profit margin increases or reduction of general expenses. Although it was not the first category introduced, one shall not underestimate its relevance since in fact, economic benefits are the major reason for a JIT implementation.

Table 8.6 shows the descriptive analysis of the elements, which were sorted in the descending order according to their median values reported or percentile 50. Based on the information in Table 8.6, it is possible to present the following conclusions:

- All elements show first quartile or 25th percentile values higher than 3 but lower than 4.00 which indicates that these benefits were at least regularly obtained in companies surveyed and according to respondents.
- All elements show 50th percentile, second quartile, or median values that are higher than 4.00 but equal or are lower than 4.25. Thus, it can be stated that, in average, all companies surveyed regularly obtain economic benefits from a JIT implementation. The range between the minimal and the maximum value of the median is only 0.25, an almost negligible value.
- The minimal third quartile or 75th percentile value observed in this category is 4.75, which indicates that almost all the companies reported obtained these benefits.

Table 8.6 Descriptive analysis of economic benefits associated with JIT

Order	JIT benefit	Percentile			IQR
		25	50	75	
1	Reduction of established costs	3.48	4.25	4.87	1.39
2	Improved competitiveness	3.44	4.23	4.86	1.42
3	Increased profit margin	3.37	4.17	4.83	1.46
4	Reduced general cost	3.32	4.10	4.77	1.45
5	Reduced production costs	3.18	4.05	4.76	1.58
6	Reduced labor costs	3.16	4.00	4.75	1.59

- *Reduced established costs*: This element was ranked first and has a median value of 4.25, which implies that companies surveyed have regularly obtained it. Moreover, it has the lowest interquartile range value for this category, indicating moderate consensus from respondents on the median value of the element. A reduction of established costs can be the result of the implementation of several elements, such as quality plans for product and process (Flynn et al. 1995; Vuppalapati et al. 1995), continuous improvement plans to solve problems in the production process (Ortiz 2010; Recht and Wilderom 1998), and plans of productive maintenance for machineries and equipment (Singh et al. 2013; Cua et al. 2001). However, these elements rely on education and training from managers and workers (The Value of JIT-Structured On-the-Job Training 2011; Deshpande and Golhar 1996; Jacobs 2010).
- *Improved competitiveness*: This element was ranked as the second most obtained economic benefit within a JIT implementation. It has a median value of 4.23, that is, two hundredths lower than the median value of the previous element, which indicates that, according to respondents, this benefit is generally obtained. Companies that rely on JIT manage to improve their competitiveness as a result of a total quality achieved in their products through quality control programs, since it is closely related to the perception of clients regarding the quality and price of the products (Bowman and Carter 1995; Li et al. 2006; Agus and Hassan 2011) and, thus, their disposition to purchase them. Similarly, a reduction of planned or established costs allows for the manufacturing of products at a minor price and can help companies reach major competitiveness in their markets.
- *Increased profit margin*: According to Table 8.6, this element has a median value of 4.17, that is, six hundredths lower than the median value of the previous element and eight hundredths lower if compared with the median value of the first element listed in the table. It can be thus stated that an increased profit margin is regularly obtained among companies surveyed. Profit margin can be referred as to the existing difference between the sale price and the production cost. This study has demonstrated that companies with JIT can reduce their production costs, but also offer better product prices. Therefore, increased profit margin is merely the result of the two previous benefits. On the one hand, profit margin can increase with a reduction of costs through the implementation of total quality control (Cua et al. 2001; Flynn et al. 1995). However, companies must also provide training to workers and demonstrate their commitment to these employees (Mokhtar and Yusof 2010; Power and Sohal 2000). Similarly, lean manufacturing techniques applied to the production system, such as SMED (Almomani et al. 2013; Ferradás and Salonitis 2013), Kaizen (Knechtges and Decker 2014; Lyu Jr 1996; Ortiz 2010), and Kanban (Brox and Fader 1997; Chan 2001; Fiscus 1987) can impact on profit margins of companies.
- *Reduced general costs*: General costs are associated to production costs and administrative cost. This element has a median value of 4.10, that is, seven

hundredths lower than the median value of the previous benefit and 15 hundredths if compared with the median value of the first element listed in the table. Reduced general costs are the result of several JIT benefits and elements previously mentioned, such as reduced inventories and their increased turnover (Brox and Fader 1997) that also help diminish maintenance costs of raw materials. Similarly, the standardization of administrative processes associated with the control and monitoring of quality plans and programs can help diminish general costs, since they are based on international norms properly established, such as ISO (Dowlatshahi and Urias 2004; Orecchini 2000). Finally, the reduction of production cost can be the result of extensive training that companies offer to their employees, who become more effective and are able to appropriately correct their few errors (Jacobs 2010; Jones 2001).

- *Reduced product cost*: This benefit has a median value of 4.05, that is, five hundredths lower than the median value of the previous element and 20 hundredths if it is compared with the median value of the first element listed in the table. This indicates that companies surveyed regularly achieve a reduction in product cost with a JIT implementation. Reduced product cost can be the result of other benefits and elements previously described, such as the implementation of a pull system, which indicates that companies request the exact amount of raw materials and thus reduce storage and inventory costs (Huq and Huq 1994; Xiong and Nyberg 2000). Similarly, product cost decreases when employees become highly qualified in their operations; thanks to the extensive training offered (Jones 2001), machinery and equipment are properly supervised by means of preventive maintenance programs (Ahmad et al. 2012; Cua et al. 2001; McCarthy and Rich 2004), and changes in product engineering are properly and efficiently approached (Chiarini 2013; Ferradás and Salonitis 2013). All these elements also ensure a continuous flow of materials along the production system that reduce technical issues.
- *Reduced labor costs*: This element has a median value of 4.00, that is, five hundredths lower than the median value of the previous immediate benefit and 25 hundredths lower if compared with the median value of the first element. It is also the last economic benefit listed, although one may hope it eventually manages to be one of the first as a result of real high expertise of workers, which is also required for the success of specialized factories in JIT systems (Jones 2001).

8.7 Conclusions

This chapter described the analysis of six categories of JIT benefits that allows for the following conclusions:

Table 8.7 Descriptive analysis of JIT benefits

Order	JIT benefit	Percentile			IQR
		25	50	75	
1	Increased efficiency	3.62	4.36	4.94	1.32
2	Reduced waste and rework	3.51	4.31	4.92	1.41
3	Increased product quality	3.45	4.26	4.89	1.44
4	Reduced established costs	3.48	4.25	4.87	1.39
5	Improved competitiveness	3.44	4.23	4.86	1.42
6	Reduced inventory	3.38	4.22	4.88	1.50

- Companies surveyed reported increased resource utilization.
- Managers surveyed informed of greater efficiency of the goals met.
- Companies have experienced improved resource utilization associated with the physical space.
- Companies surveyed improved product quality with the support and commitment of managers to quality control plans and programs.
- Companies surveyed reported decreasing inventories as a result of less administrative cost and material handling.
- Companies surveyed have reduced established and planned costs as the result of high quality products, reduced inventories, and workers expertise.

Table 8.7 shows the six main general JIT benefits regardless of their category. It can be observed that there exists no large difference between the highest and the lowest value. Data in this table allow for the following interpretations:

- Increased efficiency is reported as the first general JIT benefit obtained; it is associated with the production process.
- Reduced of waste rework is also associated with the same category and was reported as the second most obtained JIT benefit in general. Thus, it seems that JIT benefits mainly reflect on the production process.
- Product quality ranks as the third general JIT benefit obtained, indicating that it is the result of the two previous benefits.
- The fourth and fifth general JIT benefits obtained by companies surveyed are associated with the economic factor. A reduction of waste and rework and planned or established costs reflected an increased competitiveness for these companies.
- Reduced inventory along the production process was the sixth main general JIT benefit obtained by surveyed companies. This demonstrates that the main objective of JIT philosophy does not concern materials handling, which is rather viewed as a means to increase process efficiency and major product quality to improve the corporative image of companies as well as their position over their competitors by means of low prices and suitable quality.

References

Agus A, Hassan Zf (2011) Enhancing production performance and customer performance through total quality management (TQM): strategies for competitive advantage. Procedia Soc Behav Sci 24:1650–1662. doi:10.1016/j.sbspro.2011.09.019

Ahmed PK, Rafiq M (2002) Total quality management (TQM) and internal marketing (Chap. 6). In: Ahmed PK, Rafiq M (eds) Internal marketing. Butterworth-Heinemann, Oxford, pp 91–124. doi:10.1016/B978-0-7506-4838-7.50011-9

Ahmad S, Schroeder RG, Sinha KK (2003) The role of infrastructure practices in the effectiveness of JIT practices: implications for plant competitiveness. J Eng Tech Manage 20(3):161–191. doi:10.1016/S0923-4748(03)00017-1

Ahmad MF, Zakuan N, Jusoh A, Takala J (2012) Relationship of TQM and business performance with mediators of SPC, lean production and TPM. Procedia Soc Behav Sci 65:186–191. doi:10.1016/j.sbspro.2012.11.109

Aigbedo H (2007) An assessment of the effect of mass customization on suppliers' inventory levels in a JIT supply chain. Eur J. Oper Res 181(2):704–715. doi:10.1016/j.ejor.2006.06.037

Aksoy A, Öztürk N (2011) Supplier selection and performance evaluation in just-in-time production environments. Expert Syst Appl 38(5):6351–6359. doi:10.1016/j.eswa.2010.11.104

Alcaraz JLG, Maldonado AA, Iniesta AA, Robles GC, Hernández GA (2014) A systematic review/survey for JIT implementation: Mexican maquiladoras as case study. Comput Ind 65 (4):761–773. doi:10.1016/j.compind.2014.02.013

Ali J, Kumar S (2011) Information and communication technologies (ICTs) and farmers' decision-making across the agricultural supply chain. Int J Inf Manage 31(2):149–159. doi:10.1016/j.ijinfomgt.2010.07.008

Almomani MA, Aladeemy M, Abdelhadi A, Mumani A (2013) A proposed approach for setup time reduction through integrating conventional SMED method with multiple criteria decision-making techniques. Comput Ind Eng 66(2):461–469. doi:10.1016/j.cie.2013.07.011

Amasaka K (2002) "New JIT": a new management technology principle at Toyota. Int J Prod Econ 80(2):135–144. doi:10.1016/S0925-5273(02)00313-4

Amasaka K (2007) Applying new JIT—Toyota's global production strategy: epoch-making innovation of the work environment. Robot Comput Integr Manuf 23(3):285–293. doi:10.1016/j.rcim.2006.02.001

Amasaka K (2014) New JIT, new management technology principle: surpassing JIT. Procedia Technol 16:1135–1145. doi:10.1016/j.protcy.2014.10.128

Anderman EM, Gray D (2015) Motivation, learning, and instruction. In: Wright JD (ed) International encyclopedia of the social & behavioral sciences, 2nd edn. Elsevier, Oxford, pp 928–935. doi:10.1016/B978-0-08-097086-8.26041-8

Azadeh A, Bidokhti B, Sakkaki SMR (2005) Design of practical optimum JIT systems by integration of computer simulation and analysis of variance. Comput Ind Eng 49(4):504–519. doi:10.1016/j.cie.2005.03.005

Azadeh A, Motevali Haghighi S, Asadzadeh SM (2014) A novel algorithm for layout optimization of injection process with random demands and sequence dependent setup times. J Manuf Syst 33(2):287–302. doi:10.1016/j.jmsy.2013.12.008

Bayo-Moriones A, Lera-López F (2007) A firm-level analysis of determinants of ICT adoption in Spain. Technovation 27(6–7):352–366. doi:10.1016/j.technovation.2007.01.003

Bennett GF (1984) A guide to the safe handling of hazardous materials accidents: manual STP/825, American Society for testing materials, Philadelphia, PA, 1982, 53 pages. J Hazard Mater 9(3):396. doi:10.1016/0304-3894(84)87022-3

Benton WC, Shin H (1998) Manufacturing planning and control: the evolution of MRP and JIT integration. Eur J Oper Res 110(3):411–440. doi:10.1016/S0377-2217(98)00080-0

Berghman L, Matthyssens P, Vandenbempt K (2012) Value innovation, deliberate learning mechanisms and information from supply chain partners. Ind Mark Manage 41(1):27–39. doi:10.1016/j.indmarman.2011.11.014

Billesbach TJ (1991) A study of the implementation of just-in-time in the United States. Prod Inventory Manage J 32(3):1–4

Blaga P, Jozsef B (2014) Human resources, quality circles and innovation. Procedia Econ Finance 15:1458–1462. doi:10.1016/S2212-5671(14)00611-X

Bowman C, Carter S (1995) Organising for competitive advantage. Eur Manage J 13(4):423–433. doi:10.1016/0263-2373(95)00036-K

Brown JR (2014) A capacity constrained mathematical programming model for cellular manufacturing with exceptional elements. J Manuf Syst. doi:10.1016/j.jmsy.2014.09.005

Brox JA, Fader C (1997) Assessing the impact of JIT using economic theory. J Oper Manage 15 (4):371–388. doi:10.1016/S0272-6963(97)00005-3

Chan FTS (2001) Effect of kanban size on just-in-time manufacturing systems. J Mater Process Technol 116(2–3):146–160. doi:10.1016/S0924-0136(01)01022-6

Chen Y, Xiao Q, Tang X (2011) Product layout optimization and simulation model in a multi-level distribution center. Syst Eng Procedia 2:300–307. doi:10.1016/j.sepro.2011.10.042

Cheng C-F, Chang M-L, Li C-S (2013) Configural paths to successful product innovation. J Bus Res 66(12):2561–2573. doi:10.1016/j.jbusres.2012.10.006

Cheung C, Baum T, Wong A (2012) Relocating empowerment as a management concept for Asia. J Bus Res 65(1):36–41. doi:10.1016/j.jbusres.2011.07.012

Chiarini A (2013) The main methods of lean organization: Kanban, Cellular Manufacturing, SMED and TPM. In: Lean organization: from the tools of the Toyota Production System to Lean Office, vol 3. Perspectives in Business Culture. Springer, Milan, pp 81–116. doi:10.1007/978-88-470-2510-3_6

Chu KF (2003) An organizational culture and the empowerment for change in SMEs in the Hong Kong manufacturing industry. J Mater Process Technol 139(1–3):505–509. doi:10.1016/S0924-0136(03)00527-2

Colombo MG, Croce A, Grilli L (2013) ICT services and small businesses' productivity gains: an analysis of the adoption of broadband internet technology. Inf Econ Policy 25(3):171–189. doi:10.1016/j.infoecopol.2012.11.001

Cua KO, McKone KE, Schroeder RG (2001) Relationships between implementation of TQM, JIT, and TPM and manufacturing performance. J Oper Manage 19(6):675–694. doi:10.1016/S0272-6963(01)00066-3

David I, Eben-Chaime M (2003) How far should JIT vendor–buyer relationships go? Int J Prod Econ 81–82:361–368. doi:10.1016/S0925-5273(02)00274-8

de Macedo-Soares TDLvA, Lucas DC (1995) Empowerment and total quality: comparing research findings in the USA and Brazil. Technovation 15(8):475–495. doi:10.1016/0166-4972(95)96587-J

Deshpande SP, Golhar DY (1996) HRM practices of JIT firms in Canada. Prod Plan Control 7 (1):79

Dey O, Giri BC (2014) Optimal vendor investment for reducing defect rate in a vendor–buyer integrated system with imperfect production process. Int J Prod Econ 155:222–228. doi:10.1016/j.ijpe.2014.02.004

Diefendorff JM, Seaton GA (2015) Work motivation. In: Wright JD (ed) International encyclopedia of the social & behavioral sciences, 2nd edn. Elsevier, Oxford, pp 680–686. doi:10.1016/B978-0-08-097086-8.22036-9

Dowlatshahi S, Urias C (2004) An empirical study of ISO certification in the maquiladora industry. Int J Prod Econ 88(3):291–306. doi:10.1016/S0925-5273(03)00191-9

Dumond EJ, Newman W (1990) Closing the gap between buyer and vendor. Prod Inventory Manage J 31(4):13–17

Dwivedula R, Bredillet CN (2010) Profiling work motivation of project workers. Int J Project Manage 28(2):158–165. doi:10.1016/j.ijproman.2009.09.001

Evanschitzky H, Eisend M, Calantone RJ, Jiang Y (2012) Success factors of product innovation: an updated meta-analysis. J Prod Innov Manage 29:21–37. doi:10.1111/j.1540-5885.2012.00964.x

Fazel F, Fischer KP, Gilbert EW (1998) JIT purchasing vs. EOQ with a price discount: an analytical comparison of inventory costs. Int J Prod Econ 54(1):101–109. doi:10.1016/S0925-5273(97)00130-8

Ferradás PG, Salonitis K (2013) Improving changeover time: a tailored SMED approach for welding cells. Procedia CIRP 7:598–603. doi:10.1016/j.procir.2013.06.039

Ferretti S, Caputo D, Penza M, D'Addona DM (2013) Monitoring systems for zero defect manufacturing. Procedia CIRP 12:258–263. doi:10.1016/j.procir.2013.09.045

Fiscus JE (1987) Kanban and just-in-time at Toyota: edited by the Japanese management association, translated by David J. Lu. Productivity Press, Cambridge, MA, U.S.A., 1986. 192 pages, 40 figures, hardcover. ISBN 0-915299-08-0. List price: $29.95. Eng Manage Int 4 (2):157. doi:10.1016/0167-5419(87)90022-6

Flynn B, Salakibara S, Schroeder R (1995) Relationship between JIT and TQM practices and performance. Acad Manage J 38(5):1325–1360

Fong SF, Ch'ng PE, Por FP (2013) Development of ICT competency standard using the Delphi Technique. Procedia Soc Behav Sci 103:299–314. doi:10.1016/j.sbspro.2013.10.338

Forza C, Filippini R (1998) TQM impact on quality conformance and customer satisfaction: a causal model. Int J Prod Econ 55(1):1–20. doi:10.1016/S0925-5273(98)00007-3

Fullerton RR, McWatters CS (2001) The production performance benefits from JIT implementation. J Oper Manage 19(1):81–96. doi:10.1016/S0272-6963(00)00051-6

Fullerton RR, McWatters CS (2002) The role of performance measures and incentive systems in relation to the degree of JIT implementation. Acc Organ Soc 27(8):711–735. doi:10.1016/S0361-3682(02)00012-0

Fullerton RR, McWatters CS, Fawson C (2003) An examination of the relationships between JIT and financial performance. J Oper Manage 21(4):383–404. doi:10.1016/S0272-6963(03)00002-0

Gélinas R, Jacob R, Drolet J (1996) Just-in-time purchasing and the partnership strategy. Eur J Purchasing Supply Manage 2(1):39–45. doi:10.1016/0969-7012(95)00015-1

Gilbert FW, Young JA, O'Neal CR (1994) Buyer-seller relationships in just-in-time purchasing environments. J Bus Res 29(2):111–120. doi:10.1016/0148-2963(94)90015-9

González-Benito J, Suárez-González I, Spring M (2000) Complementarities between JIT purchasing practices: an economic analysis based on transaction costs. Int J Prod Econ 67 (3):279–293. doi:10.1016/S0925-5273(00)00032-3

Green Jr KW, Inman RA, Birou LM, Whitten D (2014) Total JIT (T-JIT) and its impact on supply chain competency and organizational performance. Int J Prod Econ 147(Part A):125–135. doi:10.1016/j.ijpe.2013.08.026

Gupta YP, Heragu S (1991) Implications of implementing just-in-time systems. Technovation 11 (3):143–162. doi:10.1016/0166-4972(91)90031-X

Gupta YP, Lonial SC, Glynn Mangold W (1992) Empirical assessment of the strategic orientation of JIT manufacturers versus non-JIT manufacturers. Comput Integr Manuf Syst 5(3):181–189. doi:10.1016/0951-5240(92)90029-C

Hong J-D, Hayya JC, Kim S-L (1992) JIT purchasing and setup reduction in an integrated inventory model. Int J Prod Res 30(2):255–266. doi:10.1080/00207549208942893

Huang C-H, Lin P-Y (2014) Influence of spatial layout on airflow field and particle distribution on the workspace of a factory. Build Environ 71:212–222. doi:10.1016/j.buildenv.2013.09.014

Humphreys P, Shiu WK, Lo VHY (2003) Buyer–supplier relationship: perspectives between Hong Kong and the United Kingdom. J Mater Process Technol 138(1–3):236–242. doi:10.1016/S0924-0136(03)00078-5

Huq Z (1999) Conventional shop control procedures to approximate JIT inventory performance in a job shop. J Manuf Syst 18(3):161–174. doi:10.1016/S0278-6125(99)80029-6

Huq Z, Huq F (1994) Embedding JIT in MRP: the case of job shops. J Manuf Syst 13(3):153–164. doi:10.1016/0278-6125(94)90001-9

Inman RA, Sale RS, Green KW Jr, Whitten D (2011) Agile manufacturing: relation to JIT, operational performance and firm performance. J Oper Manage 29(4):343–355. doi:10.1016/j.jom.2010.06.001

Iwase M, Ohno K (2011) The performance evaluation of a multi-stage JIT production system with stochastic demand and production capacities. Eur J Oper Res 214(2):216–222. doi:10.1016/j.ejor.2011.04.014

Iyogun P (1991) Coordination of production and distribution in JIT Systems. In: Eloranta E (ed) Advances in production management systems. Elsevier, Amsterdam, pp 515–522. doi:10.1016/B978-0-444-88919-5.50058-5

Jacobs RL (2010) Training and learning in the workplace. In: McGaw PPB (ed) International encyclopedia of education, 3rd edn. Elsevier, Oxford, pp 328–336. doi:10.1016/B978-0-08-044894-7.00773-9

Jensen JB, Malhotra MK, Philipoom PR (1996) Machine dedication and process flexibility in a group technology environment. J Oper Manage 14(1):19–39. doi:10.1016/0272-6963(95)00030-5

Jones MJ (2001) Just-in-time training. Adv Developing Human Resour 3(4):480

Jun M, Cai S, Shin H (2006) TQM practice in maquiladora: antecedents of employee satisfaction and loyalty. J Oper Manage 24(6):791–812. doi:10.1016/j.jom.2005.09.006

Kemal Karasu M, Cakmakci M, Cakiroglu MB, Ayva E, Demirel-Ortabas N (2014) Improvement of changeover times via Taguchi empowered SMED/case study on injection molding production. Measurement 47:741–748. doi:10.1016/j.measurement.2013.09.035

Kia R, Khaksar-Haghani F, Javadian N, Tavakkoli-Moghaddam R (2014) Solving a multi-floor layout design model of a dynamic cellular manufacturing system by an efficient genetic algorithm. J Manuf Syst 33(1):218–232. doi:10.1016/j.jmsy.2013.12.005

Kim S-L, Ha D (2003) A JIT lot-splitting model for supply chain management: enhancing buyer–supplier linkage. Int J Prod Econ 86(1):1–10. doi:10.1016/S0925-5273(03)00006-9

Knechtges P, Decker MC (2014) Application of Kaizen Methodology to foster departmental engagement in quality improvement. J Am Coll Radiol 11(12, Part A):1126–1130. doi:10.1016/j.jacr.2014.08.027

Kojima M, Nakashima K, Ohno K (2008) Performance evaluation of SCM in JIT environment. Int J Prod Econ 115(2):439–443. doi:10.1016/j.ijpe.2007.11.017

Krause DR, Ellram LM (1997) Success factors in supplier development. Int J Phys Distrib Logistics Manage 27(1):39–52. doi:10.1108/09600039710162277

Lawrence JJ, Hottenstein MP (1995) The relationship between JIT manufacturing and performance in Mexican plants affiliated with U.S. companies. J Oper Manage 13(1):3–18. doi:10.1016/0272-6963(95)00018-N

Li S, Ragu-Nathan B, Ragu-Nathan TS, Subba Rao S (2006) The impact of supply chain management practices on competitive advantage and organizational performance. Omega 34(2):107–124. doi:10.1016/j.omega.2004.08.002

Lim SAH, Antony J, Albliwi S (2014) Statistical process control (SPC) in the food industry—a systematic review and future research agenda. Trends Food Sci Technol 37(2):137–151. doi:10.1016/j.tifs.2014.03.010

Lortie M (2012) Analysis of the circumstances of accidents and impact of transformations on the accidents in a beverage delivery company. Saf Sci 50(9):1792–1800. doi:10.1016/j.ssci.2012.04.008

Lovell MC (2003) Optimal lot size, inventories, prices and JIT under monopolistic competition. Int J Prod Econ 81–82:59–66. doi:10.1016/S0925-5273(02)00349-3

Lyu J Jr (1996) Applying Kaizen and automation to process reengineering. J Manuf Syst 15(2):125–132. doi:10.1016/0278-6125(96)82337-5

Maiga AS, Jacobs FA (2009) JIT performance effects: a research note. Adv Account 25(2):183–189. doi:10.1016/j.adiac.2009.06.003

McCarthy D, Rich DN (2004) 2—Lean TPM. In: McCarthy D, Rich DN (eds) Lean TPM. Butterworth-Heinemann, Oxford, pp 24–48. doi:10.1016/B978-075065857-7/50005-6

McLaughlin P, Durazo-Cardenas I (2013) Cellular manufacturing applications in MRO operations. Procedia CIRP 11:254–259. doi:10.1016/j.procir.2013.07.077

Mensah P, Merkuryev Y, Longo F (2015) Using ICT in developing a resilient supply chain strategy. Procedia Comput Sci 43:101–108. doi:10.1016/j.procs.2014.12.014

Miocevic D, Crnjak-Karanovic B (2012) The mediating role of key supplier relationship management practices on supply chain orientation—the organizational buying effectiveness link. Ind Mark Manage 41(1):115–124. doi:10.1016/j.indmarman.2011.11.015
Moattar Husseini SM, O'Brien C, Hosseini ST (2006) A method to enhance volume flexibility in JIT production control. Int J Prod Econ 104(2):653–665. doi:10.1016/j.ijpe.2005.05.025
Mokhtar SSM, Yusof RZ (2010) The influence of top management commitment, process quality management and quality design on new product performance: a case of Malaysian manufacturers. Total Qual Manage Bus Excellence 21(3):291–300. doi:10.1080/14783360903553198
Monden Y (2002) The relationship between mini profit-center and JIT system. Int J Prod Econ 80 (2):145–154. doi:10.1016/S0925-5273(02)00314-6
Nassimbeni G (1996) Factors underlying operational JIT purchasing practices: results of an empirical research. Int J Prod Econ 42(3):275–288. doi:10.1016/0925-5273(95)00190-5
Niskanen T, Lauttalammi J (1989) Accidents in materials handling at building construction sites. J Occup Accidents 11(1):1–17. doi:10.1016/0376-6349(89)90002-3
Orecchini F (2000) The ISO 14001 certification of a machine-process. J Clean Prod 8(1):61–68. doi:10.1016/S0959-6526(99)00315-7
Ortiz C (2010) Kaizen vs. Lean: distinct but related. Met Finish 108(1):50–51. doi:10.1016/S0026-0576(10)80011-X
Ou-Yang C, Utamima A (2013) Hybrid estimation of distribution algorithm for solving single row facility layout problem. Comput Ind Eng 66(1):95–103. doi:10.1016/j.cie.2013.05.018
Paek J (1990) A study of an enlarged JIT and its impact on JIT implementation and performance of the production system. J Oper Manage 9(3):438–439. doi:10.1016/0272-6963(90)90188-J
Pattanaik LN, Sharma BP (2009) Implementing lean manufacturing with cellular layout: a case study. Int J Adv Manuf Technol 42(7–8):772–779. doi:10.1007/s00170-008-1629-8
Pourbabai B (1988) A just-in-time/group technology based scheduling model. In: Radharamanan R (ed) Robotics and factories of the future '87. Springer, Berlin, pp 105–110. doi:10.1007/978-3-642-73890-6_14
Power D, Sohal AS (2000) Human resource management strategies and practices in just-in-time environments: Australian case study evidence. Technovation 20(7):373–387. doi:10.1016/S0166-4972(99)00151-0
Recht R, Wilderom C (1998) Kaizen and culture: on the transferability of Japanese suggestion systems. Int Bus Rev 7(1):7–22. doi:10.1016/S0969-5931(97)00048-6
Rodrigues M, Hatakeyama K (2006) Analysis of the fall of TPM in companies. J Mater Process Technol 179(1–3):276–279. doi:10.1016/j.jmatprotec.2006.03.102
Rusu G, Avasilcai S (2014) Linking human resources motivation to organizational climate. Procedia Soc Behav Sci 124:51–58. doi:10.1016/j.sbspro.2014.02.459
Sandanayake YG, Oduoza CF, Proverbs DG (2008) A systematic modelling and simulation approach for JIT performance optimisation. Robot Comput Integr Manuf 24(6):735–743. doi:10.1016/j.rcim.2008.03.013
Seifermann S, Böllhoff J, Metternich J, Bellaghnach A (2014) Evaluation of work measurement concepts for a cellular manufacturing reference line to enable low cost automation for lean machining. Procedia CIRP 17:588–593. doi:10.1016/j.procir.2014.01.065
Shi W, Shang J, Liu Z, Zuo X (2014) Optimal design of the auto parts supply chain for JIT operations: sequential bifurcation factor screening and multi-response surface methodology. Eur J Oper Res 236(2):664–676. doi:10.1016/j.ejor.2013.11.015
Silvestre BS (2015) Sustainable supply chain management in emerging economies: environmental turbulence, institutional voids and sustainability trajectories. Int J Prod Econ 167:156–169. doi:10.1016/j.ijpe.2015.05.025
Singh S, Garg D (2011) JIT system: concepts, benefits and motivation in Indian industries. Int J Manage Bus Stud 1(1):26–30
Singh R, Gohil AM, Shah DB, Desai S (2013) Total productive maintenance (TPM) implementation in a machine shop: a case study. Procedia Eng 51:592–599. doi:10.1016/j.proeng.2013.01.084
Singh N, Jain S, Sharma P (2015) Motivations for implementing environmental management practices in Indian industries. Ecol Econ 109:1–8. doi:10.1016/j.ecolecon.2014.11.003

Spencer MS (1998) Group technology at John Deere: production planning and control issues. In: Suresh N, Kay J (eds) Group technology and cellular manufacturing. Springer, US, pp 501–516. doi:10.1007/978-1-4615-5467-7_29

Tabassi AA, Bakar AHA (2009) Training, motivation, and performance: the case of human resource management in construction projects in Mashhad, Iran. Int J Project Manage 27 (5):471–480. doi:10.1016/j.ijproman.2008.08.002

Tams S, Grover V, Thatcher J (2014) Modern information technology in an old workforce: toward a strategic research agenda. J Strateg Inf Syst 23(4):284–304. doi:10.1016/j.jsis.2014.10.001

The Value of JIT-Structured On-the-Job Training (2011) SMT: surface mount technology, vol 26 (6), pp 50–54

Tong X, Shi J, Zhou Y (2012) Greening of supply chain in developing countries: diffusion of lead (Pb)-free soldering in ICT manufacturers in China. Ecol Econ 83:174–182. doi:10.1016/j.ecolecon.2011.08.022

Vuppalapati K, Ahire SL, Gupta T (1995) JIT and TQM: a case for joint implementation. Int J Oper Prod Manage 15(5):84–94. doi:10.1108/01443579510083686

White RE, Prybutok V (2001) The relationship between JIT practices and type of production system. Omega 29(2):113–124. doi:10.1016/S0305-0483(00)00033-5

Wuttke DA, Blome C, Henke M (2013) Focusing the financial flow of supply chains: an empirical investigation of financial supply chain management. Int J Prod Econ 145(2):773–789. doi:10.1016/j.ijpe.2013.05.031

Xiong G, Nyberg TR (2000) Push/pull production plan and schedule used in modern refinery CIMS. Robot Comput Integr Manuf 16(6):397–410. doi:10.1016/S0736-5845(00)00023-5

Yasin MM, Small MH, Wafa MA (2003) Organizational modifications to support JIT implementation in manufacturing and service operations. Omega 31(3):213–226. doi:10.1016/S0305-0483(03)00024-0

Yu L, Suojapelto K, Hallikas J, Tang O (2008) Chinese ICT industry from supply chain perspective—a case study of the major Chinese ICT players. Int J Prod Econ 115(2):374–387. doi:10.1016/j.ijpe.2008.03.011

Chapter 9
Descriptive Analysis of the Causes of Slow JIT Implementation

As mentioned in Chap. 4, companies can experience different situations that may lead to slow implementation of the JIT philosophy, and which in a critical condition can lead to its definite abandonment. Therefore, it is necessary not only to become familiar with these causes but also to study them to avoid or prevent them, as it is possible.

9.1 Descriptive Analysis

Table 9.1 shows the descriptive analysis of 13 of the causes identified in Chap. 4. Similar to all descriptive analyses carried out in previous chapters, causes are ordered in descending order, according to the value obtained in every median.

Based on Table 9.1, it is possible to interpret the following results.

- As for the first quartile or 25th percentile, values obtained for all but one of the analyzed items (or causes) were lower than 3. This indicates that for most respondents these 13 problems rarely arise during the implementation of JIT in their companies.
- In the 50th percentile or second quartile, all causes of slow JIT implementation possess values lower than 4 but higher than 3. This suggests that these problems regularly arise during JIT implementation.
- As for the third quartile or 75th percentile, all the causes of slow JIT implementation possess values lower than 5 but higher than 4. This indicates that a large number of companies constantly experience these problems.
- In the fifth column is observed that all items possess moderately high interquartile ranges, since ten of them are higher than 1 and one is higher than 2. This indicates that there is a moderate consensus regarding the values of the median as the measure of central tendency.
- According to Table 9.1, two items or causes of slow JIT implementation possess the same median value of 3.86: *Lack of communication among all levels* during JIT implementation and *Lack of understanding of JIT techniques*. As far as the first element is concerned, it occupies the first place in the list, which indicates

J.L. García-Alcaraz and A.A. Maldonado-Macías, *Just-in-Time Elements and Benefits*,
Management and Industrial Engineering, DOI 10.1007/978-3-319-25919-2_9

Table 9.1 Descriptive analysis of the causes of slow JIT implementation

Order	Cause of slow JIT implementation	Percentile			IQR
		25	50	75	
1	Lack of communication among levels	2.84	3.86	4.69	1.85
2	Lack of understanding of JIT techniques	3.03	3.86	4.66	1.63
3	Lack of training	2.82	3.83	4.65	1.83
4	Lack of manager support in quality programs	2.85	3.81	4.63	1.78
5	Use of traditional methods for quality control	2.99	3.76	4.56	1.57
6	Lack of support from research development departments	2.86	3.75	4.59	1.73
7	Lack of team work	2.78	3.74	4.60	1.82
8	Poor and inadequate maintenance of machinery and equipment	2.85	3.74	4.58	1.73
9	Negative attitude of workers	2.62	3.73	4.64	2.02
10	Shortage of multifunctional workers	2.76	3.69	4.55	1.79
11	High cost of the implementation	2.88	3.66	4.46	1.58
12	Lack of consumer awareness of quality policies	2.71	3.62	4.45	1.74
13	Casual and informal quality audits	2.68	3.60	4.46	1.78

that communication issues are generally present in companies. This was already pointed out by Helms (1990) who stated that, as the title of his research, communication in and among all levels in companies was a key to JIT success; and it will always be. However, JIT requires certain organizational changes that may have a direct impact on the management of human resources, although this has not been extensively studied (Im et al. 1994). Therefore, in order to avoid communication issues among different organizational levels, senior managers should propose continuous programs to integrate all employees. For instance, (Power and Sohal 2000) proposed a series of integration strategies to ensure efficient communication. Similarly, structural equation models from recent studies have managed to relate the impact of senior management on the achievement of efficient communication among all organizational levels, which consequently had a direct and positive impact on the economic and financial performance of companies (Alcaraz et al. 2014; Green et al. 2014).

- As for the second element, a misunderstanding of JIT techniques is mainly due to the insufficient amount of training or guidance provided to operators (Jacobs 2010; Jones 2001), which is associated with a lack of support from senior management (Lemak and Reed 1997; Karatepe and Karadas 2012) and insufficient quality plans applied and their control (Vuppalapati et al. 1995; Zairi 1991).
- A lack of education and training may be the reason why JIT principles cannot be interpreted correctly; however, senior and top management departments are responsible for the execution of those training plans and programs. This means that while insufficient training can directly reflect a lack of acquaintance with JIT techniques, the main cause can also be a scarce commitment from senior management departments to preparation programs. It must be mentioned that

several maquiladoras included in this study are companies originated in other countries but established in Mexico, which often hire training services from companies abroad. As a result, communication between instructors and the operators is often challenging due to language barriers. Thus, companies must seek to create their own staff specialized in administrative tools that support JIT.

- Little training is the third major and important cause of slow JIT implementation, according to Table 9.1. It possesses an average of 3.83, which indicates a difference of 300ths if compared to the two items listed earlier. As previously stated, it is mainly caused by a lack of commitment from managers, who are responsible for providing continuous plans and programs of education and training for employees. However, it shall not be forgotten that this training must be provided at the right time, before operators begin performing their tasks; otherwise, inappropriate training would likely represent a source of many waste and reprocesses in the workplace (Globerson and Korman 2001). Therefore, it has been recommended that companies follow as much as possible the training and education programs offered by Toyota (Amasaka 2007), the company where JIT emerged. These programs have demonstrated to be the basis of growth of many performance indices in the company, and they can be adapted to the needs of every cultural environment. For instance, (Oral et al. 2003; Alcaraz et al. 2014) studied the implementation of these training programs in JIT environments in Turkey and Mexico, respectively.
- Insufficient participation from managers in quality programs is the fourth main cause of slow JIT implementation in Table 9.1. However, management departments are responsible for these plans and programs, as well as supervising and monitoring them to run and take appropriate decisions in case of deviations. This cause possesses a median of 3.81 with 200ths of difference from the previous cause and five from the first two major causes. As mentioned previously, the causes of a slow JIT implementation are mainly due to lack or absence of elements considered essential for its application. In this case, a lack of participation from managers may be due not only to their little commitment but also to the fact that they may not be familiarized enough with the mission and vision of the companies (Alcaraz et al. 2014; Lemak and Reed 1997). This happens especially in the case of maquiladoras, when high and middle managers are foreigners with little proficiency in Spanish, (Lawrence and Hottenstein 1995), and because of the cultural clash between them and engineers or other middle managers, who are generally proficient in the language (Sargent and Matthews 1997). To solve this problem, many companies train their managers at different organizational levels, not only to prepare them but also to encourage their commitment in the different programs being developed (Stank and Traichal 1998; Jun et al. 2006). Mexican maquiladoras investing in training their managers or high-level executives can differentiate them from those in other countries (Sargent and Matthews 2009), since this educational process for the highest hierarchical levels of the organization has been recognized as essential and one of the six strategies to ensure the success of maquiladoras (Hadjimarcou et al. 2013).

- The fifth major cause of slow JIT implementation concerns the use of traditional methods for quality control. This element in Table 9.1 possesses a median of 3.76, with differences of 500ths if compared with the previous immediate cause, and 1000ths in comparison with the first two major causes. This indicates that, according to the managers surveyed, it is a constant issue in the process of JIT implementation. However, this use of traditional methods to control quality can be, again, due to lack of information from senior managers and workers in terms of new techniques and methods available to control quality in processes and products (Ahmad et al. 2012), since statistical process control (SPC)—as its name suggests—requires high levels of knowledge in statistics for many of the processes, which restrains employees. They may assume that special effort is required from them. As a result, some authors have encouraged companies to implement programs of economic incentives for workers and managers along with the JIT implementation process, which would be provided according to the capabilities detected (Fullerton and McWatters 2002). This could allow organizations to maintain motivation among employees in training programs that, for some researchers (Gupta and Heragu 1991), must be executed during the process of JIT implementation to ensure part of its success.
- The sixth cause associated with a slow JIT implementation is the little amount of support obtained from the research and development departments in companies. This element possesses a median of 3.75 with 100th of difference from the previous immediate cause and 1100ths if compared with the two major causes reported in Table 9.1. The median indicates that issues related to this lack of support regularly arise in the surveyed maquiladoras, also because this environment does not count on the research and development department as such, since plans and designs of products are entirely made abroad in the parent companies and the possible changes that could arise later in these manufacturing environments are minimum. The research and development departments are located within the parent companies and the maquiladoras are only responsible for the assembling of products (Alcaraz et al. 2014). As a result, the physical distance between the headquarters of a company—in England for example—and its maquiladora industry established—in Mexico, for instance—becomes a limiting factor for top managers to provide support, not to mention the time difference between them, which can also be challenging. Therefore, when a design cannot be executed in its entirety and requires modifications, maquiladora companies must request authorization from their parent companies to perform the adjustments. However, if this occurs, for instance, at 17:00 h in Mexico, the company might have to wait to establish communication with the parent company until the following day, since it would be 01:00 h in England. This is translated as an incomplete order in process where the raw material would be obstructing the flow of materials and reducing the agility of the company (Inman et al. 2011).
- The inability to work in teams is the seventh reason that companies tend to experience for slow integration of JIT in their systems, according to the list in Table 9.1. The element possesses a median value of 3.74 with only 100th of

difference in relation to the previous immediate cause and 1200ths if compared to the two elements that hold the first place in the list. The value of the median indicates that, according to the respondents surveyed, companies regularly experience this problem. Some authors state that the lack of teamwork is mainly the result of cultural and educational beliefs instilled since childhood, where competition is encouraged and consequently favors individualism, instead of collaborative work (Fan et al. 2005; Fryxell and Judge 1995; Tong et al. 2015). Therefore, it is crucial that education plans and programs of companies emphasize on collaborative work from the beginning of their implementation. All members of the company must embrace the fact that they belong to and work together in the same company whose success also becomes theirs. Thus, companies can motivate their employees to work as a team with ten reasons (Fryxell and Judge 1995). First of all, employees must keep in mind that skills and talents of all members complement one another and the tasks assigned can be achieved faster through teamwork. Similarly, operators learn to trust their colleagues and become reliable. In addition, the ability to work as a group increases the sense of belonging, the results obtained by a team become rewarding to every participant, operators never face problems in the production system on their own, and the goals proposed become easier to achieve. Furthermore, teamwork allows members to learn more about themselves and their collaborators, since conflicts and problems encountered are more efficiently resolved. Finally, fellowship usually promotes social coexistence. However, the role of the leader of these work groups is crucial to achieve the complete integration of the team, although work meetings resulting from teamwork can take time. Thus, the commitment of top managers is essential for the success of teamwork (McLachlin 1997; Tong et al. 2015).

- Poor and inadequate maintenance of machinery was found as a main cause of slow JIT implementation. This cause holds the eighth place in Table 9.1 with a median value of 3.74 as the previous element. It is also 1200ths lower than the first two major causes of slow JIT implementation. Inappropriate maintenance is definitely the result of a lack of planning in the process of maintenance of machinery and equipment, which do not receive adequate care, begin generating delays in orders, and reduce the operating time available. Hence, it is highly recommended that companies establish preventive maintenance programs, especially corrective maintenance programs when a failure was detected (McCarthy and Rich 2004; Smith and Hawkins 2004). Moreover, preventive maintenance programs must be considered as one of the many foundations and pillars of JIT and lean manufacturing practices (McCarthy and Rich 2015), since nonoperating machines do not add value to the product. Preventive maintenance programs are usually ineffective since companies do not maintain a record of these nonoperating and operating machines or because the life span of each of their components or consumables is unknown. As a result, companies frequently experience damages or failures in their equipment and limit themselves to their correction, instead of focusing on their prevention (Rodrigues and Hatakeyama 2006). Total preventive maintenance programs guarantee that the flow of materials along the production system work simultaneously with other lean manufacturing techniques applied, such as total quality control, which requires the proper calibration of all the machinery and equipment in order to avoid errors in reprocessing work. This allows companies to integrate the Six Sigma techniques and SPC methodology (Thomas et al. 2006; Ahmad et al. 2012; Bakri et al. 2012).

- The negative attitude of the workers is the ninth cause of slow JIT implementation in a production system, according to Table 9.1. This element possesses a median value of 3.73 with a difference of 100th if compared with the two previous causes, and 1300ths if compared with the two major causes reported in the table. Attitude issues have been reported for nearly three decades in JIT environments, when the philosophy was being introduced in Western companies. For instance, (Giunipero and O'Neal 1988) analyzed the main obstacles for JIT in Western companies at the moment of its adoption and adaptation and found this issue as one of these obstacles. Similarly, (Ebrahimpour and Withers 1993) compared the attitudes of workers between companies that had applied JIT programs and those who had not. Authors noted that companies that applied JIT were better because one of the biggest differences between these was the attitude and cooperation of their employees. Additionally, (Yasin et al. 1997) mentioned that in order to secure a JIT program and its effectiveness, companies had count on the an upbeat attitude from workers toward its implementation; otherwise the results would not be positive. Therefore, companies could implement programs to foster the adequate participation and successful integration of workers in the business projects. These programs would aim to motivate and raise awareness of the benefits that both, the company and every individual as part of it, could gain from JIT implementation. This means that several organizational changes may be required inside organizations (Oral et al. 2003; Yasin et al. 2003).
- Shortage of multifunctional workers is another reason why JIT cannot offer the expected benefits, or provides them with delay or slowness. According to Table 9.1, this element possesses a median value of 3.69 that is 400ths lower than the median of previous cause and 1700ths in comparison with two major causes listed in the table. This shows that, according to the experience of the surveyed managers, companies regularly rely on an insufficient amount of multifunctional workers. However, in order to understand this problem, one must become familiar with the environment of the maquiladoras surveyed for this research, where unions can ban employees from performing the tasks or activities of others (Teagarden et al. 1992). However, lack of multifunctional workers in companies can also be the result of insufficient educational programs, especially those that are quality oriented, and time as an essential component of this concept which demands on-time deliveries to customers (Zairi 1991; Flynn et al. 1995; Vuppalapati et al. 1995). Another common source of this problem appears when employees refuse to learn additional skills and abilities to become multifunctional operatives. Many of them prefer specializing in a particular discipline or area that allows them to stand out from their peers but encourages

competition and, therefore, low levels of collaborative work in the company (McLachlin 1997; Tong et al. 2015).

- The high cost of JIT implementation is also considered as another cause of its slow implementation in production systems. The element is placed eleven in Table 9.1 and possesses a median value of 3.66, which differs in 300ths if it is compared to the previous immediate cause and 2000ths in comparison with the first two major causes listed. The value of the median indicates that, for the managers surveyed, the cost of implementing JIT is regularly a barrier for its implementation. However, one could have expected from this cause to be the major obstacle, but in the discussions and interviews performed, all managers surveyed for this research recognized that JIT programs were highly profitable investments from an economic point of view. Thus, it may seem that the costs of the implementation are not a major problem, unlike communication issues which hold the first place (Fullerton et al. 2003; Maiga and Jacobs 2009). Many recent studies have determined the direct relationship between JIT and the economic and financial benefits obtained in companies (Sen et al. 2015; Alcaraz et al. 2014) to demonstrate the economic viability of the philosophy. However, the reason why JIT may represent a major cost is attributable to the senior management department, who determines the plans and programs to run. Therefore, it is very likely that without the commitment of managers, not only a JIT program but also any other program sought to be implemented may fail.
- Table 9.1 listed little consumer awareness of quality policies as the penultimate issue that can cause slow JIT implementation in production systems. The element possesses an average value of 3.62 that is 400ths lower than the previous immediate cause, but 2400ths in comparison with the two major causes listed in the table. This indicates that companies of this research regularly experience this problem. Thus, although customers tend to consider the price as one of the few attributes of a product, companies must convey their outstanding product to justify the selling price. They may draw upon advertising campaigns to highlight those characteristics that make their product different from the rest in the market. Moreover, a high-quality product may not be sold, since it might offer the same general functionalities as traditional ones at a similar price. In these cases, companies can also rely on advertising campaigns to stand out the product. Thus, several authors have proposed models that demonstrate the economic benefit that companies could obtain when they managed to convince customers of the features of their products (Forza and Filippini 1998; Zhang et al. 2003). Similarly, companies can promote quality awareness by informing costumers of the type and versatility of the product purchased (Pan and Nguyen 2015; Zhang et al. 2009), since the relationship between a company and its costumer does not end with the purchase, it merely changes.
- According to Table 9.1, casual and informal quality audits also generate slow JIT implementation in production systems. This is the last element listed in the table and possesses a median value of 3.60, with a difference of 200ths if it is compared to the previous immediate cause, and 26 in comparison with the two major causes listed in the table. Unsuitable quality audits are the result of little

quality planning and/or inappropriate follow-up and control, since effective quality plans become useless when they are not provided with the adequate follow-up to take corrective actions when deviations occur. For instance, if a company relies on a suitable quality plan but does not provide it with adequate follow-up, the state of quality in the production system and the flow of materials would become unknown, and corrective actions cannot be undertaken. Thus, uncertainty should not govern quality control; quality must rather be completely planned, and audits ought to be performed regularly, if the process requires so, (Ball et al. 2015) and preferably by external personnel (Kouaib and Jarboui 2014). Furthermore, audits should be run not only in the production process but also with the suppliers of raw material, since quality issues in products may arise from that stage (Chen and Jeter 2008). Finally, it shall not be forgotten that despite the efficiency of plans and quality programs and appropriate audits, the quality of a product is not completely guaranteed because in production process the human resources have an important role (Powell et al. 2013).

9.2 Conclusions

This chapter described 13 factors that cause slow JIT implementation in production systems. According to the managers surveyed, it can be concluded that both insufficient communication and little understanding of JIT techniques are the major obstacles that prevent a successful implementation of the philosophy. This is extremely important, since they depict two major problems that most organizations experience: lack of communication and insufficient training (Jacobs 2010; Jones 2001).

Finally, had these 13 causes been grouped, it could be noticed that most of them relate to human resources and education and training. Thus, management commitment and leadership from supervisors are key factors to predict the success or failure of the implementation of this type of lean manufacturing philosophy.

References

Ahmad MF, Zakuan N, Jusoh A, Takala J (2012) Relationship of TQM and business performance with mediators of SPC, lean production and TPM. Proc Soc Behav Sci 65:186–191. doi:10.1016/j.sbspro.2012.11.109

Alcaraz JLG, Maldonado AA, Iniesta AA, Robles GC, Hernández GA (2014) A systematic review/survey for JIT implementation: Mexican maquiladoras as case study. Comput Ind 65 (4):761–773. doi:10.1016/j.compind.2014.02.013

Amasaka K (2007) Applying New JIT—Toyota's global production strategy: epoch-making innovation of the work environment. Robot Comput Integr Manuf 23(3):285–293. doi:10.1016/j.rcim.2006.02.001

Bakri AH, Rahim ARA, Yusof NM, Ahmad R (2012) Boosting lean production via TPM. Proc Soc Behav Sci 65:485–491. doi:10.1016/j.sbspro.2012.11.153
Ball F, Tyler J, Wells P (2015) Is audit quality impacted by auditor relationships? J Contemp Acc Econ 11(2):166–181. doi:10.1016/j.jcae.2015.05.002
Chen H, Jeter D (2008) The role of auditing in buyer-supplier relations. J Contemp Acc Econ 4(1):1–17. doi:10.1016/S1815-5669(10)70027-8
Ebrahimpour M, Withers BE (1993) A comparison of manufacturing management in JIT and non-JIT firms. Int J Prod Econ 32(3):355–364. doi:10.1016/0925-5273(93)90048-P
Fan X, Yen J, Volz RA (2005) A theoretical framework on proactive information exchange in agent teamwork. Artif Intell 169(1):23–97. doi:10.1016/j.artint.2005.06.005
Flynn B, Salakibara S, Schroeder R (1995) Relationship between JIT and TQM practices and performance. Acad Manage J 38(5):1325–1360
Forza C, Filippini R (1998) TQM impact on quality conformance and customer satisfaction: a causal model. Int J Prod Econ 55(1):1–20. doi:10.1016/S0925-5273(98)00007-3
Fryxell GE, Judge WQ (1995) Individualism and teamwork: synthesizing the dialectic in the management of R&D in biotechnology. J High Technol Manage Res 6(1):33–54. doi:10.1016/1047-8310(95)90005-5
Fullerton RR, McWatters CS (2002) The role of performance measures and incentive systems in relation to the degree of JIT implementation. Acc Organ Soc 27(8):711–735. doi:10.1016/S0361-3682(02)00012-0
Fullerton RR, McWatters CS, Fawson C (2003) An examination of the relationships between JIT and financial performance. J Oper Manage 21(4):383–404. doi:10.1016/S0272-6963(03)00002-0
Giunipero LC, O'Neal C (1988) Obstacles to JIT procurement. Ind Mark Manage 17(1):35–41. doi:10.1016/0019-8501(88)90024-7
Globerson S, Korman A (2001) The use of just-in-time training in a project environment. Int J Proj Manage 19(5):279–285. doi:10.1016/S0263-7863(00)00012-0
Green KW Jr, Inman RA, Birou LM, Whitten D (2014) Total JIT (T-JIT) and its impact on supply chain competency and organizational performance. Int J Prod Econ Part A 147:125–135. doi:10.1016/j.ijpe.2013.08.026
Gupta YP, Heragu S (1991) Implications of implementing just-in-time systems. Technovation 11 (3):143–162. doi:10.1016/0166-4972(91)90031-X
Hadjimarcou J, Brouthers LE, McNicol JP, Michie DE (2013) Maquiladoras in the 21st century: six strategies for success. Bus Horiz 56(2):207–217. doi:10.1016/j.bushor.2012.11.005
Helms MM (1990) Communication: the key to jit success. Prod Inventory Manage J 31(2):18–21
Im JH, Hartman SJ, Bondi PJ (1994) How do jit systems affect human resource management? Prod Inventory Manage J 35(1):1–4
Inman RA, Sale RS, Green KW Jr, Whitten D (2011) Agile manufacturing: relation to JIT, operational performance and firm performance. J Oper Manage 29(4):343–355. doi:10.1016/j.jom.2010.06.001
Jacobs RL (2010) Training and learning in the workplace. In: McGaw PPB (ed) International encyclopedia of education, 3rd edn. Elsevier, Oxford, pp 328–336. doi:10.1016/B978-0-08-044894-7.00773-9
Jones MJ (2001) Just-in-time training. Adv Dev Human Resour 3(4):480
Jun M, Cai S, Shin H (2006) TQM practice in maquiladora: antecedents of employee satisfaction and loyalty. J Oper Manage 24(6):791–812. doi:10.1016/j.jom.2005.09.006
Karatepe OM, Karadas G (2012) The effect of management commitment to service quality on job embeddedness and performance outcomes. J Bus Econ Manage 13(4):614–636. doi:10.3846/16111699.2011.620159
Kouaib A, Jarboui A (2014) External audit quality and ownership structure: interaction and impact on earnings management of industrial and commercial Tunisian sectors. J Econ Finance Adm Sci 19(37):78–89. doi:10.1016/j.jefas.2014.10.001

Lawrence JJ, Hottenstein MP (1995) The relationship between JIT manufacturing and performance in Mexican plants affiliated with U.S. companies. J Oper Manage 13(1):3–18. doi:10.1016/0272-6963(95)00018-N

Lemak DJ, Reed R (1997) Commitment to total quality management: is there a relationship with firm performance? J Qual Manage 2(1):67

Maiga AS, Jacobs FA (2009) JIT performance effects: a research note. Adv Acc 25(2):183–189. doi:10.1016/j.adiac.2009.06.003

McCarthy D, Rich DN (2004) 2—lean TPM. In: McCarthy D, Rich DN (eds) Lean TPM. Butterworth-Heinemann, Oxford, pp 24–48. doi:10.1016/B978-075065857-7/50005-6

McCarthy D, Rich N (2015) Chapter two—the lean TPM master plan. In: McCarthy D, Rich N (eds) Lean TPM, 2nd edn. Butterworth-Heinemann, Oxford, pp 27–54. doi:10.1016/B978-0-08-100090-8.00002-1

McLachlin R (1997) Management initiatives and just-in-time manufacturing. J Oper Manage 15(4):271–292. doi:10.1016/S0272-6963(97)00010-7

Oral EL, Mıstıkoglu G, Erdis E (2003) JIT in developing countries—a case study of the Turkish prefabrication sector. Build Environ 38(6):853–860. doi:10.1016/S0360-1323(03)00021-0

Pan J-N, Nguyen HTN (2015) Achieving customer satisfaction through product–service systems. Eur J Oper Res 247(1):179–190. doi:10.1016/j.ejor.2015.05.018

Powell DA, Erdozain S, Dodd C, Costa R, Morley K, Chapman BJ (2013) Audits and inspections are never enough: a critique to enhance food safety. Food Control 30(2):686–691. doi:10.1016/j.foodcont.2012.07.044

Power D, Sohal AS (2000) Human resource management strategies and practices in just-in-time environments: Australian case study evidence. Technovation 20(7):373–387. doi:10.1016/S0166-4972(99)00151-0

Rodrigues M, Hatakeyama K (2006) Analysis of the fall of TPM in companies. J Mater Process Technol 179(1–3):276–279. doi:10.1016/j.jmatprotec.2006.03.102

Sargent J, Matthews L (1997) Skill development and integrated manufacturing in Mexico. World Dev 25(10):1669–1681. doi:10.1016/S0305-750X(97)00056-9

Sargent J, Matthews L (2009) China versus Mexico in the global EPZ industry: Maquiladoras, FDI quality, and plant mortality. World Dev 37(6):1069–1082. doi:10.1016/j.worlddev.2008.10.002

Sen P, Roy M, Pal P (2015) Exploring role of environmental proactivity in financial performance of manufacturing enterprises: a structural modelling approach. J Cleaner Prod (0). doi:10.1016/j.jclepro.2015.05.076

Smith R, Hawkins B (2004) 3—total productive maintenance (TPM). In: Smith R, Hawkins B (eds) Lean maintenance. Butterworth-Heinemann, Burlington, pp 55–104. doi:10.1016/B978-075067779-0/50003-0

Stank TP, Traichal PA (1998) Logistics strategy, organizational design, and performance in a cross-border environment. Transp Res Part E: Logistics and Transp Rev 34(1):75–86. doi:10.1016/S1366-5545(97)00040-9

Teagarden MB, Butler MC, Von Glinow MA (1992) Mexico's maquiladora industry: where strategic human resource management makes a difference. Organ Dyn 20(3):34–47. doi:10.1016/0090-2616(92)90023-G

Thomas AJ, Jones GR, Vidales P (2006) An integrated approach to TPM and six sigma development in the castings industry. In: Pham DT, Eldukhri EE, Soroka AJ (eds) Intelligent production machines and systems. Elsevier Science Ltd, Oxford, pp 620–625. doi:10.1016/B978-008045157-2/50108-5

Tong DYK, Rasiah D, Tong XF, Lai KP (2015) Leadership empowerment behaviour on safety officer and safety teamwork in manufacturing industry. Safety Sci 72:190–198. doi:10.1016/j.ssci.2014.09.009

Vuppalapati K, Ahire SL, Gupta T (1995) JIT and TQM: a case for joint implementation. Int J Oper Prod Manage 15(5):84–94. doi:10.1108/01443579510083686

Yasin MM, Small M, Wafa MA (1997) An empirical investigation of JIT effectiveness: an organizational perspective. Omega 25(4):461–471. doi:10.1016/S0305-0483(97)00005-4

Yasin MM, Small MH, Wafa MA (2003) Organizational modifications to support JIT implementation in manufacturing and service operations. Omega 31(3):213–226. doi:10.1016/S0305-0483(03)00024-0

Zairi M (1991) 5—Defining TQM: Tools and techniques. In: Zairi M (ed) Total quality management for engineers. Woodhead Publishing, pp 87–123. doi:10.1533/9781845698911.87

Zhang Q, Vonderembse MA, Lim J-S (2003) Manufacturing flexibility: defining and analyzing relationships among competence, capability, and customer satisfaction. J Oper Manage 21(2):173–191. doi:10.1016/S0272-6963(02)00067-0

Zhang Q, Vonderembse MA, Cao M (2009) Product concept and prototype flexibility in manufacturing: implications for customer satisfaction. Eur J Oper Res 194(1):143–154. doi:10.1016/j.ejor.2007.12.013

Part V
Structural Equation Models

Chapter 10
Causal Models of JIT Elements Associated with Human Resources and Obtained Benefits

This chapter introduces three structural equation models that associate JIT elements previously reported concerning human resources with the benefits also previously reported. Readers may refer to methodology addressed in Chap. 5 for a full comprehension of the procedures here described concerning the planning, development, and execution of models.

Since JIT benefits were previously classified into five categories, a factor analysis was carried out before the proposal of models in order to identify the number of latent variables associated to human resources category as well as the items or variables that should be comprised in each of these latent variables.

10.1 Factor Analysis of JIT Elements Associated with Human Resources

A factor analysis was carried out according to methodology in Chap. 5 for the generation of latent variables that contain the 12 elements or items related to human resources category. According to the KMO index, the measure of sampling adequacy indicated that dataset was factorable; however, it could be reduced to fewer dimensions, since the index reported was higher than 0.800 units (the minimum acceptable value). This was also corroborated by the Bartlett's sphericity test, since chi-squared test is statistically significant when P-value is lower than 0.05—the maximum value allowed for a 95 % confidence level (Table 10.1).

Table 10.1 KMO and Barlett's tests

Kaiser-Meyer-Olkin measure of sampling adequacy		0.866
Bartlett's sphericity test	Approximate chi-square	664.867
	gl	66
	Sig.	0.000

J.L. García-Alcaraz and A.A. Maldonado-Macías, *Just-in-Time Elements and Benefits*,
Management and Industrial Engineering, DOI 10.1007/978-3-319-25919-2_10

Table 10.2 Total variance explained

Component	Initial eigenvalues	Extraction sums of squared loadings			Rotation sums of squared loadings		
	Total	Percentage of variance	Cumulative percentage	Total	Percentage of variance	Cumulative percentage	Total
1	4.970	41.419	41.419	4.970	41.419	41.419	4.472
2	1.550	12.913	54.332	1.550	12.913	54.332	3.388
3	1.057	8.809	63.141	1.057	8.809	63.141	1.717

Table 10.3 Structure matrix

Items	Name of factor	
Multifunctional workers	0.829	Communication and education
Effective communication	0.779	
Error prevention	0.764	
Employment empowerment in QC	0.748	
Workers motivation	0.725	
Self-correction of errors	0.712	
Flexibility oriented training	0.660	
Adequate evaluation and selection of suppliers	0.835	Suppliers
Zero deviation in the production program	0.804	
Supplier quality certification	0.791	
Short delivery times	0.571	Delivery times
Long-term contracts	0.911	

Table 10.2 shows that three new factors or dimensions, also called latent variables, were sufficient to explain 63.14 % of the information from this category of human resources which contains the 12 items initially analyzed. Table 10.3 depicts the distribution of the items among the three factors or latent variables identified. These latent variables were named in the proposed structural equations models. The first factor or latent variable is composed of seven observed variables or items associated with communication and education provided to operators. Similarly, the second factor or latent variable consists of three items or observed variables mostly related with suppliers. Finally, the third factor or latent variable is associated with delivery times both requested to the suppliers and offered to customers.

10.2 Model 1 Elements: Human Resources-JIT Benefits: Human Resources and Production Process

This model associates four latent variables; two of them concern items related to human resources and the remaining concern benefits reflected on the production process and human resources. Figure 10.1 illustrates six hypotheses to be tested in this first model.

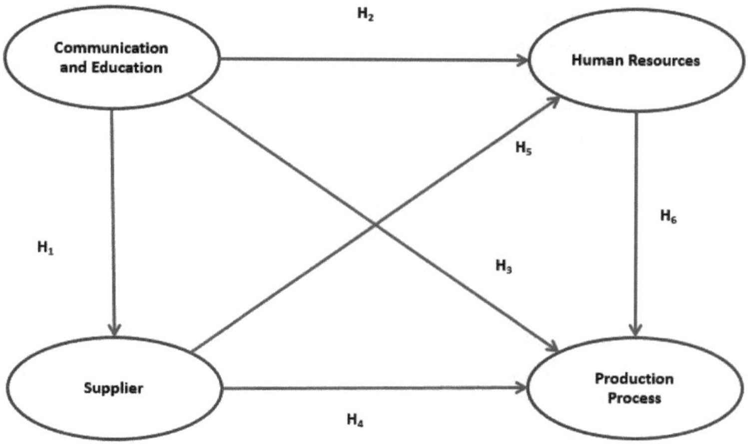

Fig. 10.1 Model 1 Elements: Human resources-benefits: production process and human resources

10.2.1 Hypotheses of Proposed Model 1

High levels of communication with suppliers are considered as an effective start of a supply chain (Mohan and Singh 1995; Gilbert et al. 1994). For instance, educational level of managers or personnel responsible for the procurement departments can influence the relationship that companies maintain with their suppliers (Dumond and Newman 1990; Macbeth et al. 1988; O' Neal 1987). Similarly, some authors have recommended considering suppliers as partners in the production systems (Chicksand), which thus requires high levels of communication. Based on this information, it is possible to propose the following hypothesis:

H1: *Communication and educational level of managers and employees have a direct and positive impact on the administrative processes of manufacturing companies established with suppliers.*

On the other hand, it has been mentioned that communication and education are the basis of all progress and development, including JIT implementation (Helms 1990). For instance, some causal models have allowed to associate communication with a number of benefits that companies could obtain (Alcaraz et al. 2014), since when communication is efficient, collaborative work in the supply chain is achieved (Huang et al. 2009). Therefore, since it is supposed that communication impacts on human resources and it is believed that companies should rely on new strategies to promote such an efficient interaction (Im et al. 1994), the following hypothesis for the model can be proposed:

H2: *Proper communication built and education provided during the process of JIT implementation have a direct and positive impact on the human resources of companies.*

Several theories consider that businesses are composed of three aspects: people, the production processes, and products. This became a point of reference for the development of this research, since it has been pointed out that human resources and communication among them are the true elements running a production system in a JIT environment (Schneider and Leatherman 1992). Furthermore, workers use their knowledge to transform raw materials into finished products (Valmohammadi and Roshanzamir 2015; Ooi 2014; Ahmad et al. 2012), provide preventive maintenance to machinery and equipment (Thomas et al. 2006; Cua et al. 2001), and calibrate changes and adjustment for the process of a new product (Chiarini 2013; Almomani et al. 2013). Therefore, the following hypothesis could be proposed:

H3: *Communication at different structural levels and high levels of education provided to workers during a JIT implementation have a direct and positive impact on indexes of the production process.*

If suppliers are responsible for providing companies with the raw material required for a production process, they should certainly impact this process. Thus, if poor quality materials are integrated into the production system, companies would be unable to offer quality products in spite of the reputation of the selected supplier. Thus, companies must rely on strict supplier selection processes (Shaw et al. 2012; Lin 2012) based on rules that do not merely evaluate economic features, but also assess delivery times, quality, ability to rapidly respond to uncertainties in demand (Wang et al. 2009; Lee et al. 2009), and ecological aspects (Villanueva-Ponce et al. 2015; Gurel et al. 2015; Galankashi et al. 2015), which have recently become an important part in the assessment.

Hence, the following hypothesis was proposed to find the possible relationship between the selection of suppliers and the process of production:

H4: *Proper selection of suppliers has a direct and positive effect on the production process in a JIT environment.*

In addition, supplier selection processes may also impact on different aspects of human resources, such as employee motivation, increased teamwork, and better resource utilization. For instance, employees receiving defective raw material may not feel that motivated to generate new products, since they may be aware of the fact that components do not meet quality standards (De Toni and Nassimbeni 2000). Unfortunately, this often becomes an obstacle for the appropriate functioning of a JIT system (Giunipero and O'Neal 1988). Therefore, in order to prove that adequate selection processes of suppliers impacts on human resources, the following hypothesis can be proposed.

H5: *Adequate selection of suppliers of raw materials in a production system has a direct and positive impact on the benefits associated with human resources in a JIT environment.*

Since human resources are accountable for the production process, it is assumed that the former have an impact on the latter (García-Alcaraz et al. 2015). Thus, human resources are responsible for making decisions to approach plan deviations or errors in the production process associated to product quality, which they could resolve by proposing solutions in quality circles (Blaga and Jozsef 2014). Moreover, human resources are in charge of executing plans and programs of preventive and corrective maintenance (Cua et al. 2001), and they are the difference between the success or failure of companies, which reflects on the performance indices of the production process (Kesti 2012). Hence, in order to test the relationship between human resources and the production process, the following hypothesis is proposed.

H6: *The benefits obtained in human resources during the process of JIT implementation have a direct positive impact on the production process.*

10.2.2 Evaluation of Model 1

The model presented in Fig. 10.1 was tested by means of the structural equation modeling (SEM) technique described in the methodology. Figure 10.2 shows the direct effects that exist in the relations between latent variables. The figure merely shows statistically significant relationships; however, note that the relationship between the latent variable Communication and Education and Production Processes was eliminated since the test reported it as statistically not significant.

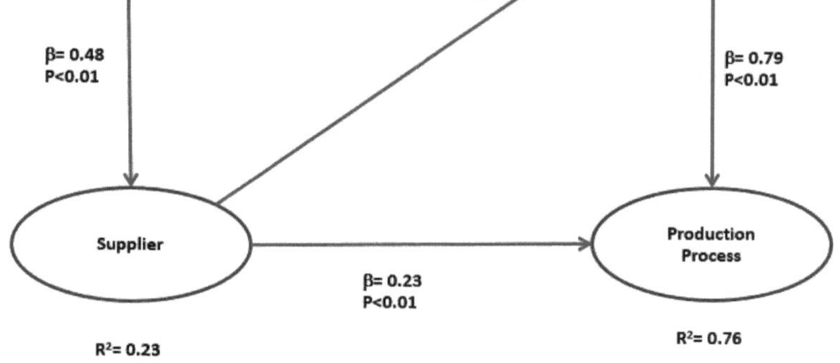

Fig. 10.2 Evaluated Model 1

10.2.2.1 Efficiency Indexes of Model 1

The efficiency indexes of the model tested are:

1. Average path coefficient (APC) = 0.434, $P < 0.001$
2. Average R-squared (ARS) = 0.459, $P < 0.001$
3. Average adjusted R-squared (AARS) = 0.453, $P < 0.001$
4. Average block VIF (AVIF) = 1.135, acceptable if ≤ 5, ideally ≤ 3.3
5. Average full collinearity VIF (AFVIF) = 2.923, acceptable if ≤ 5, ideally ≤ 3.3
6. Tenenhaus GoF (GoF) = 0.536, small ≥ 0.1, medium ≥ 0.25, large ≥ 0.36
7. Sympson's paradox ratio (SPR) = 1.000, acceptable if ≥ 0.7, ideally = 1
8. R-squared contribution ratio (RSCR) = 1.000, acceptable if ≥ 0.9, ideally = 1
9. Statistical suppression ratio (SSR) = 1.000, acceptable if ≥ 0.7
10. Nonlinear bivariate causality direction ratio (NLBCDR) = 1.000, acceptable if ≥ 0.7

Based on the ARS, APC, and AARS indexes with a P-value that makes it possible to determine statistical significance, it is concluded that the model is statistically significant, since the P-value of all these parameters is lower than 0.05—the maximum value allowed—with a 95 % confidence level. Similarly, according to AFVIF and AVIF values, it can be inferred that there is no collinearity among latent variables. In both values the parameter is lower than 3.3, which is the maximum value established. Therefore, it can be concluded that the model is efficient and its relationships could be interpreted.

10.2.2.2 Coefficients of Latent Variables of Model 1

Coefficients of latent variables are important since they contain information on their validity. Table 10.4 shows that Model 1 has predictive validity, since all values of dependent latent variables are higher than 0.02—the minimum acceptable cutoff

Table 10.4 Coefficient of latent variables

	Communication and education	Supplier	Human resources	Production process
R-squared		0.227	0.356	0.795
Adjusted R-squared		0.222	0.347	0.792
Composite reliability	0.896	0.871	0.887	0.916
Cronbach's alpha	0.86	0.778	0.841	0.892
AVE	0.59	0.693	0.612	0.61
FVIF	1.532	1.478	2.213	2.468
Q-squared		0.228	0.357	0.795

value. Similarly, adjusted *R*-squared and *Q*-squared are similar to *R*-squared values, which indicate that both parametric and nonparametric tests of predictive validity tests are statistically significant. Furthermore, it is observed that composite and internal validity measured by the Cronbach's alpha coefficient is appropriate, since all latent variables show values higher than 0.7—the minimum acceptable value—in these indexes.

10.2.2.3 Direct Effects of Model 1: Hypotheses Validation

Direct effects depicted in Fig. 10.2 serve to accept or reject the initial hypotheses proposed in the model. According to this information reported in the figure, it is possible to state the following conclusions:

- H1. There is enough statistical evidence to state that communication and the level of education of managers and employees have a direct and positive impact on the administrative processes established with suppliers, since when the first variable increases its standard deviation by one unit, the standard deviation of the second latent variable also increases by 0.48 units.
- H2. There is sufficient statistical evidence to point out that proper communication established and education offered during the implementation process of JIT have a direct and positive impact on the human resources of a company. When the first variable increases its standard deviation by one unit, then the standard deviation of the second latent variable increases by 0.53 units.
- H3. There is not enough statistical evidence to support that communication at different structural levels of the company and a high level of education provided to workers during the process of JIT implementation have a direct and positive impact on the indexes generated in production process, since the *P*-value for this hypothesis is higher than 0.05, which is the accepted cutoff value.
- H4. There is necessary statistical evidence to declare that adequate selection of suppliers has a direct and positive effect on production process during the implementation of JIT philosophy, since when the first latent variable increases its standard deviation by one unit, the standard deviation of the second latent variable also rises by 0.23 units.
- H5. There is enough statistical evidence to state that the proper selection of raw material suppliers in a production system has a direct and positive impact on the benefits of human resources during JIT implementation. When the first latent variable increases its standard deviation by one unit, then the standard deviation of the second latent variable increases by 0.14 units.
- H6. There is sufficient statistical evidence to declare that benefits of human resources during JIT implementation have a direct and positive impact on the production process where they work, since when the first latent variable increases its standard deviation by one unit, the standard deviation of the second latent variable rises 0.79 units.

It is also important to consider the effects sizes of independent variables over dependent latent variables. For instance, human resources (see Fig. 10.2) is 36 % explained by communication and education and providers. However, 30.6 % of this effect originates from the first latent variable and only 4.9 % emerges from the second latent variable. Therefore, it can be concluded that to obtain the benefits associated with human resources, adequate communication and education is more significant than the selection of suppliers, since the percentage of variability explained is greatly enhanced by the first latent variable. Similarly, dependent latent variable named Production Processes is 76 % explained by Communication and Education and Benefits Obtained in Human Resources. Since 11.1 % of this effect originates from the first latent variable and 68.4 % corresponds to the second latent variable, it can be concluded that striving to achieve the benefits of human resources is more important for production processes than education and communication.

Finally, the sole hypothesis rejected was hypothesis 3 (H3), concerning the relationship between communication and education and production processes. The relation between these two latent variables is indirect and given through suppliers and mostly human resources.

10.2.2.4 Sum of Indirect Effects of Model 1

Indirect effects occur between two latent variables through one or more mediating variables, for what they are often called effects across segments. Table 10.5 shows the indirect effects of Model 1; the table also depicts the effect of every latent variable over others, the *P*-value of the statistic test that shows that indirect effects in all cases are significant, and the effect size.

Table 10.5 shows that indirect effect between communication and education and production process is given through suppliers and human resources. According to the table, the effect is high since it has a value of 0.582. Therefore, when the first variable—Communication and Education—increases its standard deviation by one unit, the standard deviation of the second latent variable—Production Process—increases by 0.582 units. This proves the strong indirect impact between these variables. However, note that the direct effect analyzed between these latent variables proved to be not significant, which indicted that effective communication and education should be achieved in order to hold positive relations with suppliers, select them appropriately, and improve workers' motivation, and thus collaborative

Table 10.5 Sum of indirect effects

	Communication and education	Supplier
Human resources	0.069 ($P < 0.01$) ES = 0.040	
Production process	0.582 ($P < 0.01$) ES = 0.279	0.144 ($P = 0.008$) ES = 0.055

Table 10.6 Total effects of Model 1

	Communication and education	Supplier	Human resources
Supplier	0.477 ($P < 0.01$) ES = 0.227		
Human resources	0.596 ($P < 0.01$) ES = 0.346	0.144 ($P = 0.01$) ES = 0.049	
Production process	0.582 ($P < 0.01$) ES = 0.279	0.346 ($P < 0.01$) ES = 0.165	0.792 ($P < 0.01$) ES = 0.684

work and better resource utilization. Finally, it must also be noted that this indirect relationship between the variables explains production process in 27.9 %, since the effect size is 0.279. Similar interpretations can be inferred for other indirect effects.

10.2.2.5 Total Effects of Model 1

Table 10.6 introduces the total effects (sum of direct and indirect effects) of Model 1. The table also shows the *P*-values that tested statistical significance of the effects and their size to indicate the percentage of variance that can be explained by the relationship between these two latent variables.

According to Table 10.6, all total effects are statistically significant since in all cases the *P*-value is less than 0.05. As for the effect size, the relation between human resources holds the highest value among all relations; that is, 68.4 %, which is only caused by a direct effect.

10.2.3 Conclusions of Model 1

Six hypotheses were tested in Model 1 that integrated the four latent variables: Communication and Education, Suppliers, Human Resources, and Production Processes. While the first two latent variables are considered as required elements for a successful implementation of JIT program, the latter two represent benefits obtained after its implementation. In this model, it has been assumed that communication and education are independent latent variables and from it, the other three variables are depending, but that dependent latent variables are related to benefits that can be obtained from these activities previously performed in the production process.

One hypothesis was rejected, since it has been found that it was statistically not significant. This rejected hypothesis concerned the relation between Communication and Education and Production Processes; however, it has been also found that these two variables are highly dependent on each other, although the indirect effect is caused through the other two mediating variables. In fact, communication must be promoted and education must be offered to human resources who eventually demonstrate their abilities to perform their activities in the production process.

10.3 Model 2 Elements: Human Resources-JIT Benefits: Quality and Economic Benefits

This model (Model 2) seeks to associate three latent variables concerning human resources and two latent variables concerning quality in the production process and the economic benefits obtained. It is assumed that Education and Communication is the initial independent latent variable on which the remaining variables depend. Similarly, in the model structure shown in Fig. 10.3, it is assumed that economic benefits are the dependent latent variable. Therefore, economic benefits depend on all the JIT elements considered in the model but also on the quality obtained in the production process.

In order to relate the variables of this model, ten working hypothesis were proposed and expected to be statistically proved. However, please note that hypothesis 1 (H1) concerning the relation between communication and education with suppliers was already explained and analyzed in Model 1; thus, this model will not analyze it once more since the effect of this relationship is already known.

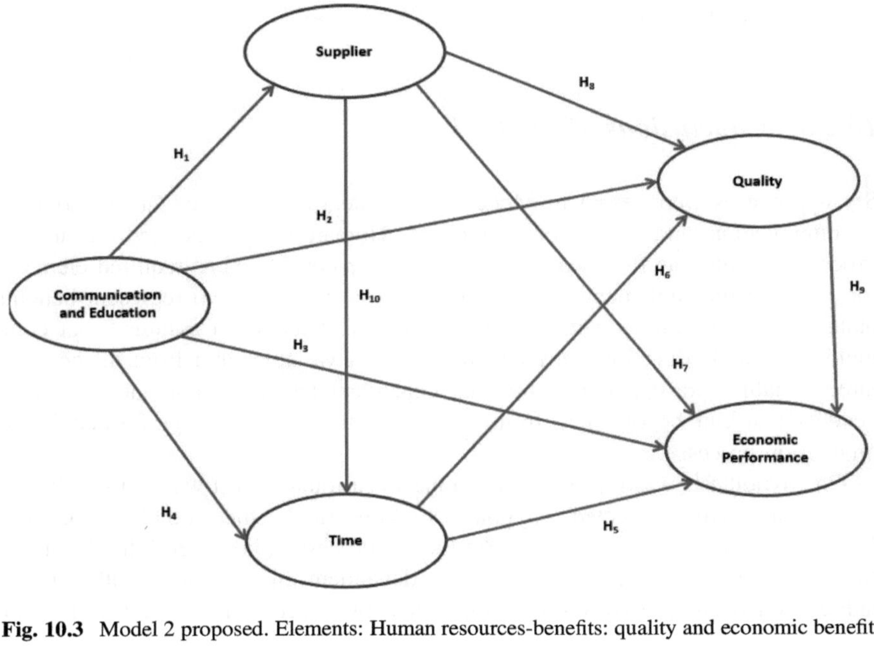

Fig. 10.3 Model 2 proposed. Elements: Human resources-benefits: quality and economic benefits

10.3.1 Hypotheses

Quality obtained in the production process is the result of human performance; workers must be properly trained to carry out their activities with the minimum possible errors (Ooi 2014; Jun et al. 2006). Nevertheless, communication, especially concerning quality of the product, must also be effective among the different departments of the company, which must jointly interfere in the processes of decision-making when it is required (Huang et al. 2009; Helms 1990). Furthermore, it is well known that the training and education provided to employees, is crucial for the prevention of errors and rework in the production process. Therefore, to prove the relation between education and communication over product quality, the following hypothesis was proposed.

H2: *Communication and education provided during the process of JIT implementation have a direct and positive impact on product quality obtained.*

All communication plans and protocols and education programs established within a company seek to improve its performance (Fullerton et al. 2003). However, this relationship might be promoted through the human resources (mediator variable) who provide such training (Globerson and Korman 2001; Deshpande and Golhar 1996), although in other industrial sectors, some authors have managed to find a relationship between education and the economic performance of companies. Therefore, the following hypothesis was proposed to prove that more education and communication will ensure greater economic benefits.

H3: *Communication and education provided to operators have a direct and positive impact on the economic indices of companies in a JIT environment.*

Similarly, communication inside and outside of the companies may affect the punctuality of its commitments (Gilbert et al. 1994; O' Neal 1987). That is, effective communication with suppliers can allow companies to meet time deliveries to customers and hold long-term contracts with workers (Kwak et al. 2006; Chen and Jeter 2008). When suppliers are properly integrated into the production system and companies consider them as partners in the production (De Toni and Nassimbeni 2000; David and Eben-Chaime 2003); they are likely to become reliable and maintain an effective communication due to the level of trust established between them and the customer company, which also reduces the amount of administrative work. Therefore, these reliable suppliers will enable to allow the company to build or commit to short delivery times with the client and within it, may have long-term contracts with employees. Therefore, in order to test the relationship between communication and education and these variables, the following hypothesis is proposed.

H4: *Communication and education provided to workers in a JIT environment have a direct and positive effect on on-time commitments, such as delivery times to customer and long-term contracts with employees.*

Quality is defined as the process by which companies generate a product in the right amount, with the appropriate requirements or specifications, and that is delivered on time to the customer (Schniederjans and Schniederjans 2015). Therefore, time is a crucial dimension of quality, and customers usually request short-time deliveries (Pan and Nguyen 2015), which implies saving costs since raw materials would not remain long in warehouses and the production process, and finished goods will be stored for little time (Avelar-Sosa et al. 2015; Khan and Sarker 2002). Therefore, it is suitable for manufacturer companies to meet short delivery times agreed with customers. However, it must be noted that this requires skilled workers with great knowledge of the operations to perform, who preferably work under long-term contracts, since a company cannot afford to regularly unemploy operators that have become fully acquainted with the activities of the production process and are integrated in the company. Therefore, in order to prove the relation between short delivery times and long contracts on the performance of a company, the following hypothesis was proposed.

H5: *Short delivery times of finished products and long-term contracts established with operators, have a direct and positive impact on the economic performance companies.*

As mentioned above, customers are more satisfied with delivery times of products (Zhang et al. 2009; Chen et al. 2014; Colledani et al. 2014). Nevertheless, another advantage of quality is that operators may be motivated to produce quality products when they work under long-term contracts. One must remember those Eastern companies mentioned previously, who mostly rely on lifetime contracts with their employees, who have demonstrated motivation to improve the production process where they work (Singh et al. 2015). However, it is common for Western companies to draw upon short-term contracts usually of less than a year or, as it is the case in the United States, to remunerate workers depending on the hours worked in a day. In these cases workers cannot be sure whether they will be employed to work a complete day, and it often promotes an unfavorable work environment (Rusu and Avasilcai 2014; Singh and Garg 2011). Hence, the following hypothesis was proposed in order to demonstrate the relation existing between long-term contracts and short-time deliveries and product quality.

H6: *Long-term contracts with operators and short delivery times offered to customers have a direct and positive impact on quality product.*

When companies select and assess their suppliers appropriately and successfully integrate these suppliers into their production process, they can rely on certain economic benefits (Kumar et al. 2014). That is, if companies trust their providers who are successfully integrated into the production process (Humphreys et al. 2003), few of them are required for raw materials supply. Therefore, administrative work and management processes reduce and costs for companies consequently diminish (Araz and Ozkarahan 2007); and it is found that raw material costs represent up to 70 % of the cost of production in some industrial sectors (Ngai et al. 2011). However, in order to maintain high integration, companies must regularly run audits and assess their suppliers (Chen and Jeter 2008; Dey et al. 2015). Based on the previous statements, the following hypothesis was proposed to prove the existing relation between the selection and evaluation of suppliers and the economic performance of companies.

H7: *Selection and evaluation of suppliers and their certifications have a direct and positive impact on the economic performance of companies.*

Product quality begins with the design of the product as well as with the technical specifications that must be met to ensure the objective of the product (Yoo et al. 2015). Therefore, raw material suppliers required to convert that prototype into a product play an important role to meet these specifications, and their selection and evaluation must be based on their compliance (Kim and Ha 2003). The successful integration of suppliers as partners in the production system can promote greater agility of the supply chain and higher inventory turnover (Inman et al. 2011), which would consequently impact on the reduction of costs involved (Sueyoshi and Goto 2010). Therefore, the following hypothesis is proposed as a means to prove the relation that exists between the selection and evaluation of suppliers on quality product and process.

H8: *The selection and evaluation of suppliers have a direct positive impact on process and product quality.*

The ultimate goal of a business organization is generating wealth for shareholders, which suggests that all lean manufacturing techniques implemented during the production process are simply tools to reach that ultimate goal (Prajogo and Sohal 2006). Thus, high-quality products are not the final goal of the company but the means to generate these economic benefits. However, all this does not simply emerge; it is rather the result of several factors, such as quality of raw materials, training plans and programs, maintenance plans and programs, and above all, human resources (Kassicieh and Yourstone 1998). Some authors state that the relation between product quality and financial performance of companies is difficult to understand and unclear due to the large number of mediating factors (Corredor and Goñi 2011; Calvo-Mora et al. 2014; Ooi 2014). Hence, in order to contribute to this discussion, the following working hypothesis is proposed.

H9: *Product quality has a direct and positive impact on the economic performance of companies during the implementation process and execution of a JIT program.*

The selection and evaluation of suppliers cannot be considered a luxury for companies, since it actually represents expenses and costs to assume (Esmaeili Aliabadi et al. 2013; Chai et al. 2013). However, organizations rely on this in order to identify reliable suppliers that can be successfully integrated in the production process as partners (Kumar et al. 2013). This would guarantee the reduction of administrative costs or management process and the punctual arrivals of raw materials requested. Consequently, costs of management and storage of raw materials can be deminished. Therefore, the following hypothesis was proposed based on this information.

H10: *The selection and evaluation of suppliers have a direct and positive impact on established delivery times to final customers and long-term contracts with providers.*

10.3.2 Evaluation of Model 2

Model 2 was evaluated according to the methodology described in Chap. 5 based on a structural equation modeling technique and with the use of WarpPLS 4 software. Results from the analysis are shown in Fig. 10.4 for direct effects, indirect effects, and total effects.

10.3.2.1 Efficiency Indexes of Model 2

Before describing the results obtained from the analysis, it is suitable to define the efficiency levels reached, which are defined as follows:

1. Average path coefficient (APC) = 0.285, $P < 0.001$
2. Average R-squared (ARS) = 0.374, $P < 0.001$
3. Average adjusted R-squared (AARS) = 0.364, $P < 0.001$
4. Average block VIF (AVIF) = 1.308, acceptable if ≤ 5, ideally ≤ 3.3
5. Average full collinearity VIF (AFVIF) = 1.858, acceptable if ≤ 5, ideally ≤ 3.3

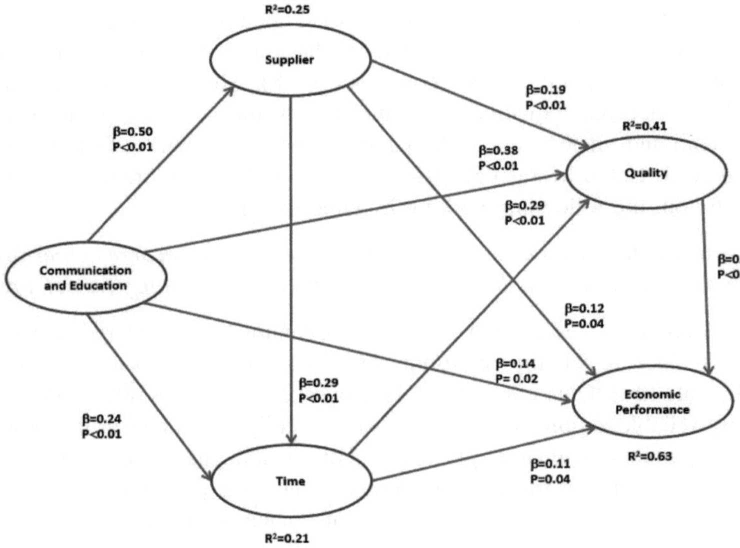

Fig. 10.4 Evaluated Model 2

6. Tenenhaus GoF (GoF) = 0.493, small > = 0.1, medium > = 0.25, large > = 0.36
7. Sympson's paradox ratio (SPR) = 1.000, acceptable if > = 0.7, ideally = 1
8. *R*-squared contribution ratio (RSCR) = 1.000, acceptable if > = 0.9, ideally = 1
9. Statistical suppression ratio (SSR) = 1.000, acceptable if > = 0.7
10. Nonlinear bivariate causality direction ratio (NLBCDR) = 1.000, acceptable if > = 0.7

According to APC, AARS, and ARS indexes, the model has acceptable validity since *P*-value in all latent variables is lower than 0.05. Similarly, based on the AVIF, collinearity was not reported in any latent variable. Finally, since the values of the remaining indexes were below the limit values, the model can be analyzed and discussed.

10.3.2.2 Coefficients of Latent Variables of Model 2

Table 10.7 shows the indexes associated with quality and validity of every latent variable analyzed in Model 2. All latent variables dependent on others show *R*-squared and adjusted *R*-squared values above 0.02—the lower limit of cutoff value. Therefore, it can be concluded that latent variables analyzed have sufficient predictive validity from a parametric view.

Likewise, *Q*-squared has positive and close values to *R*-squared in all latent variables, which indicates that, from a nonparametric perspective, all latent variables have predictive validity. Similarly, all values show Cronbach's alpha and composite reliability indexes below 0.7 (the minimum cutoff value previously established), which indicates that the latent variables analyzed in Model 2 have internal adequate reliability. As for AVES, all variables show values above 0.5 in all variables, which demonstrates that they all have convergent validity.

Finally, according to AFVIF and AVIF values, it can be inferred that there is no collinearity among latent variables, since the parameter on both values is lower than

Table 10.7 Coefficients of latent variables for Model 2

Coefficient	Communication and education	Supplier	Time	Quality	Economic performance
R-squared		0.251	0.207	0.408	0.628
Adjusted *R*-squared		0.245	0.196	0.395	0.617
Composite reliability	0.899	0.871	0.791	0.885	0.911
Cronbach's alpha	0.869	0.778	0.706	0.805	0.882
AVE	0.561	0.693	0.654	0.720	0.631
FVIF	1.546	1.495	1.306	2.462	2.481
Q-squared		0.252	0.211	0.418	0.628

3.3, which is the maximum cutoff value allowed. Therefore, based on the foregoing information, it can be concluded that Model 2 is valid in their variables which show no collinearity problems. The effects can now be analyzed.

10.3.2.3 Direct Effects of Model 2—Hypotheses Validation

Direct effects are shown in Fig. 10.4, which further indicates values of regression as well as *P*-values of the statistical significance tested. All *P*-values concluded that all latent variables were valid and should remain in the model. Therefore, based on these values, the hypotheses initially proposed were tested, and it was possible to state the following occlusions. However, note that H1 will not be stated once more since the same relation was tested and proved in the previous model.

- H2. There is sufficient statistical evidence to declare that effective communication and education provided during JIT implementation in a production system have a direct and positive impact on quality product, since when the first latent variable increases its standard deviation by one unit, the standard deviation of the second latent variable also increases by 0.38 units.
- H3. There is enough statistical evidence to state that effective communication and education provided to operators have a direct and positive effect on the economic performance indexes of companies in a JIT environment. When the first variable latent increases its standard deviation by one unit, the standard deviation of the second latent variable also increases by 0.14 units.
- H4. Enough statistical evidence points out that effective communication and education provided to operators on a JIT environment have a direct effect on punctual commitments of companies, such as onetime deliveries and long-term contracts with employers, since when the first latent variable increases its standard deviation by one unit, the standard deviation of the second latent variable also increases by 0.24 units.
- H5. There is enough statistical evidence to declare that short delivery times of finished products and long-term contracts signed with operators, have a direct and positive impact on the economic performance of the company, since when the first latent variable increases its deviation standard by one unit, the standard deviation of the second latent variable also increases by 0.11 units.
- H6. Statistical evidence is sufficient to declare that long-term contracts with operators and short delivery times with customers have a direct and positive impact on the quality of products manufactured, since when the first latent variable increases its standard deviation by one unit, the standard deviation of the second latent variable also increases by 0.29 units.
- H7. Statistical evidence is enough to declare that selection and evaluation of suppliers and their certifications have a direct positive impact on the economic performance companies, since when the first latent variable increases its standard deviation by one unit, the standard deviation of the second latent variable also increases by 0.12 units.

- H8. There is sufficient statistical evidence to declare that the selection of suppliers has a direct and positive impact on the quality obtained in the production process and the product generated, since when the first latent variable increases its standard deviation by one unit, the standard deviation of the second latent variable also increases by 0.19 units.
- H9. There is sufficient statistical evidence to point out that product quality has a direct and positive impact on the economic performance companies during a JIT implementation process, since when the first latent variable increases its standard deviation in one unit, the standard deviation of the second latent variable also increases by 0.59 units.
- H10. There is sufficient statistical evidence to declare that selection and evaluation of suppliers have a direct and positive impact on delivery times established with customers and long-term contracts with suppliers, since when the first latent variable increases its deviation standard by one unit, the standard deviation of the second latent variable also increases by 0.29 units.

Four dependent latent variables in this model depend on one or some others that have their *R*-squared value. However, for those variables explained through more than one latent variable, it is difficult to identify the exact percentage variance explained through every latent variable. Thus, in order to facilitate the comprehension of these relations, Table 10.8 shows the effect sizes.

According to Table 10.8, it is possible to state the following conclusions:

1. The Suppliers latent variable is explained by only one independent latent variable. Therefore, the total direct effect—25.1 %—is due to that single latent variable.
2. The latent variable called Time is explained by two latent variables. The total direct effect is 21 %, which originates 9 % from Communication and Education and 11.8 % from Suppliers.
3. The latent variable Quality is explained by three latent variables. The total direct effect of the variable is 41 %, which is 20.1 % explained by Communication and Education, 7.9 % by Suppliers and 12.8 % by the latent variable Time. It is therefore concluded that the latent variable Communication and Education is the most significant variable to Quality.
4. The latent variable Economic Performance is explained by four latent variables and has a total direct effect of 63 %, from which 7.3 % is explained through

Table 10.8 Effect size of direct effects of Model 2

Communication and education	Communication and education	Supplier	Time	Quality	*R*-squared
Supplier	0.251				0.251
Time	0.090	0.118			0.21
Quality	0.201	0.079	0.128		0.41
Economic performance	0.073	0.052	0.050	0.543	0.63

Communication and Education, 5.2 % by Suppliers, 5.0 % through the latent variable Time, and 54.3 % through Quality. It is thus concluded the Quality is the most significant variable to improve economic performance in companies.

10.3.2.4 Sum of Indirect Effect of Model 2

As previously mentioned, indirect effects occur between two latent variables through a third latent variable named Mediator. Table 10.9 shows the sum of indirect effects between latent variables evaluated in Model 2. The table also shows the tested statistical significance of *P*-value, the effect size for every variable. Observe that all indirect effects, according to the *P*-value, are statistically significant since they are lower than 0.05, the maximum cutoff value.

The relation between Communication and Education and Economic Performance shows the largest indirect effect among all the relations analyzed. When the first latent variable increases its standard deviation by one unit, the standard deviation of the second latent variable also rises by 0.45 units, which is a relatively high value. This indirect effect can also explain in 23.7 % the variance of the dependent variable. Moreover, the direct effect between these same two variables was only 0.14, indicating that the greatest effect is actually indirect and occurs through the appropriate selection of suppliers, the quality obtained in the production process, and the long-term contracts of companies with workers.

Similarly, the relation between Communication and Education and Product Quality has the second highest indirect effect. When the first latent variable increases it standard deviation by one unit, the standard deviation of the second latent variable also increases by 0.205 units. The independent variable can account for up to 10.8 % of the dependent variable. Similar conclusions can be inferred from the remaining indirect relationships between the latent variables, which even though they are small, they remain statistically significant.

Table 10.9 Sum of indirect effects of Model 2

To	From		
	Communication and education	Supplier	Time
Time	0.146 (*P* = 0.001) ES = 0.055		
Quality	0.205 (*P* = 0.001) ES = 0.108	0.085 (*P* = 0.037) ES = 0.036	
Economic performance	0.45 (*P* < 0.001) ES = 0.237	0.195 (*P* = 0.002) ES = 0.087	0.17 2 (*P* < 0.001) ES = 0.076

10.3.2.5 Sum of Total Effects of Model 2

Table 10.10 introduces the total effects of the relationships of Model 2. The table also shows the *P*-value from the statistical significance test for every estimated latent variable or parameter and the effect size or percentage of variance that explains the effect of one independent variable on one dependent variable. According to *P*-values obtained, the sum of the effects in all cases is statistically significant, since they were lower than 0.05. Note that total effect can also be equal to the sum of indirect effects or to the direct effect between the variables.

According to its size, it is observed that the largest effect occurs between latent variables Quality and Economic Performance. The value of these total effects is also equal to the value of their direct effect, since no indirect effects were reported between these latent variables. This indicates that when the first variable—that is Quality—increases its standard deviation by one unit, the standard deviation of the second latent variable also increases by 0.594 units.

Also, the relation between Communication and Education and Economic Performance shows the second largest total effect, according to its size, since when the former increases its standard deviation by one unit, the standard deviation of the second latent variable also increases by 0.589 units. However, great part of this effect is indirect, and the independent latent variable can explain up to 45.3 % of the variability of the dependent latent variable.

Furthermore, Communication and Education and Quality have also a significant large total effect; when the first latent variable increases its standard deviation by one unit, the standard deviation of the second latent variable also increases by 0.585. The independent variable can explain 30.9 % of the variability of the dependent latent variable. However, it can be noted that Communication and Education also impact on Providers, since when the first latent variable increases its standard deviation by one unit, the standard deviation of the second variable

Table 10.10 Total effects of Model 2

To	From			
	Communication and education	Supplier	Time	Quality
Supplier	0.501 ($P < 0.001$) ES = 0.251			
Time	0.385 ($P < 0.001$) ES = 0.144	0.292 ($P < 0.001$) ES = 0.118		
Quality	0.585 ($P < 0.001$) ES = 0.309	0.271 ($P < 0.001$) ES = 0.115	0.29 ($P < 0.001$) ES = 0.128	
Economic performance	0.589 ($P < 0.001$) ES = 0.310	0.31 ($P < 0.001$) ES = 0.139	0.287 ($P < 0.001$) ES = 0.126	0.594 ($P < 0.001$) ES = 0.453

increases by 0.501 units. Moreover, the first latent variable can explain up to 31 % of the variability. Similar interpretations can be stated for the remaining relationships analyzed.

10.3.3 Conclusions of Model 2

This model initially proposed 10 working hypotheses, which all proved to be statistically significant and whose results were analyzed in the previous sections. The conclusions that can be derived from the generation of model are:

- Communication and education is one of the most important variables in the management and implementation process of JIT, since it strongly impacts on quality, suppliers, times established, and the economic performance of companies.
- Product quality is one of the most important factors that explain the economic performance of companies, since customers tend to seek for quality products.
- Companies must strive to rely on long-term contracts with their employees and establish short delivery times and short cycles with their suppliers. This could motivate workers to find methods for a fast flow of materials throughout the production system.

10.4 Model 3 Elements: Human Resources-JIT Benefits: Materials Handling and Engineering Process

This causal model aims to identify the impact of human resources, comprising latent variables associated with suppliers and time, on the JIT benefits obtained for materials handling and the engineering process. Six hypotheses were established in this model, from that already proposed and tested in Model 2. This hypothesis concerns the relation between Suppliers and Time and will thus not be discussed in this section. However, this model considers it as the first hypothesis proposed (H1), although it was addressed last in the previous section.

Figure 10.5 illustrates the third causal model proposed in this section, where assumptions and the direction of the relationships between variables can be appreciated.

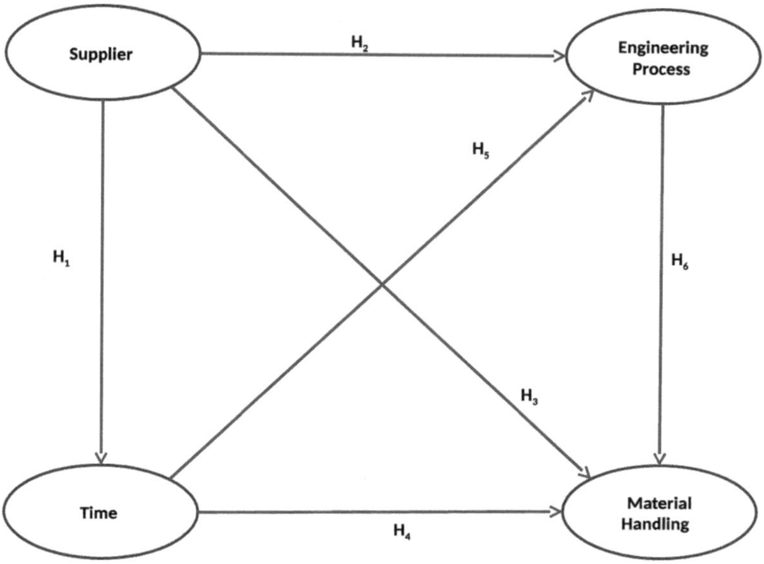

Fig. 10.5 Proposed Model 3

10.4.1 Hypotheses

Companies must consider several attributes and characteristics when evaluating and selecting their suppliers, and for this reason many techniques rely on multi-attribute or multi-criteria assessments. In fact, several of these have been widely and deeply discussed (Govindan et al. 2015; Igarashi et al. 2013; Chai et al. 2013; Lima Junior et al. 2014; Aksoy et al. 2014). Therefore, the quality of the benefits that suppliers can offer in the engineering process of a product will surely be taken into account in their selection and evaluation. Manufacturing companies will then purchase raw material from suppliers who can favor changes of product design in the production line (Ferradás and Salonitis 2013) and contribute to a reduction of travel distances. However, companies above all will consider a supplier as suitable when raw material offered does not require large spaces for its storage. For companies, this concerns the type of machinery and equipment required to add value to raw material, and the ability to offer these machines proper maintenance (Chiarini 2013). In order to understand and contribute to the analysis of this issue, the following hypothesis is proposed.

H2: *The selection and evaluation of suppliers and the relationship between manufacturing companies and them have a direct positive impact on engineering process associated with space expansion requirements, movement distances, and rapid changes during the JIT implementation process.*

Additional attributes to select and evaluate suppliers are based on the ability of manufacturing companies to handle the raw material provided (Karsak and Dursun 2015; Du et al. 2015; Awasthi et al. 2009), the minimum lot size that suppliers offer for its purchase (Mazdeh et al. 2015; Cárdenas-Barrón et al. 2015; Choudhary and Shankar 2013, 2014; Lee et al. 2013), the extent to which they favor inventory turnover in the warehouses of manufacturers (Scott et al. 2015; Sarkis and Dhavale 2015; Jadidi et al. 2015). Therefore, providers that offer purchases of raw material in small sizes will be ideal to companies, since this will reduce time of storage of material and save administrative costs. Companies can also rely on regional suppliers, because the relationships with these can be easy to manage (Kumar et al. 2014). Hence it is recommended that manufacturing companies purchase raw material from a distant supplier that allows for or facilitate a proper relationship (Humphreys et al. 2003; David and Eben-Chaime 2003; Kaufmann et al. 2012). In order to support the impact of these selection criteria of suppliers over materials handling, the following hypothesis is proposed.

H3: *The selection and evaluation of suppliers, as well as the relationship between them and manufacturing companies, have a direct positive impact on the efficiency indexes associated with materials handling.*

When companies receive a production order from customers, they must be specific with delivery times. Likewise, suppliers must be punctual and provide all information concerning the delivery time of raw materials to manufacturers (Chakravorty 2009). These delivery times of materials or finished products, when they are short, force companies to generate strategies to increase inventory turnovers, although if production orders occur very often, it is more likely that companies rely on small-sized orders to minimize the costs of administration and material management (Parvini 2011; Rossi et al. 2013). Furthermore, short delivery times of raw materials make suppliers integrate into the production system of the manufacturing company, which promotes a much closer relationship (Sánchez 2015; Zhang et al. 2015). Moreover, when operators work under long-term contracts, they feel integrated into the environment and strive to support inventory reduction and handle materials more appropriately (Crumpton 2015). Therefore, the following hypothesis was proposed in order to contribute to this relationship between delivery times and materials handling.

H4: *Commitments established with suppliers concerning delivery time and long-term contracts with operators have a direct positive impact on efficiency indexes associated with materials handling.*

An agile production system requires short delivery times of raw material from suppliers (Krause and Ellram 1997; Chan 2001). However, involves manufacturers in the implementation and execution of plans and programs for better distribution of machines and equipment responsible for adding value to raw materials (Ahmad et al. 2012; Thomas et al. 2006). It is intended that these materials move as little as it is possible, since when they move, they do not receive any added value.

Similarly, short delivery times imply that small-sized lots were purchased, which lead to reduced administrative and management costs (Teksan and Geunes 2015; Cárdenas-Barrón et al. 2015) and to higher inventory turnover. Moreover, it must be remembered that accidents in production process occur during materials handling, causing absenteeism of workers, and an increase of insurance premiums (Azevedo et al. 2014; Niskanen and Lauttalammi 1989; Bennett 1984). The following hypothesis is proposed to support the aforementioned discussion.

H5: *Commitment to delivery times from suppliers and with customers, as well as contracts for operators has a direct and positive impact on efficiency indexes of engineering process associated with space requirements, movement distances, and rapid changes.*

Manufacturing cells group a series of activities to generate a product in a small space (Marodin et al. 2015; Li et al. 2015). Therefore, greater efficiency of available resources reduce the movement of materials and consequently avoids possible accidents (Niskanen and Lauttalammi 1989). These manufacturing cells are usually small and allow for reduced inventories, leading to have a high turnover of them (Lee et al. 2014; Chen et al. 2011).

Similarly, manufacturing cells are highly specialized in making a series of works and they are generally an automated system with information integrated, which allows for monitoring and controlling the status of raw material along the cell. This helps identify the parts or product components that are in process, to process, or missing, among others (Mensah et al. 2015; Tams et al. 2014; den Butter et al. 2012; Dias et al. 2009; Boersma and Kingma 2005).

Finally, manufacturing cells also require high integration from manufacturers and suppliers, since raw material orders are small and remain short times in the cells, and a lack of coordination between these two organizations could produce a work stoppage due to lack of components. Therefore, the sixth hypothesis is proposed.

H6: *Engineering process associated with requirements, movement distances, and rapid changes has a direct and positive impact on benefits associated with materials handling.*

10.4.2 Evaluation of Model 3

This model focused on the management of materials from human resources. Figure 10.6 introduces the results of the model after its evaluation according to methodology described in Chap. 5. Similarly, the figure shows for all parameters or latent variables their estimated values, the P-values that determined their statistical significance, and the R-squared values that indicated their percentage of variance.

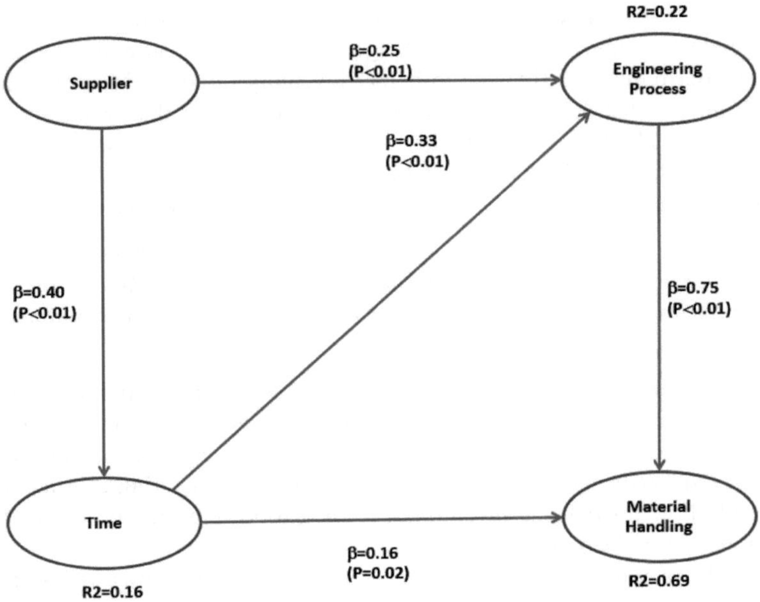

Fig. 10.6 Evaluated Model 3

Note that Fig. 10.6 merely illustrates five segments from the six hypothesis previously proposed. The relationship between suppliers and materials handling was not statistically significant and was removed.

10.4.2.1 Efficiency Indexes of Model 3

The indices obtained for this model are illustrated below.

1. Average path coefficient (APC) = 0.380, $P < 0.001$
2. Average R-squared (ARS) = 0.357, $P < 0.001$
3. Average adjusted R-squared (AARS) = 0.350, $P < 0.001$
4. Average block VIF (AVIF) = 1.137, acceptable if ≤5, ideally ≤3.3
5. Average full collinearity VIF (AFVIF) = 2.113, acceptable if ≤5, ideally ≤3.3
6. Tenenhaus GoF (GoF) = 0.489, small ≥ 0.1, medium ≥ 0.25, large ≥ 0.36
7. Sympson's paradox ratio (SPR) = 1.000, acceptable if ≥0.7, ideally = 1
8. R-squared contribution ratio (RSCR) = 1.000, acceptable if ≥0.9, ideally = 1
9. Statistical suppression ratio (SSR) = 1.000, acceptable if ≥0.7
10. Nonlinear bivariate causality direction ratio (NLBCDR) = 1.000, acceptable if ≥0.7

According to APC, ARS, and AARS indexes, the model is acceptable at 95 % confidence level, since the values of all these indexes are lower than 0.05. Similarly, AVIF and AVFIF indexes reported values below 3.3, indicating that no collinearity problems exist between latent variables. Finally, the remaining efficiency indexes reported met the values required, which lead to the conclusion that the model is adjusted, according to the parameters analyzed.

10.4.2.2 Latent Variable Coefficients in Model 3

Table 10.11 reports validity of latent variables, which allows for the interpretation of the following:

- All values of *R*-squared coefficient are greater than 0.02, which was established as the minimum acceptable value. Therefore, it is concluded that all latent-dependent variables have sufficient predictive validity to be analyzed by the statistic test.
- As for the adjusted *R*-squared, all coefficients are below A.02, indicating that latent-dependent variables have predictive validity and can be analyzed.
- *Q*-squared coefficient values of all latent variables analyzed in this model are positive or higher than zero and close to the *R*-squared values. Thus, validity of latent variables is confirmed.
- Both composite reliability and Cronbach's alpha coefficients show values higher than 0.7—the minimum acceptable value—for all latent variables observed. Hence, it is concluded that there is sufficient validity and internal consistency in the variables analyzed.
- As for variance inflation, all latent variables show values below, 3.3, the maximum allowed value. Therefore, collinearity problems did not occur between variables.

10.4.2.3 Direct Effects—Validation of Raised Hypotheses

Since hypothesis (H1) of this model was already proposed, tested, and discussed in the previous model as hypothesis 10 (H10), it will be not addressed in this section.

Table 10.11 Coefficients of latent variables in Model 3

	Latent variables			
	Supplier	Time	Materials handling	Engineering process
R-squared coefficients		0.163	0.685	0.224
Adjusted *R*-squared coefficients		0.157	0.681	0.213

- H2. There is sufficient statistical evidence to declare that the selection and evaluation of suppliers, as well as their relation with the manufacturing companies, have a direct and positive impact on engineering process associated with the space expansion requirements, movement distances, and rapid changes during the JIT implementation process, since when the first latent variable increases its standard deviation by one unit, the standard deviation of the second latent variable also increases by 0.25 units.
- H3. There is not enough statistical evidence to declare that the selection and evaluation of suppliers and their relationship with manufacturing companies have a direct and positive impact on efficiency indexes associated with materials handling, since the test hypothesis test for the parameter showed a value above 0.05—the maximum allowed value—with 95 % of confidence level.
- H4. There is sufficient statistical evidence to declare that the commitments from suppliers to delivery times of raw materials and contracts with operators have a direct and positive impact on efficiency indexes associated with materials handling. When the first latent variable increases its standard deviation by one unit, the standard deviation of the second variable rises by 0.16 units.
- H5. There is sufficient statistical evidence to declare that commitment to delivery times with suppliers and customers, as well as contracts of operators, has a direct and positive impact on efficiency indexes of engineering process related to space requirements, movement distances, and rapid changes, since when the first latent variable increases its standard deviation by one unit, the standard of the second latent variable rises by 0.33 units.
- H6. There is enough statistical evidence to declare that the engineering process associated with the movement distances, requirements, and rapid changes has a direct and positive impact on the benefits associated with materials handling, since when the first latent variable increases its standard deviation by one unit, the standard deviation of the second also increases by 0.75 units.

One may observe that latent-dependent variables, which in this case concern JIT benefits—have values of R-squared coefficient explained by two independent latent variables. Thus, it is suitable to decompose the total of the coefficient of determination explained by every independent variable.

- Dependent latent variable Time is 16 % explained by the independent latent variable Suppliers.
- The dependent latent variable Engineering Process is explained in 22 %. However, 8.9 % originates from the independent latent variable Suppliers and 13.4 % is explained through the latent variable Time. Hence, Time variable plays a key role to obtain benefits in the Engineering Process.
- The dependent latent variable Materials Handling is explained in 69 %, from which 7.2 % is due to independent latent variable Time, and 61. 4 % of the variability originates from latent variable Engineering Processes, which included space requirements, movement distances, and rapid changes of engineering. Therefore, Engineering Processes holds the most significant relation with Materials Handling.

Table 10.12 Sum of indirect effects of Model 3

To	From	
	Supplier	Time
Materials handling	0.356 ($P < 0.001$) ES = 0.127	0.249 ($P < 0.001$) ES = 0.112
Engineering process	0.133 ($P = 0.011$) Es = 0.047	

10.4.2.4 Sum of Indirect Effects of Model 3

Table 10.12 shows the indirect effects that latent variables have over other latent variables. The table also allows for the interpretation of data as it follows.

- All indirect effects between latent variables are statistically significant with a 95 % confidence level, since their *P*-values are below 0.05, maximum value allowed.
- According to its size, the largest indirect effect occurs between the selection of Suppliers and Materials Handling, since when the first latent variable increases its standard deviation by one unit, the standard deviation of the second latent variable rises by 0.356 units. Note that this same relation had a direct effect statistically not significant, which implies that the impact of suppliers over materials handling occurs through the mediating variables Time and Engineering Process. That is, the selection and evaluation of suppliers do not directly affect the management of materials. Companies must first rely on proper space requirements, systems to reduce movement distances, and the ability to respond quickly to engineering changes. It is also important to mention that providers can explain 12.7 % of the variability of latent variable named Materials Handling.
- The second largest effect, according to its size, concerns Time and Materials Handling, because when the first latent variable increases its standard deviation by one unit, the standard deviation of the second latent variable also increases by 0.249 units. Therefore, it seems that short delivery of raw materials from suppliers become useless when the production process is unable to change from one product to another, or when raw material is moved constantly and excessively. Only when production systems respond rapidly to engineering changes and material movements are reduced, it may be possible to minimize inventories and increase their turnover.

10.4.2.5 Sum of Total Effects of Model 3

Table 10.13 introduces the total effects (sum of direct and indirect effects) for every relation between variables.

- All total effects of Model 3 are statistically significant, since the *P*-value for every relation between variables is below 0.05, which is the highest value allowed for a 95 % confidence level.
- Based on its size, the relation between Engineering Process and Materials Handling shows the largest total effect. When the first latent variable increases

Table 10.13 Sum of total effects of Model 3

To	From		
	Supplier	Time	Engineering process
Time	0.403 ($P < 0.001$) ES = 0.163		
Materials handling	0.356 ($P < 0.001$) ES = 0.127	0.408 ($P < 0.001$) ES = 0.183	0.753 ($P < 0.001$) ES = 0.614
Engineering process	0.387 ($P < 0.001$) ES = 0.136	0.33 ($P < 0.001$) ES = 0.134	

its standard by one unit, the standard deviation of the second latent variable also increases by 0.753 units. Moreover, the first independent latent variable explains 61.4 % of the variability of the dependent variable. It must also be noted that indirect effects do not occur between these two variables; thus, the value of the total effect reported refers to the value of the direct effect.

- The second largest total effect was found in the relation between latent variable Time and Materials Handling. When the first latent variable increases its standard deviation by one unit, the standard deviation of the second latent variable rises by 0.408 units. Moreover, Time explains 18.4 % of variability of Materials Handling and the indirect effect in this relation originates from Engineering Process.
- The total effect that latent variable Providers causes over Engineering Process is reported as the third largest total effect, since an indirect effect occurs between them through Time. Therefore, when the first latent variable increases its standard deviation by one unit, the standard deviation of the second latent variable increases by 0.387 units. Moreover, Providers explains 13.6 % of the variability of Engineering Process.

10.4.3 Conclusions of Model 3

Model 3 considered four latent variables to be analyzed and proposed six hypotheses or direct relations between them, from which five were completely new and one hypothesis had already been analyzed and discussed in Model 2. From the results of the evaluation of the model, the following conclusions can be provided.

- The selection and evaluation of suppliers is a variable that must be considered to achieve the objectives of a JIT program concerning engineering and materials handling. However, it must be remembered that the relationship between these variables can often be indirect.
- Time concerning reception of raw materials from suppliers, deliveries of final products, and long-term contracts offered to workers has both direct and indirect effects over the engineering process and materials handling.

– To achieve the objectives and obtain benefits for materials handling in a JIT environment, companies must first strive and manage to obtain benefits in the engineering process of products, since the latter precedes the first latent variable.

References

Ahmad MF, Zakuan N, Jusoh A, Takala J (2012) Relationship of TQM and business performance with mediators of SPC, lean production and TPM. Proc Soc Behav Sci 65:186–191. doi:10.1016/j.sbspro.2012.11.109

Aksoy A, Sucky E, Öztürk N (2014) Dynamic strategic supplier selection system with fuzzy logic. Proc Soc Behav Sci 109:1059–1063. doi:10.1016/j.sbspro.2013.12.588

Alcaraz JLG, Maldonado AA, Iniesta AA, Robles GC, Hernández GA (2014) A systematic review/survey for JIT implementation: Mexican maquiladoras as case study. Comput Ind 65(4):761–773. doi:10.1016/j.compind.2014.02.013

Almomani MA, Aladeemy M, Abdelhadi A, Mumani A (2013) A proposed approach for setup time reduction through integrating conventional SMED method with multiple criteria decision-making techniques. Comput Ind Eng 66(2):461–469. doi:10.1016/j.cie.2013.07.011

Araz C, Ozkarahan I (2007) Supplier evaluation and management system for strategic sourcing based on a new multicriteria sorting procedure. Int J Prod Econ 106(2):585–606. doi:10.1016/j.ijpe.2006.08.008

Avelar-Sosa L, García-Alcaraz J, Vergara-Villegas O, Maldonado-Macías A, Alor-Hernández G (2015) Impact of traditional and international logistic policies in supply chain performance. Int J Adv Manuf Technol 76(5–8):913–925. doi:10.1007/s00170-014-6308-3

Awasthi A, Chauhan SS, Goyal SK, Proth J-M (2009) Supplier selection problem for a single manufacturing unit under stochastic demand. Int J Prod Econ 117(1):229–233. doi:10.1016/j.ijpe.2008.10.012

Azevedo R, Martins C, Teixeira JC, Barroso M (2014) Obstacle clearance while performing manual material handling tasks in construction sites. Safety Sci 62:205–213. doi:10.1016/j.ssci.2013.08.016

Bennett GF (1984) A guide to the safe handling of hazardous materials accidents: manual STP/825, American society for testing materials. J Hazard Materials 9(3):396. doi:10.1016/0304-3894(84)87022-3

Blaga P, Jozsef B (2014) Human resources, quality circles and innovation. Proc Econ Finance 15:1458–1462. doi:10.1016/S2212-5671(14)00611-X

Boersma K, Kingma S (2005) From means to ends: the transformation of ERP in a manufacturing company. J Strateg Inf Syst 14(2):197–219. doi:10.1016/j.jsis.2005.04.003

Calvo-Mora A, Ruiz-Moreno C, Picón-Berjoyo A, Cauzo-Bottala L (2014) Mediation effect of TQM technical factors in excellence management systems. J Bus Res 67(5):769–774. doi:10.1016/j.jbusres.2013.11.042

Cárdenas-Barrón LE, González-Velarde JL, Treviño-Garza G (2015) A new approach to solve the multi-product multi-period inventory lot sizing with supplier selection problem. Comput Oper Res 64:225–232. doi:10.1016/j.cor.2015.06.008

Chai J, Liu JNK, Ngai EWT (2013) Application of decision-making techniques in supplier selection: a systematic review of literature. Expert Syst Appl 40(10):3872–3885. doi:10.1016/j.eswa.2012.12.040

Chakravorty SS (2009) Improving distribution operations: implementation of material handling systems. Int J Prod Econ 122(1):89–106. doi:10.1016/j.ijpe.2008.12.026

Chan FTS (2001) Effect of kanban size on just-in-time manufacturing systems. J Mater Process Technol 116(2–3):146–160. doi:10.1016/S0924-0136(01)01022-6

Chen H, Jeter D (2008) The role of auditing in buyer-supplier relations. J Contemp Acc Econ 4 (1):1–17. doi:10.1016/S1815-5669(10)70027-8
Chen Y, Xiao Q, Tang X (2011) Product layout optimization and simulation model in a multi-level distribution center. Syst Eng Proc 2:300–307. doi:10.1016/j.sepro.2011.10.042
Chen CX, Martin M, Merchant KA (2014) The effect of measurement timing on the information content of customer satisfaction measures. Manage Acc Res 25(3):187–205. doi:10.1016/j.mar.2013.12.003
Chiarini A (2013) The main methods of lean organization: Kanban, cellular manufacturing, SMED and TPM. In: Lean organization: from the tools of the Toyota production system to lean office, vol 3. Perspectives in business culture. Springer, Milan, pp 81–116. doi:10.1007/978-88-470-2510-3_6
Choudhary D, Shankar R (2013) Joint decision of procurement lot-size, supplier selection, and carrier selection. J Purchasing Supply Manage 19(1):16–26. doi:10.1016/j.pursup.2012.08.002
Choudhary D, Shankar R (2014) A goal programming model for joint decision making of inventory lot-size, supplier selection and carrier selection. Comput Ind Eng 71:1–9. doi:10.1016/j.cie.2014.02.003
Colledani M, Tolio T, Fischer A, Iung B, Lanza G, Schmitt R, Váncza J (2014) Design and management of manufacturing systems for production quality. CIRP Ann Manuf Technol 63 (2):773–796. doi:10.1016/j.cirp.2014.05.002
Corredor P, Goñi S (2011) TQM and performance: is the relationship so obvious? J Bus Res 64 (8):830–838. doi:10.1016/j.jbusres.2010.10.002
Crumpton MA (2015) 3—human resources as a strategic partner. In: Crumpton MA (ed) Strategic human resources planning for academic libraries. Chandos Publishing, Oxford, pp 23–29. doi:10.1016/B978-1-84334-764-4.00003-4
Cua KO, McKone KE, Schroeder RG (2001) Relationships between implementation of TQM, JIT, and TPM and manufacturing performance. J Oper Manage 19(6):675–694. doi:10.1016/S0272-6963(01)00066-3
David I, Eben-Chaime M (2003) How far should JIT vendor–buyer relationships go? Int J Prod Econ 81–82:361–368. doi:10.1016/S0925-5273(02)00274-8
De Toni A, Nassimbeni G (2000) Just-in-time purchasing: an empirical study of operational practices, supplier development and performance. Omega 28(6):631–651. doi:10.1016/S0305-0483(00)00016-5
den Butter FAG, Liu J, Tan Y-H (2012) Using IT to engender trust in government-to-business relationships: the authorized economic operator (AEO) as an example. Gov Inf Q 29(2):261–274. doi:10.1016/j.giq.2011.05.004
Deshpande SP, Golhar DY (1996) HRM practices of JIT firms in Canada. Prod Plann Control 7 (1):79
Dey PK, Bhattacharya A, Ho W (2015) Strategic supplier performance evaluation: a case-based action research of a UK manufacturing organisation. Int J Prod Econ 166:192–214. doi:10.1016/j.ijpe.2014.09.021
Dias JCQ, Calado JMF, Osório AL, Morgado LF (2009) RFID together with multi-agent systems to control global value chains. Ann Rev Control 33(2):185–195. doi:10.1016/j.arcontrol.2009.03.005
Du B, Guo S, Huang X, Li Y, Guo J (2015) A pareto supplier selection algorithm for minimum the life cycle cost of complex product system. Expert Syst Appl 42(9):4253–4264. doi:10.1016/j.eswa.2015.01.056
Dumond EJ, Newman W (1990) Closing the gap between buyer and vendor. Prod Inventory Manage J 31(4):13–17
Esmaeili Aliabadi D, Kaazemi A, Pourghannad B (2013) A two-level GA to solve an integrated multi-item supplier selection model. Appl Math Comput 219(14):7600–7615. doi:10.1016/j.amc.2013.01.046
Ferradás PG, Salonitis K (2013) Improving changeover time: a tailored SMED approach for welding cells. Proc CIRP 7:598–603. doi:10.1016/j.procir.2013.06.039

Fullerton RR, McWatters CS, Fawson C (2003) An examination of the relationships between JIT and financial performance. J Oper Manage 21(4):383–404. doi:10.1016/S0272-6963(03)00002-0

Galankashi MR, Chegeni A, Soleimanynanadegany A, Memari A, Anjomshoae A, Helmi SA, Dargi A (2015) Prioritizing green supplier selection criteria using fuzzy analytical network process. Proc CIRP 26:689–694. doi:10.1016/j.procir.2014.07.044

García-Alcaraz J, Prieto-Luevano D, Maldonado-Macías A, Blanco-Fernández J, Jiménez-Macías E, Moreno-Jiménez J (2015) Structural equation modeling to identify the human resource value in the JIT implementation: case maquiladora sector. Int J Adv Manuf Technol 77(5–8):1483–1497. doi:10.1007/s00170-014-6561-5

Gilbert FW, Young JA, O'Neal CR (1994) Buyer-seller relationships in just-in-time purchasing environments. J Bus Res 29(2):111–120. doi:10.1016/0148-2963(94)90015-9

Giunipero LC, O'Neal C (1988) Obstacles to JIT procurement. Ind Mark Manage 17(1):35–41. doi:10.1016/0019-8501(88)90024-7

Globerson S, Korman A (2001) The use of just-in-time training in a project environment. Int J Proj Manage 19(5):279–285. doi:10.1016/S0263-7863(00)00012-0

Govindan K, Rajendran S, Sarkis J, Murugesan P (2015) Multi criteria decision making approaches for green supplier evaluation and selection: a literature review. J Cleaner Prod 98:66–83. doi:10.1016/j.jclepro.2013.06.046

Gurel O, Acar AZ, Onden I, Gumus I (2015) Determinants of the green supplier selection. Proc Soc Behav Sci 181:131–139. doi:10.1016/j.sbspro.2015.04.874

Helms MM (1990) Communication: the key to JIT success. Prod Inventory Manage J 31(2):18–21

Huang C-Y, Huang T-S, Chen W-L (2009) Communication protocols for order management in collaborative manufacturing. Int J Prod Econ 122(1):257–268. doi:10.1016/j.ijpe.2009.06.004

Humphreys P, Shiu WK, Lo VHY (2003) Buyer–supplier relationship: perspectives between Hong Kong and the United Kingdom. J Mater Process Technol 138(1–3):236–242. doi:10.1016/S0924-0136(03)00078-5

Igarashi M, de Boer L, Fet AM (2013) What is required for greener supplier selection? A literature review and conceptual model development. J Purchasing Supply Manage 19(4):247–263. doi:10.1016/j.pursup.2013.06.001

Im JH, Hartman SJ, Bondi PJ (1994) How do jit systems affect human resource management? Prod Inventory Manage J 35(1):1–4

Inman RA, Sale RS, Green KW Jr, Whitten D (2011) Agile manufacturing: relation to JIT, operational performance and firm performance. J Oper Manage 29(4):343–355. doi:10.1016/j.jom.2010.06.001

Jadidi O, Cavalieri S, Zolfaghari S (2015) An improved multi-choice goal programming approach for supplier selection problems. Appl Math Model 39(14):4213–4222. doi:10.1016/j.apm.2014.12.022

Jun M, Cai S, Shin H (2006) TQM practice in maquiladora: antecedents of employee satisfaction and loyalty. J Oper Manage 24(6):791–812. doi:10.1016/j.jom.2005.09.006

Karsak EE, Dursun M (2015) An integrated fuzzy MCDM approach for supplier evaluation and selection. Comput Ind Eng 82:82–93. doi:10.1016/j.cie.2015.01.019

Kassicieh SK, Yourstone SA (1998) Training, performance evaluation, rewards, and TQM implementation success. J Qual Manage 3(1):25–38. doi:10.1016/S1084-8568(99)80102-3

Kaufmann L, Kreft S, Ehrgott M, Reimann F (2012) Rationality in supplier selection decisions: the effect of the buyer's national task environment. J Purchasing Supply Manage 18(2):76–91. doi:10.1016/j.pursup.2012.04.004

Kesti M (2012) Organization human resources development connection to business performance. Proc Econ Finance 2:257–264. doi:10.1016/S2212-5671(12)00086-X

Khan LR, Sarker RA (2002) An optimal batch size for a JIT manufacturing system. Comput Ind Eng 42(2–4):127–136. doi:10.1016/S0360-8352(02)00009-8

Kim S-L, Ha D (2003) A JIT lot-splitting model for supply chain management: enhancing buyer–supplier linkage. Int J Prod Econ 86(1):1–10. doi:10.1016/S0925-5273(03)00006-9

Krause DR, Ellram LM (1997) Success factors in supplier development. Int J Phys Distrib Logistics Manage 27(1):39–52. doi:10.1108/09600039710162277

Kumar D, Singh J, Singh OP, Seema (2013) A fuzzy logic based decision support system for evaluation of suppliers in supply chain management practices. Math Comput Model 57(11–12):2945–2960. doi:10.1016/j.mcm.2013.03.002

Kumar A, Jain V, Kumar S (2014) A comprehensive environment friendly approach for supplier selection. Omega 42(1):109–123. doi:10.1016/j.omega.2013.04.003

Kwak TC, Kim JS, Moon C (2006) Supplier–buyer models for the bargaining process over a long-term replenishment contract. Comput Ind Eng 51(2):219–228. doi:10.1016/j.cie.2006.02.004

Lee AHI, Kang H-Y, Hsu C-F, Hung H-C (2009) A green supplier selection model for high-tech industry. Expert Syst Appl 36(4):7917–7927. doi:10.1016/j.eswa.2008.11.052

Lee AHI, Kang H-Y, Lai C-M, Hong W-Y (2013) An integrated model for lot sizing with supplier selection and quantity discounts. Appl Math Model 37(7):4733–4746. doi:10.1016/j.apm.2012.09.056

Lee H-H, Zhou J, Hsu P-H (2014) The role of innovation in inventory turnover performance. Decis Support Syst (0). doi:10.1016/j.dss.2015.02.010

Li Y, Li X, Gupta JND (2015) Solving the multi-objective flowline manufacturing cell scheduling problem by hybrid harmony search. Expert Syst Appl 42(3):1409–1417. doi:10.1016/j.eswa.2014.09.007

Lima Junior FR, Osiro L, Carpinetti LCR (2014) A comparison between fuzzy AHP and Fuzzy TOPSIS methods to supplier selection. Appl Soft Comput 21:194–209. doi:10.1016/j.asoc.2014.03.014

Lin R-H (2012) An integrated model for supplier selection under a fuzzy situation. Int J Prod Econ 138(1):55–61. doi:10.1016/j.ijpe.2012.02.024

Macbeth DK, Baxter L, Farguson N, Neil G (1988) Buyer-vendor relationship with just-in-time: lessons from US multinationals. Ind Eng 20(9):38–41

Marodin G, Saurin T, Tortorella G, Denicol J (2015) How context factors influence lean production practices in manufacturing cells. Int J Adv Manuf Technol 79(5–8):1389–1399. doi:10.1007/s00170-015-6944-2

Mazdeh MM, Emadikhiav M, Parsa I (2015) A heuristic to solve the dynamic lot sizing problem with supplier selection and quantity discounts. Comput Ind Eng 85:33–43. doi:10.1016/j.cie.2015.02.027

Mensah P, Merkuryev Y, Longo F (2015) Using ICT in developing a resilient supply chain strategy. Proc Comput Sci 43:101–108. doi:10.1016/j.procs.2014.12.014

Mohan MTR, Singh NP (1995) Buyer-supplier relationships: the case from Indian diary co-operatives. Prod J 35(4):643–647

Ngai EWT, Chau DCK, Chan TLA (2011) Information technology, operational, and management competencies for supply chain agility: findings from case studies. J Strateg Inf Syst 20(3):232–249. doi:10.1016/j.jsis.2010.11.002

Niskanen T, Lauttalammi J (1989) Accidents in materials handling at building construction sites. J Occup Accid 11(1):1–17. doi:10.1016/0376-6349(89)90002-3

O' Neal C (1987) The buyer seller linkage in a JIT environment. J Purchasing Mater Manage (Spring, pp 7–13)

Ooi K-B (2014) TQM: A facilitator to enhance knowledge management? A structural analysis. Expert Syst Appl 41(11):5167–5179. doi:10.1016/j.eswa.2014.03.013

Pan J-N, Nguyen HTN (2015) Achieving customer satisfaction through product–service systems. Eur J Oper Res 247(1):179–190. doi:10.1016/j.ejor.2015.05.018

Parvini M (2011) 9—packaging and material handling. In: Kardar RZFR (ed) Logistics operations and management. Elsevier, London, pp 155–180. doi:10.1016/B978-0-12-385202-1.00009-8

Prajogo DI, Sohal AS (2006) The relationship between organization strategy, total quality management (TQM), and organization performance—the mediating role of TQM. Eur J Oper Res 168(1):35–50. doi:10.1016/j.ejor.2004.03.033

Rossi D, Bertoloni E, Fenaroli M, Marciano F, Alberti M (2013) A multi-criteria ergonomic and performance methodology for evaluating alternatives in "manuable" material handling. Int J Ind Ergon 43(4):314–327. doi:10.1016/j.ergon.2013.04.009

Rusu G, Avasilcai S (2014) Linking human resources motivation to organizational climate. Proc Soc Behav Sci 124:51–58. doi:10.1016/j.sbspro.2014.02.459

Sánchez AJ (2015) Chapter 27—handling for micro-manufacturing. In: Qin Y (ed) Micromanufacturing engineering and technology, 2nd edn. William Andrew Publishing, Boston, pp 637–659. doi:10.1016/B978-0-323-31149-6.00027-X

Sarkis J, Dhavale DG (2015) Supplier selection for sustainable operations: a triple-bottom-line approach using a Bayesian framework. Int J Prod Econ 166:177–191. doi:10.1016/j.ijpe.2014.11.007

Schneider JD, Leatherman MA (1992) Integrated just-in-time: a total business approach. Prod Inventory Manage J 33(1):78–82

Schniederjans D, Schniederjans M (2015) Quality management and innovation: new insights on a structural contingency framework. Int J Qual Innov 1(1):1–20. doi:10.1186/s40887-015-0004-8

Scott J, Ho W, Dey PK, Talluri S (2015) A decision support system for supplier selection and order allocation in stochastic, multi-stakeholder and multi-criteria environments. Int J Prod Econ 166:226–237. doi:10.1016/j.ijpe.2014.11.008

Shaw K, Shankar R, Yadav SS, Thakur LS (2012) Supplier selection using fuzzy AHP and fuzzy multi-objective linear programming for developing low carbon supply chain. Expert Syst Appl 39(9):8182–8192. doi:10.1016/j.eswa.2012.01.149

Singh S, Garg D (2011) JIT system: concepts, benefis and motivation in Indian industries. Int J Manage Bus Stud 1(1):26–30

Singh N, Jain S, Sharma P (2015) Motivations for implementing environmental management practices in Indian industries. Ecol Econ 109:1–8. doi:10.1016/j.ecolecon.2014.11.003

Sueyoshi T, Goto M (2010) Measurement of a linkage among environmental, operational, and financial performance in Japanese manufacturing firms: a use of data envelopment analysis with strong complementary slackness condition. Eur J Oper Res 207(3):1742–1753. doi:10.1016/j.ejor.2010.07.024

Tams S, Grover V, Thatcher J (2014) Modern information technology in an old workforce: toward a strategic research agenda. J Strateg Inf Syst 23(4):284–304. doi:10.1016/j.jsis.2014.10.001

Teksan ZM, Geunes J (2015) A polynomial time algorithm for convex cost lot-sizing problems. Oper Res Lett 43(4):359–364. doi:10.1016/j.orl.2015.03.009

Thomas AJ, Jones GR, Vidales P (2006) An integrated approach to TPM and six sigma development in the castings industry. In: Pham DT, Eldukhri EE, Soroka AJ (eds) Intelligent production machines and systems. Elsevier Science Ltd, Oxford, pp 620–625. doi:10.1016/B978-008045157-2/50108-5

Valmohammadi C, Roshanzamir S (2015) The guidelines of improvement: relations among organizational culture, TQM and performance. Int J Prod Econ 164:167–178. doi:10.1016/j.ijpe.2014.12.028

Villanueva-Ponce R, Garcia-Alcaraz J, Cortes-Robles G, Romero-Gonzalez J, Jiménez-Macías E, Blanco-Fernández J (2015) Impact of suppliers' green attributes in corporate image and financial profit: case maquiladora industry. Int J Adv Manuf Technol. pp 1–20. doi:10.1007/s00170-015-7082-6

Wang J-W, Cheng C-H, Huang K-C (2009) Fuzzy hierarchical TOPSIS for supplier selection. Appl Soft Comput 9(1):377–386. doi:10.1016/j.asoc.2008.04.014

Yoo SH, Shin H, Park M-S (2015) New product development and the effect of supplier involvement. Omega 51:107–120. doi:10.1016/j.omega.2014.09.005

Zhang Q, Vonderembse MA, Cao M (2009) Product concept and prototype flexibility in manufacturing: implications for customer satisfaction. Eur J Oper Res 194(1):143–154. doi:10.1016/j.ejor.2007.12.013

Zhang Y, Zhang G, Du W, Wang J, Ali E, Sun S (2015) An optimization method for shopfloor material handling based on real-time and multi-source manufacturing data. Int J Prod Econ 165:282–292. doi:10.1016/j.ijpe.2014.12.029

Chapter 11
Causal Models of JIT Elements Associated with Production Process and the Obtained Benefits

This chapter addresses three models to associate JIT elements from production process with JIT benefits obtained. Readers may once more refer to methodology addressed in Chap. 5 for a full comprehension of the procedures here described concerning the planning, development, and execution of models. Moreover, since the category of JIT elements associated with the production process comprises 18 items required for the successful JIT production process, it was necessary to perform a factor analysis to identify how many latent variables were comprised in this category.

11.1 Factor Analysis of JIT Elements Associated with Production Process

The objective of this analysis is to identify the latent dependent variables to be integrated into the structural equation model according to methodology described in Chap. 5. Table 11.1 shows the factorability result of the analysis of these 18 items. It is observed that according to the median or adequacy of the sample or KMO index, it is possible to carry out an actuarial analysis, since the value obtained for that index showed to be higher than the minimum accepted value of 0.8. Moreover results of the Bartlett's sphericity test showed a *P*-value of 0.000, which is definitely lower than the maximum accepted value for a 95 % confidence level. Therefore, it can be concluded that correlation matrix of these 18 items is different from an identity matrix. It is thus inferred that correlations of that of correlation matrix are different from 0, indicating that variables are highly correlated.

Table 11.2 illustrates the total explained variability, which explains each of these elements based on the eigenvalues of the correlation matrix. It is observed that only three items have all values higher than the unit. They also explain 59.301 % of the variability contained in the 18 variables or initial items. Furthermore, Table 11.2 also shows that the first factor or found latent variable explains 42.79 % of the variability contained in the initial 18 items, while the second variable explains it in 9.48 %. Finally, the third latent variable found explains 7.03 % of the variability,

J.L. García-Alcaraz and A.A. Maldonado-Macías, *Just-in-Time Elements and Benefits*, Management and Industrial Engineering, DOI 10.1007/978-3-319-25919-2_11

Table 11.1 KMO and Bartlett's test

Measure of sampling adequacy Kaiser-Meyer-Olkin		0.885
Bartlett's sphericity test	Approximate Chi-squared	1326.772
	gl	153
	Sig.	0

Table 11.2 Total variance explained

Factors	Initial Eigenvalues	Extraction sums of squared loadings			Rotation sums of squared loadings		
	Total	Percentage of variance	Percentage accumulated	Total	Percentage of the variance	Percentage accumulated	Total
1	7.702	42.790	42.790	7.702	42.790	42.790	6.163
2	1.707	9.484	52.274	1.707	9.484	52.274	6.262
3	1.266	7.032	59.306	1.266	7.032	59.306	3.507

Table 11.3 Structure matrix for elements in production process

Item	Factor loading	Factor name
Process flexibility	0.875	Lean techniques
Reduced work in process	0.762	
Small lot sizes	0.611	
Preparation time reduction (SMED)	0.599	
Just-in-time purchases	0.595	
Pull system	0.572	
Total preventive maintenance (TPM)	0.535	
Improved plant layout	0.501	
Kanban system	0.833	Production organization and material flow
Cellular manufacturing	0.765	
Group technology	0.736	
Kaizen system	0.720	
Use of robots	0.702	
Scheduling production below capacity	0.689	Capacity and inventory management
Safety stock	0.689	
Specialized factories	0.671	
Standardized containers	0.612	

which combined with the previous two latent variables, accumulates 59.306 % of the total variability explained.

Table 11.3 illustrates how items or elements analyzed are associated with each of the latent variables found. It is important to mention that the element or variable named Implementation of Process Control was eliminated, since it showed little association with latent variables and had no greater load factor than 0.5. Hence, while the first factor or latent variable named Lean Techniques is composed of eight items, the second factor or latent variable includes five items, all related to distribution of machinery and equipment, flow of materials, and continuous improvement of production process. Finally, the third factor or latent variable comprises four items related to aspects of plant capacity and inventory management.

11.2 Model 1. JIT Elements: Lean Techniques and Production Organization and Flow of Materials. Benefits: Production Process and Process Engineering

This model associates four latent variables. Two of them concern JIT elements of Lean Techniques, which includes eight items, and Production Organization and Material Flow, which comprises five items. On the other hand, the remaining two latent variables concern benefits of Production Process with seven items included, and Engineering Process with four items or variables included. Twenty-four variables were analyzed in total. Figure 11.1 depicts the model with the relationships proposed and the hypothesis to be tested for every relationship. None of these hypotheses has been previously tested and discussed.

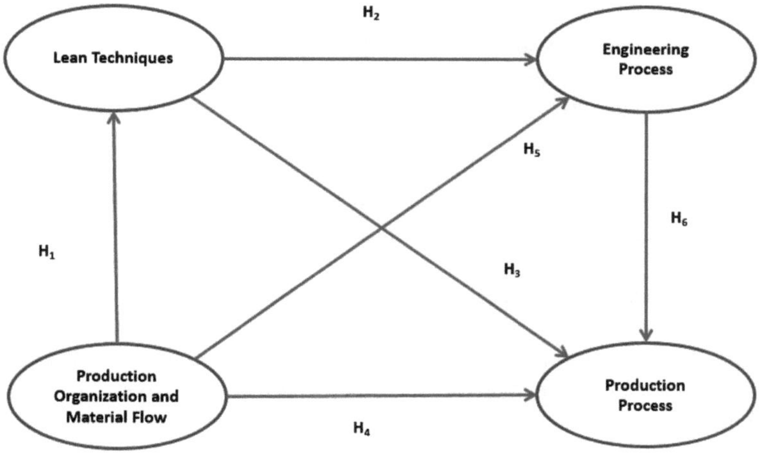

Fig. 11.1 Proposed model 1

11.2.1 Hypotheses Proposed of Model 1

The model associates two latent variables from category of JIT elements of the production process with two latent variables associated with the benefits received in the engineering and production processes. As a result of this association, six hypotheses were proposed and eventually tested.

Companies plan their plant distribution of machinery and equipment depending on the activities that these machines perform for raw materials processing. (Kia et al. 2014; Chen et al. 2011). For instance, while cellular manufacturing distribution groups activities associated with the same product family (Pattanaik and Sharma 2009; Ertay et al. 2006), and the same raw material is processed there from start to finish, technology groups integrate machines that develop similar activities (Jensen et al. 1996; Pourbabai 1988). However, despite their different approach, these forms of plan distribution rely on high technologies such as robots and computers that focus on minimizing the number of accidents and material handling. Moreover, companies support the efficiency of their production process with the implementation of lean manufacturing techniques, such as continuous improvement or Kaizen and Kanban system (Oropesa-Vento et al. 2015; Recht and Wilderom 1998; Radharamanan et al. 1996), which allows for the development of a pull system that can be effective but requires extensive knowledge of machinery maintenance and the ability of machinery to respond rapidly to changes from one product design to another. In order to contribute to this discussion, the following hypothesis is proposed.

H1: *There is a direct and positive effect between production organization and material flow and lean manufacturing techniques that can be applied in a production system in a JIT environment.*

Appropriate plant layout and pull system can contribute to minimize the movement distances of raw materials throughout the production system (Sundar et al. 2014; Selçuk 2013) and ensure better use of available space. Moreover, SMED technique should lead to rapid changes in engineering (Chiarini 2013; Almomani et al. 2013) and machinery and equipment to adapt to new designs and processes, which is closely associated with a suitable program of total preventive maintenance (Kumar et al. 2014). Similarly, just-in-time purchases indicate that companies rely on pull production systems (Al-Tahat and Mukattash 2006) that produce the exact quantity desired, which also results in a reduction of movement distances of raw materials, since companies are familiar with the exact process of that order. This means that since they produce what has been ordered, they purchase only the exact amount of material required to fulfill that order. Therefore, as a contribution to the present discussion, the following hypothesis is proposed.

H2: *There is a direct and positive relationship between lean manufacturing techniques applied in a production system during the implementation process of JIT and the benefits obtained in the engineering process.*

Just-in-time purchases and production process that follow the established methods and standards may certainly help achieve short deliveries to customers, which it now stands as one of the most significant metrics of efficiency in manufacturing (Selçuk 2013; Pragman 1996). However, total productive maintenance programs and rapid changes in the production process can also facilitate such short delivery times, since broken and damaged machines would certainly cause delays (Chan 2001; Kim and Tang 1997; Chiarini 2013; Cua et al. 2001). Similarly, the ability to make quick modifications in the production process from one design or prototype to another allows companies to accept several different orders and increase its production flexibility (Kumar et al. 2014; Kemal Karasu et al. 2014). This agility and flexibility allow manufacturers to cater for larger quantities of orders that will result in economic benefits for them (Ragin-Skorecka 2014; Huumonen 2011). Finally, high levels of expertise, education, and training of workers ensure a reduction of errors in the production process or waste, which also improves efficiency and increases productivity (Lee and Johnson 2012). Therefore, the following hypothesis was proposed as a contribution to this discussion.

H3: *Lean manufacturing techniques applied to the production process during the implementation of JIT have a direct and positive impact on efficiency indexes of production process.*

Competitive advantage can be measured with productivity indexes, a reduction of waste (Marodin et al. 2015; Deep and Singh 2015) and rework, and according to the integration of common or singular production activities (Calvo-Mora et al. 2014) that allow for the employment of highly trained workers. Therefore, machines and equipment used for the transformation of raw materials are distributed in such a way that they increase this competitive advantage of companies.

Similarly, techniques that support continuous flow of materials, such as Kanban, can reduce delivery times of finished products (Rahman et al. 2013a, b; Hou and Hu 2011; Lage Junior et al. 2010; Chan 2001). However, production processes also often require the implementation of continuous improvement programs, such as Kaizen, whose popularity is due to the fact that it enables the integration of operators into the improvement of production processes (Oropesa-Vento et al. 2015; Recht and Wilderom 1998; Radharamanan et al. 1996), since they are the most familiarized with this process, run the machines, and ensure the flow of materials. Therefore, as a means to contribute to the discussion, it was possible to propose the following hypothesis.

H4: *Production organization and material flow have a direct and positive impact on efficiency indexes achieved in the production process in a JIT environment.*

The use of Kanban system in a production system will result in a number of benefits associated with a reduced space requirements for the production process and should also help minimize movement distances of raw materials (Sylvain and Duguay 1997; Lummus 1995; Fukukawa and Hong 1993). Nevertheless, if this card system is implemented in group technology or manufacturing cells, profits could be much higher, since these centers are organized to bring together a certain number of similar activities (Spencer 1998; Jensen et al. 1996; Pourbabai 1988) or group of families of products, where operators become highly skilled in their activities (Tabassi and Abu Bakar 2009; Kassicieh and Yourstone 1998) and continuous improvement is achieved through innovative proposals that enable better production process, products, and distribution of machinery and equipment (Farris et al. 2009; Kumiega and Van Vliet 2008). Therefore, since physical organization of machines and equipment used in the production system equipment can generate a number of benefits reflected in the engineering process and its indexes, it is possible to propose the fifth hypothesis of this model.

H5: *Production organization and flow of materials in a production process has a direct and positive impact on efficiency indexes of engineering process in a JIT environment.*

An efficient engineering process reflects on the reduction of space requirements (Huang and Lin 2014), movement distances of raw materials (Ohno 2011; Chakravorty 2009), and rapid changes (Chiarini 2013; Almomani et al. 2013). Therefore, these indications would measure productivity of the production process based on a minimum of waste (Inman et al. 2011; Huq 1999; Huq and Huq 1994), an increased efficiency and flexibility, and thus short delivery times. However, for every action there is a reaction, and in order to statistically demonstrate the meaning of these facts, the following hypothesis is proposed.

H6: *Indexes of an engineering process have a direct and positive impact on the indexes associated with the production process in a JIT environment.*

11.2.2 Results of Model 1

Results of the model evaluation are illustrated in Fig. 11.2. Every segment includes a value or parameter that measures the relation between the two latent variables involved in that segment. Similarly, *P*-values are also included in segments as a measure of statistical significance (the level of significance in all models executed in this book is 0.05). Finally, every latent dependent variable shows an R-squared value as a measure of its explained variance.

The segment colored in red indicates that the direct effect for this relation cannot be considered significant with a 95 % confidence level but with a 90 % confidence level. Moreover, unlike models in Chap. 10, this relation was not eliminated since indirect and total effects that occur through this segment are significant.

11.2.2.1 Efficiency Indices of Model 1

Before describing the direct and indirect effects of this first model, it is necessary to determine whether it meets certain indexes of goodness of fit that permit drawing

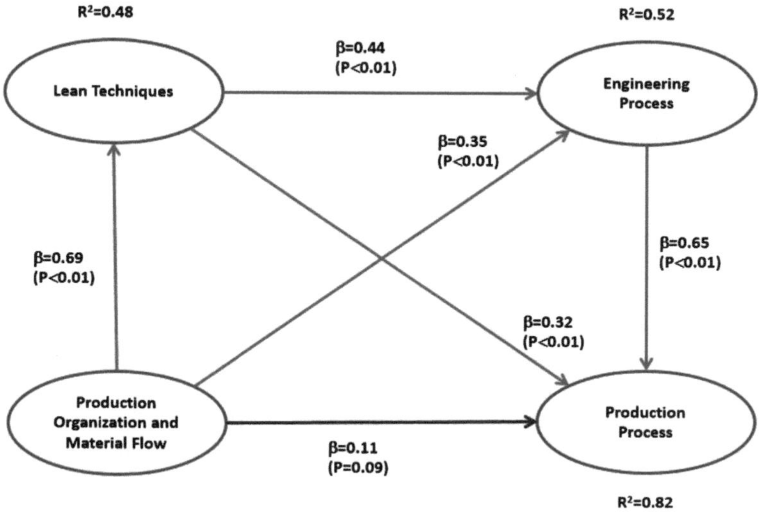

Fig. 11.2 Evaluated model 1—validating hypotheses

conclusions from it, since the objective of the model is to be as predictive as it is possible to determine the existing relations between variables. Efficiency indexes of this model are listed below.

1. Average path coefficient (APC) = 0.419, $P < 0.001$
2. Average R-squared (ARS) = 0.608, $P < 0.001$
3. Average adjusted R-squared (AARS) = 0.603, $P < 0.001$
4. Average block VIF (AVIF) = 2.080, acceptable if ≤5, ideally ≤3.3
5. Average full collinearity VIF (AFVIF) = 3.092, acceptable if ≤5, ideally ≤3.3
6. Tenenhaus GoF (GoF) = 0.609, small ≥0.1, medium ≥0.25, large ≥0.36
7. Sympson's paradox ratio (SPR) = 1.000, acceptable if ≥0.7, ideally = 1
8. R-squared contribution ratio (RSCR) = 1.000, acceptable if ≥0.9, ideally = 1
9. Statistical suppression ratio (SSR) = 1.000, acceptable if ≥0.7
10. Nonlinear bivariate causality direction radio (NLBCDR) = 1.000, acceptable if ≥0.7

According to the APC, AARS, and AARS indexes that show *P*-values to measure statistical significance, it can be concluded that the model is efficient, has predictive capacity, and, in average, all parameters measuring the relations between variables latent are statistically significant. It is important to be clear with the concept of "average" here addressed. It refers to the overall value of all relationships in the model, since as shown in Fig. 11.2, one of this relation is not statistically significant. Similarly, when the variance inflation factor was analyzed, no collinearity problems were observed among these latent variables studied in general or average terms. Similar interpretations can be proposed for the remaining indexes shown. However, note that GoF index, which is recommended to be greater than

0.36, shows a value of 0.609 in this model; this indicates that it is as an appropriate model fit.

11.2.2.2 Coefficients of Latent Variables of Model 1

Indices reported above indicate that the model is in general efficient and can be used to interpret the relations between latent variables. However, it is also important to analyze each of these latent variables. Table 11.4 depicts validity indexes for each of these variables.

The first two rows of Table 11.4 show that values of the R-squared and the adjusted R-squared are higher than 0.02. Therefore, all latent variables that are dependent or explained by independent latent variables have sufficient predictive validity from a parametric view. Moreover, Q-squared values of all latent variables are higher than 0 and close to their R-squared and adjusted R-squared values.

As for internal validity, Cronbach's alpha and composite reliability show values higher than 0.7 in all latent variables, which is the minimum accepted value. Therefore, all latent variables have sufficient internal validity. Similarly, AVE indexes have values greater than 0.5 in all latent variables, leading to the conclusion that they have by far convergent validity. Finally, all variables have adequate collinearity, since none of them have values greater than 3.3, the maximum value admitted. Therefore, based on the indexes above, it is concluded that the structural equations model is suitable and can be analyzed.

11.2.2.3 Direct Effect of Model 1

According to direct effect, the hypotheses can be tested and the following conclusions are given:

- H1: There is sufficient statistical evidence to declare that organization of production and material flow have a direct and positive effect on lean

Table 11.4 Coefficients of latent variables of model 1

	Lean techniques	Production organization and material flow	Production process	Engineering process
R-squared	0.481		0.820	0.522
Adj. R-squared	0.477		0.816	0.515
Composite reliability	0.907	0.886	0.916	0.891
Cronbach's alpha	0.880	0.838	0.892	0.836
AVE	0.553	0.609	0.610	0.672
FVIF	2.678	2.164	3.141	3.086
Q-squared	0.485		0.747	0.521

manufacturing techniques applied in a JIT production system, since when the first latent variable increases its standard deviation by one unit, the standard deviation of the second latent variable increases by 0.69 units.

- H2: There is enough statistical evidence to declare that lean manufacturing techniques applied in a JIT production system have a direct and positive impact on the benefits obtained in the engineering process. When the first latent variable increases its standard deviation by one unit, the standard deviation of the second latent variable also increases by 0.44 units
- H3: There is sufficient statistical evidence to declare that lean manufacturing techniques implemented in a JIT production process have a direct and positive impact on its efficiency indexes, because when the first latent variable increases its standard deviation by one unit, the standard deviation of the second latent variable also increases by 0.32 units.
- H4: There is no enough statistical evidence to declare that production organization and material flow have a direct and positive impact on the efficiency indexes achieved in the production process in a JIT environment, since the *P*-value obtained from the significance test of the hypothesis showed a value above 0.05. Thus, with a 95 % confidence established, this hypothesis was rejected.
- H5: There is enough statistical evidence to declare that production organization and flow of materials in a JIT production process have a direct and positive impact on efficiency indexes of engineering process, since when the first latent variable increases its standard deviation by one unit, the standard deviation of the second latent variable also rises by 0.35 units.
- H6: There is sufficient statistical evidence to declare that efficiency indexes of engineering process have a direct and positive impact on indexes associated with the production process in JIT environment, because when the first latent variable increases its standard deviation by one unit, the standard deviation of the second latent variable increases by 0.65 units.

For dependent variables with R-squared values, it is important to apportion their total direct effect and discuss the percentage of direct effects that is explained through the different independent latent variables. Table 11.5 shows the effect sizes that independent latent variables have on dependent latent variables.

- Dependent latent variable Lean Techniques is explained in 48.4 % by latent variable Production Organization and Material Flow, since the R-squared value is 0.484. Note this variable is only explained by the independent variable.

Table 11.5 Effect sizes of model 1

	Lean techniques	Production organization and material flow	Engineering process
Lean techniques		0.484	0.546
Production process	0.238	0.032	
Engineering process	0.29	0.225	

- Latent dependent variable Engineering Processes is explained in 52 %, from which 29 % can be explained by independent latent variable Lean Techniques and 22.5 % by independent latent variable Production Organization and Material Flow.
- Latent dependent variable Production Process is explained in 82 % by three latent variables. Thus, 23.6 % of variability is given through Lean Techniques, 3.2 % originates from Production Organization and Material Flow (although the direct effect between these two variables was statistically not significant), and 54.6 % of variability originates from Engineering Process. Therefore, it can be stated that process engineering is the most important element to improve production process indexes, since it explains the greatest percentage of production process variable.

11.2.2.4 Sum of Indirect Effects

Table 11.6 depicts the sum of indirect effects between two variables given through a mediator variable. However, it shall not be forgotten that the same variable may have an effect on several variables, and such effects may occur through different other mediator variables. Note that the *P*-values for statistical significance of this effects were lower than 0.05; thus, indirect effects are considered statistically significant.

Based on the information shown in Table 11.6, it is possible to conclude the following:

- Latent variable Lean Techniques has an indirect effect on dependent latent variable Production Process, since when the first increases its standard deviation by one unit, the standard deviation of the second latent variable increases by 0.286 units. Besides, this independent variable can explain up to 21.2 % of the variability of the latent dependent variable, since the R-squared value is 0.212.
- The independent latent variable Production Organization and Material Flow has a strong impact on dependent latent variable Production Process, since when the first increases its standard deviation by one unit, the standard deviation of the second rises by 0.649 units. Moreover, the independent latent variable can explain up to 38.1 % of the variability of the dependent latent variable, since the R-squared value is 0.381.

Table 11.6 Sum of indirect effects of model 1

To	From	
	Lean techniques	Production organization and material flow
Production process	0.286 ($P < 0.01$) ES = 0.212	0.649 ($P < 0.01$) ES = 0.381
Engineering process		0.305 ($P < 0.01$) ES = 0.197

- Independent latent variable Production Organization and Material Flow also has an indirect effect on dependent latent variable Engineering Process, since when the former increases its standard deviation by one unit, the standard deviation of the latter also increases by 0.305 units. The independent variable also accounts for up to 19.7 % of the variability of the dependent latent variable, since the R-squared value is 0.197.

As a conclusion, it can be stated that independent latent variable Production Organization and Material Flow is crucial when implementing a JIT system or, since several elements and benefits that companies can obtain depend on it, such as the implementation of Kanban system and Kaizen. The former is directly associated with the flow of materials, while the latter supports a continuous improvement of the entire production process.

11.2.2.5 Total Effects of Model 1

The sum of direct and indirect effects for every relation between variables is depicted in Table 11.7. Likewise, every parameter that measures the total effect among a relation includes a *P*-value associated with its statistical significance. The same table also introduces the size of the effects and allows for the conclusions offered below.

- All total effects are statistically significant, since the *P*-value of the statistical significance test for every relation was lower than 0.05, the maximum value accepted for a 95 % confidence level.
- One of the total effects has a value lower than 0.6, while the remaining five effects show values below 0.6. This demonstrates the importance of the relations of these latent variables.
- The relationship between Production Organization and Material Flow and Production Process shows the largest total effect; when the former latent variable increases its standard deviation by one unit, the standard deviation of the latter increases by 0.710 units. Moreover, the first latent variable explains 41.6 % of

Table 11.7 Total effects of model 1

To	From		
	Lean techniques	Production organization and material flow	Engineering process
Lean techniques		0.693 ($P < 0.01$) ES = 0.481	
Production process	0.609 ($P < 0.01$) ES = 0.450	0.710 ($P < 0.01$) ES = 0.416	0.652 ($P < 0.01$) ES = 0.546
Engineering process	0.440 ($P < 0.01$) ES = 0.297	0.653 ($P < 0.01$) ES = 0.422	

the variability of the latter, since the effect size equals 0.416. Also, the direct relation between these latent variables analyzed by H4 was not significant; thus, the indirect effects between them occur through latent variables Lean Techniques and Engineering Process.

- The relation between latent variables Production Organization and Material Flow and Lean Techniques was reported as the second relation with the largest size of total effects. When the standard deviation of the first variable increases by one unit, the standard deviation of the second also increases by 0.693 units. Also, the independent latent variable explains up to 48.1 % of the variability of the dependent latent variable, since the size effect is 0.481.
- The total effect from the relation between latent variable Production Organization and Material Flow and Engineering Process is also considerably high and reported as the third largest total effect. When the independent latent variable increases its standard deviation by one unit, the standard deviation of the second latent variable increases by 0.653 units. Moreover, the independent variable explains up to 42.2 % of the variability of the dependent latent variable, since the effect size is 0.422.
- The relation between latent variable Lean Techniques and Production Process shows the fourth relation with the highest total effect size. It refers to the implementation of lean manufacturing techniques in the production line. When the former latent variable increases its standard deviation by one unit, the standard deviation of the latter rises 0.609 units. Moreover, the first latent variable explains 45 % of variability of the second latent variable, since the effect size is 0.45 %
- Similar interpretations were obtained for the remaining two relations between latent variables.

11.2.2.6 Conclusions of Model 1

This model associated four latent variables, from which two concerned JIT elements from production processes and two referred to benefits and production process. A total of 24 items or variables were analyzed within these four latent variables. Moreover, six hypotheses were proposed and analyzed to indicate the direct effect between latent variables; five of them were accepted since their *P*-values were lower than 0.05, the maximum allowed for a 95 % confidence level.

The latent variable Production Organization and Material Flow reported the highest total effect over the latent variable Production Process, although their direct effect was statistically not significant. Therefore, the total effect is mostly due from indirect effects occurring through latent variables Lean Techniques and Engineering Process. Moreover, latent variable Production Process is explained in 82 % by the latent variables that influence on it, which is one of the highest values obtained in the models proposed so far. Industrial implications of this model are stated as it follows:

- The implementation of a JIT program must be based on the physical distribution and organization of machinery and equipment within a plant, since many lean manufacturing techniques depend on it.
- Lean manufacturing techniques alone do not provide great benefits to the production process; they must be supported by effective engineering processes. Note that the direct effect that causes latent variable Engineering Process over Production Process is much higher than the direct effect between Lean and Production Organization and Material flow; however, their direct effect is not statistically significant.

11.3 Model 2. Elements: Lean Techniques and Production Organization and Material Flow. Benefits: Human Resources and Quality

This model associates four latent variables; two of them concern JIT elements of Production Process, which are: Production Organization and Material Flow and Lean Techniques. The remaining two refer to benefits obtained for human resources and quality in the production process. The former two latent variables together comprise 13 items or variables, while the latter two latent variables include eight items or variables. Thus, 21 items were integrated into the four latent variables of this second model. Similar to the previous model, latent variable Production Organization and Material Flow here is considered independent latent variable, while Quality is considered as dependent variable from all others.

Figure 11.3 shows the relationship between the four latent variables and displays six working hypotheses. Five of them are recent, while (H1) was previously tested and analyzed. It concerns the relation between Production Organization and Material Flow and Lean Techniques.

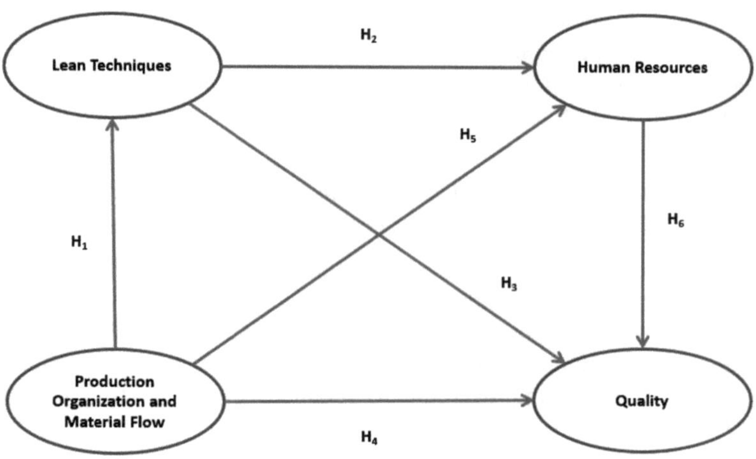

Fig. 11.3 Proposed model 2

11.3.1 Hypotheses Proposed of Model 2

This model seeks to associate two latent variables of production processes with two latent variables associated with the benefits gained after a JIT implementation. As mentioned above, one hypothesis was previously tested and analyzed in model 1, but is depicted in this model to provide the complete proposal. However, it will not be discussed once more.

Companies exist and grow as a result of human resources and their skills and abilities; thus, resources and techniques to bring greater motivation among them will be constantly sought and implemented. Besides, this motivation and encouragement will be reflected on improved work performance. For instance, flexibility depends on the capacity of human resources to perform diverse tasks (Chlivickas 2014), which impacts on a continuous material flow and can provide employees with satisfaction once they become multifunctional (Chlivickas 2014; Nen 2015); moreover, they feel more integrated. Several authors nowadays consider that human resources are top partners of any organization before suppliers (Crumpton 2015).

Another factor with a positive effect on workers motivation is their opportunity to rely on machinery and equipment that are properly calibrated by suitable programs of total productive maintenance. It results in satisfying work performance and a reduction of errors along the production process (Chiarini 2013; Thomas et al. 2006). However, companies must seek to implement other programs and plans associated with total productive maintenance, such as those that enable to perform quick changes between product designs. These other programs can also improve morale of workers and promote collaborative work as a result of communication (Chiarini 2013; Huang et al. 2009).

Another JIT crucial element is just-in-time purchases to highly reliable suppliers who become efficiently and highly integrated into the production system as a result of the communication skills of human resources (Gilbert et al. 1994; Macbeth et al. 1988). The importance of such suppliers is that they become the beginning of a supply chain, and without their ability to deliver raw materials on time, manufacturing companies cannot guarantee punctual deliveries of finished products to customers. This could significantly affect the motivation of employees when they consider they cannot achieve the goals of the company and theirs. Therefore, as a means to contribute to this discussion, the second hypothesis proposed states as it follows:

H2: *The implementation of lean manufacturing techniques along with a JIT philosophy has a direct and positive impact on the human resources of companies.*

Lean manufacturing techniques applied in the production process must—to any extent—impact on product quality. For instance, when companies produce in small lots, they force suppliers to deliver raw material in similar small quantities. This facilitates the inspection and audit of this material reaching the warehouses and consequently can lead to improved quality product (Absi et al. 2012; Lovell 2003; Kim and Ha 2003).

Similarly, total preventive maintenance programs can improve product quality, since properly calibrated and adjusted machines can reduce errors in the production process (Ahmad et al. 2012; Cua et al. 2001). However, although many authors discuss whether maintenance and quality are issues to consider separately, for others it is more convincing to believe that they are closely related aspects, since quality is the result of appropriate maintenance (Konecny and Thun 2011).

A reduction of work in process is also a JIT element considered by many as crucial to improve production processes. Companies can often struggle with great amounts of raw material being processed somewhere in the production process. For instance, feedstock is usually placed somewhere close to the machinery and equipment where it will be processed, resulting in obstructed aisles and hampered visibility that bring the impression of factories with little organization and planning. Moreover, this disorganization would affect quality indices due to an increasing handling of materials (Ohno et al. 2015; Bettayeb et al. 2014). Therefore, in order to support the relation between lean manufacturing techniques and quality product, the following hypothesis is proposed.

H3: *Lean manufacturing techniques applied in a JIT production process have a direct and positive impact on product quality.*

The organization for the improvement of material flows can group a product family in what might be called manufacturing cells (Pattanaik and Sharma 2009; Williams and David 1991), or a set of similar activities in what might be named group technology (Spencer 1998; Pourbabai 1988). However, companies must also seek to implement additional techniques that favor the flow of materials under such layout. Kanban, also known as the cards system (Sylvain and Duguay 1997; Reda 1987; Fiscus 1987), is a widely used technique that provides with exact instructions to operators on what must be performed with a certain component or subassembly. This can reduce errors within manufacturing cells and technology groups and thus improve quality of the final product (Liberopoulos and Koukoumialos 2005; Chan 2001). However, when problems or errors do arise within these plants, operators can be integrated and encourage to propose solutions to these issues. This is also a form of improving the production process and therefore the quality of the final product. Based on all information previously stated, it is possible to formulate the following hypothesis.

H4: *Production organization and material flow have a direct and positive effect on product quality within a JIT environment.*

Machinery and equipment of maquiladoras are usually designed and purchased overseas, and when these types of companies establish in the Mexican territory, some of their organization systems also arrive with a prearranged distribution of the physical space. This prearranged distribution was conceived according to the type of product to be manufactured. Nevertheless, the range of products often increases and several adjustments in the plant layout must be executed, which also involve a number of changes in human resources (Caggiano and Teti 2012; Yalcin 2004; Williams and David 1991).

Several techniques applied in a production process are executed simultaneously and can have an impact on product quality. For instance, as it was previously mentioned, Kanban system allows for a reduction of errors in the production process, and this can bring greater motivation to employees when they realize they are able to meet the goal of their companies (Liu et al. 2009; Ruckart and Burgess 2007). Similarly, Kaizen can improve product quality and material flow, since it aims to organize and integrate workers in the process of solving a specific problem of the production system. This promotes high levels of collaborative work and thus supports effective communication among operators, which often translates into a greater social coexistence (Knechtges and Decker 2014; Glover et al. 2011; Farris et al. 2009). Therefore, since it is considered that production organization can impact on human resources, the following hypothesis was proposed.

H5: *Production organization and material flow have a direct and positive impact on human resources within a JIT environment.*

Companies emerge and remain existent as the result of human resources and their skills and abilities, which is a reason that companies must strive to implement programs and techniques to integrate their workforces and make them feel an integral part of the organization (Rusu and Avasilcai 2014; Martínez-Jurado et al. 2014). Similarly, human resources are responsible for the flexibility and agility that companies wish to accomplish in their supply chains, since they can hold several positions and perform several different tasks so that the flow of material does not stop (Ragin-Skorecka 2014) and companies reach on-time deliveries. Moreover, authors have associated the capacity of innovation of companies with the quality circles established in them (Blaga and Jozsef 2014a), which generally results in an increased efficiency (Blaga and Jozsef 2014b; Lengnick-Hall et al. 2013).

As a conclusion, it has been stated that the success and performance of companies are largely due to the different roles of human resources (Farris et al. 2009; Power and Sohal 2000; Jayaram et al. 1999). Thus, the following hypothesis can be formulated.

H6: *The levels of satisfaction, motivation, and communication of human resources have a direct and positive effect on the quality indices of the product that is manufactured within a JIT environment.*

11.3.2 Results of Model 2

Figure 11.4 illustrates the results from the evaluation of this model according to methodology stated in Chap. 5. Every relation or hypothesis in the model shows a value of the beta parameter and a *P*-value from the statistical significance test to determine whether these hypotheses are accepted or rejected. Similarly, every dependent latent variable in the figure shows an R-squared value that measures its variability explained by independent latent variables.

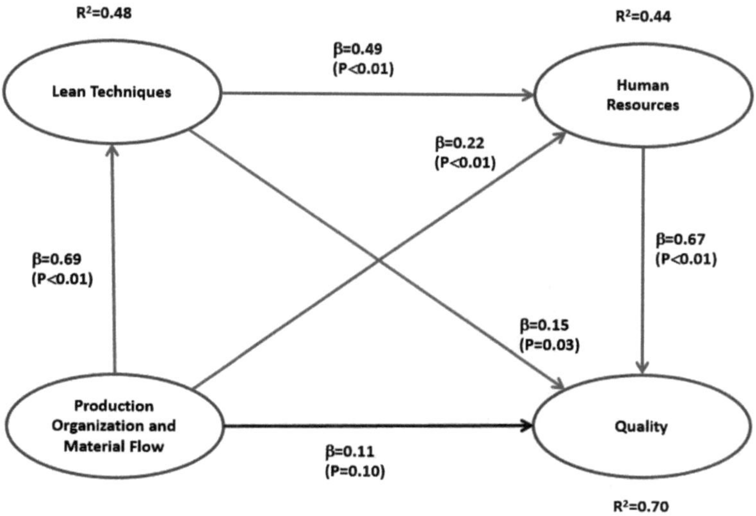

Fig. 11.4 Evaluated model 2—validation of hypotheses

The segment or hypothesis in red indicates that this relation between variables is not statistically significant at a 95 % confidence level. Its *P*-value is considerably high, while the beta value is visibly low. On the other hand, the remaining five relations were statistically significant, according to their *P*-value.

11.3.2.1 Efficiency Indices of Model 2

Before drawing conclusions from this second model, it is necessary to determine whether it meets certain indexes of goodness to accurately predict relations between variables.

1. Average path coefficient (APC) = 0.388, $P < 0.001$
2. Average R-squared (ARS) = 0.540, $P < 0.001$
3. Average adjusted R-squared (AARS) = 0.534, $P < 0.001$
4. Average block VIF (AVIF) = 1.935, acceptable if ≤5, ideally ≤3.3
5. Average full collinearity VIF (AFVIF) = 2.702, acceptable if ≤5, ideally ≤3.3
6. Tenenhaus GoF (GoF) = 0.580, small ≥0.1, medium ≥0.25, large ≥0.36
7. Sympson's paradox ratio (SPR) = 1.000, acceptable if ≥0.7, ideally = 1
8. R-squared contribution ratio (RSCR) = 1.000, acceptable if ≥0.9, ideally = 1
9. Statistical suppression ratio (SSR) = 1.000, acceptable if ≥0.7
10. Nonlinear bivariate causality direction ratio (NLBCDR) = 1.000, acceptable if ≥0.7

In APC, ARS, and AARS indexes their *P*-values to determine statistical significance are all lower than 0.05, which is the maximum value allowed for a 95 % confidence level. Therefore, relationships between variables are on average statistically significant and dependent latent variables have sufficient predictive validity.

Similarly, AFVIF and AVIF indexes demonstrate that there are no collinearity problems in latent variables analyzed and the model can be eventually interpreted. The remaining indexes are also suitable for the model. However, the Tenenhaus index in this model is 0.58, while the minimum value suggested is 0.36. This value is positively high and suitable for the model.

11.3.2.2 Coefficients of Latent Variables

Indices reported above indicate that the model is in general efficient and can be used to interpret the relations between variables. However, it is also necessary to analyze latent variables independently. Table 11.8 shows validity indexes for each of these variables.

- All R-squared and adjusted R-squared values are higher than 0.2; thus, all dependent latent variables have predictive validity from a parametric point of view. Similarly, Q-squared values of all variables are higher than 0 and close to R-squared and adjusted R-squared values. From a nonparametric point of view, all dependent latent variables have predictive validity.
- As for internal validity and reliability, it is observed that Cronbach's alpha and composite reliability indexes are higher than 0.7 in all variables analyzed. This indicates that all latent variables are properly integrated with their items.
- Value of AVES index for every latent variable is higher than 0.5—the minimum acceptable cutoff value. Therefore, it is that all latent variables analyzed have sufficient convergent validity and can be discussed.
- According to the index of variance inflation in all variables, there are no collinearity problems among them.

11.3.2.3 Direct Effects—Validation of Hypotheses of Model 2

Hypotheses initially raised in Fig. 11.3 were evaluated according to the established methodology, resulting in the values shown in Fig. 11.4 for the analyzed direct effects. The following conclusions can be stated from this last figure. Hypothesis 1 (H1) was discussed in the previous model.

Table 11.8 Coefficients of latent variables of model 2

To	From			
	Lean techniques	Production organization and material flow	Quality	Human resources
R-squared	0.481		0.702	0.437
Adj. R-squared	0.477		0.696	0.429
Composite reliability	0.907	0.886	0.885	0.887
Cronbach's alpha	0.88	0.838	0.805	0.841
AVE	0.553	0.609	0.72	0.612
FVIF	2.298	1.997	3.326	3.184
Q-squared	0.485		0.705	0.439

- H2: There is enough statistical evidence to declare that lean manufacturing techniques used along a JIT program have a direct and positive impact on human resources, because when the independent latent variable increases its standard deviation by one unit, the standard deviation of the dependent latent variable increases by 0.49 units.
- H3: There is sufficient statistical evidence to state that lean manufacturing techniques applied simultaneously with a JIT production process have a direct and positive impact on the quality of finished products, since when the independent latent variable increases its standard deviation by one unit, the standard deviation of the dependent latent variable increases by 0.15 units.
- H4: There is not enough statistical evidence to declare that production organization and material flow have a direct and positive effect on product quality in a JIT environment. The *P*-value obtained for the statistical test of the estimated parameter is greater than 0.05, the maximum acceptable value for a 95 % confidence level.
- H5: There is sufficient statistical evidence to point out that production organization and material flow have a direct and positive impact on human resources in a JIT environment. When the first latent variable increases its standard deviation by one unit, the standard deviation of the second latent variable rises by 0.22 units.
- H6: There is enough statistical evidence to declare that levels of satisfaction, motivation, and communication among human resources have a direct and positive effect on product quality indexes in a JIT environment, since when the first latent variable increases its standard deviation by one unit, the standard deviation of the second variable increases by 0.67 units.

All latent dependent variables have a R-squared value that indicates the percentage of their variability explained by independent latent variables. However, some of these dependent latent variables are explained by more than one independent latent variable. Thus, it is important to apportion their total direct effect and discuss the percentage of direct effect that is explained through the different independent latent variables. Table 11.9 shows the effect sizes that independent latent variables have on dependent latent variables.

Figure 11.4 and data from Table 11.6 allow for the statement of the following conclusions:

- Latent variable Lean Techniques is explained in 48.1 % by latent variable Production Organization and Material Flow.

Table 11.9 Effect sizes of model 2

	Lean techniques	Production organization and material flow	Human resources
Lean techniques		0.481	
Quality	0.098	0.059	0.545
Human resources	0.316	0.12	

- Motivation, increased teamwork, and communication among human resources are explained in 44 % by two latent variables, since the effect size is 0.44. Therefore, 32.6 % of this variability is explained by latent variable Lean Techniques, while 12.4 % originates from Production Organization and Material Flow. Thus, it seems that lean techniques are more significant to increase workers motivation and collaborative work among human resources.
- Quality of a product is explained in 70 % by three latent variables, since the size effect is 0.70. Thus, Human Resources explains 54.5 % of this variability, 5.9 % originates from Production Organization and Material Flow, and 9.8 % of this variability can be explained by Lean Techniques employed in companies. Hence, it can be stated that human resources are the most significant factor to achieve quality in a production system.

11.3.2.4 Sum of Indirect Effects of Model 2

As shown in Fig. 11.4, it is possible that latent variables associated with JIT elements have an effect on latent variables with benefits through mediator variables. Table 11.10 illustrates the sum of the indirect effects between latent variables, the *P*-value of the statistical significance test of the parameter, and the effect size or variability that could be explained.

According to Table 11.10, all indirect effects are statistically significant, since the *P*-value for every one is lower than 0.05, the maximum accepted value for a 95 % confidence level. Moreover, it is possible to draw the following conclusions:

- Lean Techniques has an indirect effect on quality obtained in a product, since when the first latent variable increases its standard deviation by one unit, the standard deviation of the second latent variable increases by 0.328 units. Moreover, the first latent variable can explain 21.4 % of the variability of the second latent variable, since the effect size is 0.214.
- Latent variable Production Organization and Material Flow has a total indirect effect on quality of products, since when the first latent variable increases its standard deviation by one unit, the standard deviation of the second latent variable increases by 0.477 units. Moreover, the first latent variable can explain up to 26.8 % of variability of quality, since the effect size is 0.268.

Table 11.10 Sum of indirect effects of model 2

To	From	
	Lean techniques	Production organization and material flow
Lean techniques		
Quality	0.328 ($P < 0.001$) ES = 0.214	0.477 ($P < 0.001$) ES = 0.268
Human resources		0.342 ($P < 0.001$) ES = 0.189

11.3.2.5 Total Effects of Model 2

Table 11.11 introduces the total reported in this second model. Similar to previous analysis, every total effect shows the *P*-value of the statistical significance test, and the effect sizes are also shown to demonstrate the variability of a dependent latent variable explained from an independent latent variable. The same table also allows for the following interpretation of data.

- All total effects are statistically significant, since in all cases *P*-value is lower than 0.05, the maximum allowed value for a 95 % confidence level of 95 %.
- Four total effects are relatively high, since their values are higher than 0.5. Three of these concerned effects of latent variable Production Organization and Material Flow over the remaining latent variables, while one total effect refers to latent variable Human Resources over quality obtained in the final product.
- Latent independent variable Production Organization and Material Flow has the largest total effect over Lean Techniques. When the first latent variable increases its standard deviation by one unit, the standard deviation of the second latent variable increases by 0.693 units. Moreover, the former explains up to 48.1 % of the variability of the latter. The effect size is 0.481 units.
- The second largest total effect is caused by independent latent variable Human Resources over latent variable Quality of the product, since when the former increases its standard deviation by one unit, the standard deviation of the latter increases 0.666 units. Moreover, the first latent variable explains up to 54.5 % of variability of the second latent variable, since the effect size is 0.545 units.
- The third most significant total effect is caused by independent latent variable Production Organization and Material Flow over latent variable Quality. When the first latent variable increases its standard deviation by one unit, the standard deviation of the second latent variable increases by 0.582 units. Moreover, the former explains up to 32.8 % of the variability of the latter, since the value of the effect size is 0.328.
- The fourth most significant total effect is caused by the same independent latent variable Organization Production and Material Flow over latent dependent variable Human Resources. When the former increases its standard deviation by one unit, the standard deviation of the latter increases by 0.56 units. Also, the first latent variable explains 30.9 % of variability of the second latent variable, since the value of the effect is 0.309. Similar interpretations can be concluded for the remaining relations between variables.

Table 11.11 Total effects of model 2

	Lean techniques	Production organization and material flow	Human resources
Lean techniques		0.693 ($P < 0.001$) ES = 0.481	
Quality	0.479 ($P < 0.001$) ES = 0.312	0.582 ($P < 0.001$) ES = 0.328	0.666 ($P < 0.001$) ES = 0.545
Human resources	0.493 ($P < 0.001$) ES = 0.316	0.56 ($P < 0.001$) ES = 0.309	

11.3.2.6 Conclusions of Model 2

This model associated four latent variables, from which two of them belonged to JIT elements for production process, while the other two concerned benefits obtained after the implementation of JIT in production lines. It is assumed that Production Organization and Material Flow is the independent latent variable, while Quality of the final product was considered the dependent latent variable, since most latent variables had an effect on it. Moreover, from the six hypotheses initially raised, five were new and one (H1) had already been tested and discussed in the previous model. Finally, four from these new hypotheses were statistically significant and accepted, while one was statistically not significant and rejected.

Based on the results, it was proved that latent variable Production Organization and Material Flow is one of the most significant from a statistical point of view, since it has an effect on all others. However, its direct relation with Quality was not statistically significant, but indirect effects occurred between them through latent variable Lean Techniques and Human Resources. Therefore, production organization and material flow can have an impact on product quality if human resources properly implement and apply lean manufacturing techniques. Human resources here are crucial variable, since workforce is responsible for generating quality.

Finally, it was also observed that lean manufacturing techniques have a slightly low direct impact on quality; however, the total effect that the former causes on the latter is visibly high due to an indirect effect caused through human resources. Therefore, it can be stated that human resources are crucial to achieve the objectives of product quality.

11.4 Model 3: JIT Elements: Production Organization and Material Flow and Production Capacity and Inventory Management. Benefits: Material Handling and Economic Performance

This model also integrates four latent variables, from which two relate to JIT elements for the production process, while the other two latent variables concern benefits after a JIT implementation in production lines. Latent variables associated with JIT elements are:

- Production Organization and Material Flow
- Production Capacity and Inventory Management

On the other hand, latent variables associated with JIT benefits are:

- Material Handling
- Economic Performance

Six working hypotheses were proposed from the relation of these four latent variables. None of them has been previously analyzed. Moreover, this model

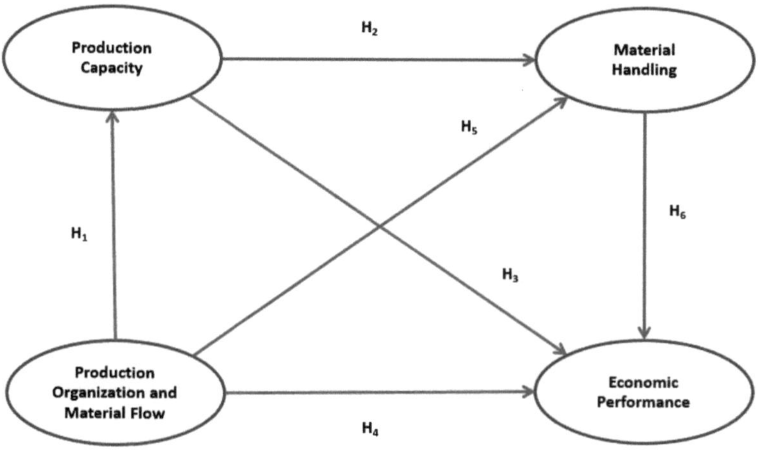

Fig. 11.5 Proposed model 3

assumes that surveyed maquiladoras have a defined and previously established organization of production and material handling and, as a consequence, all applied techniques and production schedules are based on production capacity of companies. Therefore, this latent variable is regarded as independent with an impact on all the others. Economic Performance is then considered the dependent variable since all other latent variables have a direct or indirect effect on it. Figure 11.5 depicts the model proposed with its six working hypotheses.

11.4.1 Proposed Model 3

Production planning is based on the production capacity and plant layout of every company, since when manufacturers produce above their capacity, they can compromise on-time delivery dates (Špicar and Januška 2015; Dong et al. 2015), or must outsource a portion of the amount order requested by the client and assume the risks it may entails (Gyulai et al. 2014). Thus, several manufacturing companies prefer scheduling production below their capacity to be able to fulfill the requested orders and avoid possible dissatisfaction from customers. However, since changes in the production system from one model to another can be expensive, companies must rely on high runs to prevent regular design changes, reduce the cost of production, and offer customers attractive discounts (Wu and Wu 2015; Jiang and Seidmann 2014).

Maquiladoras surveyed for this research are characterized by being highly specialized factories (Alcaraz et al. 2014; Sargent and Matthews 2009) that rely on advanced manufacturing technologies, either hard or soft. For instance, robots are employed for those movements of materials that can be dangerous or nonoptimal for people, while as for soft technologies, these companies generally implement systems of continuous improvement or Kaizen (Oropesa-Vento et al. 2015) and card systems or Kanban (Panayiotou and Cassandras 1999), which help improve the management of inventory in process and streamline the flow of materials. The implementation of these lean manufacturing techniques to support production processes is based on their production organization. Therefore, to support the relation between production organization and inventory management, it is possible to propose the following hypothesis.

H1: *Production organization and material flow have a positive and direct effect on production capacity and inventory management in a JIT environment.*

In addition to ensuring on-time deliveries of final product to customers, (Lagemann and Meier 2014; van der Laan et al. 1999), scheduling below production capacity can help reduce inventories and administrative costs due to high levels of inventory turnover (Kim and Ha 2003). Similarly, it allows companies to purchase small lots, which provides a direct—and perhaps closer—relationship with suppliers who must be integrated into the production system (David and Eben-Chaime 2003).

Moreover, although just-in-time philosophy focuses on minimizing inventories along the entire production system, one of its important elements is maintaining a safety stock that would help companies cope with possible sudden fluctuations in the demand (Amit et al. 2015; Hong et al. 2015) and prevent them from losing or rejecting an order due to missing parts or components. Similarly, the use of standardized containers for raw material components is crucial for JIT programs (Ji et al. 2015; Dong et al. 2015; Myung and Moon 2014), since it aims to minimize materials handling and accidents that might occur (Lortie 2012; Niskanen and Lauttalammi 1989), especially from mismanagement of raw materials or when they were not properly packaged. Therefore, in order to contribute to this discussion, the following working hypothesis was formulated.

H2: *Production capacity and inventory management have a direct and positive effect on benefits associated with materials handling in a successful JIT environment.*

It is traditionally stated that one of the major advantages of scheduling below installed capacity is that companies do not lose costumers due to late deliveries. However, the economic impact of this approach has been rarely analyzed, although it may be important (Pan and Nguyen 2015). That is, research has emphasized on the fact that companies do not lose costumers but little has been discussed about the economic costs of losing these costumers. Moreover, when the impact of safety stock on sudden changes in demand is analyzed, research tends to study the number of accepted orders rather than those rejected due to a lack of raw materials. Therefore, it may be suitable to direct research toward an analysis of the marginal profit of companies when they refuse these orders (Liberopoulos and Koukoumialos 2005; Lemak and Reed 1997).

The level of expertise of companies is also another important element to determine the production capacity to establish. Highly specialized industries are also high-technology industries that are suitable for mass production or to manufacture large lots or products with little variation (Teagarden et al. 1992). This favors the reduction of operating, production, and labor costs mainly due to automated machines found in the production line (Villa and Taurino 2013; Amasaka 2007), which also help minimize accidents of material handling, since robots become responsible for its management (Niskanen and Lauttalammi 1989). Based on this discussion, it is possible to propose the following hypothesis.

H3: *Production below capacity installed and inventory management have a direct and positive impact on the economic performance of the company in a successful JIT environment.*

Organizations seek to reduce material handling since it is a major source of accidents and fatal casualties (Nenonen 2011; Argilés-Bosch et al. 2014). Moreover, disability payments to workers can be high (Hajakbari and Minaei-Bidgoli 2014; Chinniah 2015; Fernández and Pérez 2015). Therefore, the planned flow of material must have an economic impact on the company; otherwise, it would not be justified unless it had implications associated with the health and safety of workers. Similarly, the cards system or Kanban can have an economic impact on companies, since it supports the flow of materials and thus increases inventory turnover. This allows for the proposal of two hypotheses.

H4: *Production organization and material flow have a direct and positive impact on the indexes of economic performance indices of companies within successful JIT environment.*
H5: *Production organization and material flow have a direct and positive effect on the reduction of inventories and accidents associated with materials handling. Moreover, they strengthen relationships between suppliers and the manufacturing companies and increases inventory turnover.*

Policies for inventory reduction that minimize material handling, favor the purchase of small lots, and encourage close relationships with suppliers, are also implemented since companies can obtain certain economic benefits. For instance, authors have widely discussed the economic impact of transmission and distribution systems and have pointed out that appropriate deliveries of raw materials and finished products can bring economic advantages (Arıkan et al. 2014; Schaefer and Konur 2015).

Inside companies, inventory reduction decreases the costs of its maintenance (Kouki and Jouini 2015; Avelar-Sosa et al. 2015). Similarly, effective relationships with suppliers may encourage collaboration to face uncertainty in demand (Chen and Jeter 2008; Humphreys et al. 2003; David and Eben-Chaime 2003), which can represent larger amounts of production orders accepted and not rejected due to missing components, which also increases economic performance. Likewise, successful relationships with suppliers allow for the procurement and purchase of small-sized lots (Kwak et al. 2006; David and Eben-Chaime 2003; Dong et al. 2001;

De Toni et al. 2000), which represent low inventory levels and a reduction in costs. Finally, little material handling implies that raw material is transported and moved only when it is necessary, since this transportation of raw materials does not add any value to the final product. Thus, raw materials should be moved only when they must be incorporated into a previously established activity. Kanban can certainly be useful in these cases. As a means to contribute to this discussion, the following hypothesis is proposed.

H6: *Benefits obtained in material handling in a JIT environment have a direct and positive impact on the economic performance of companies.*

11.4.2 Results of Model 3

Figure 11.6 introduces the results obtained after the evaluation of the model according to methodology described in Chap. 5. The two red segments indicate that these hypotheses or relations are statistically not significant, since their *P*-value of the statistical significance test was higher than 0.05, the maximum acceptable value for a 95 % confidence level.

11.4.2.1 Efficiency Indexes of Model 3

Before describing the direct and indirect effects of this model, it is necessary to determine whether it meets certain indexes of goodness of fit to accurately determine the existing relations between variables. Indexes are described below.

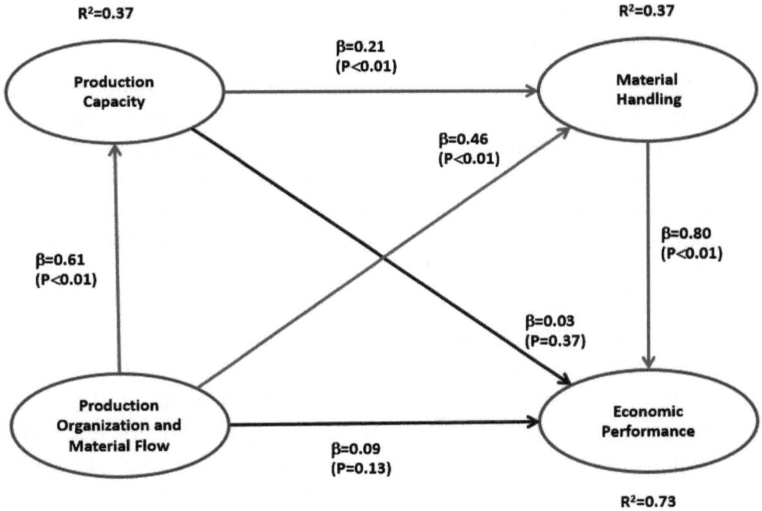

Fig. 11.6 Evaluated model 3—validation of hypotheses

1. Average path coefficient (APC) = 0.366, $P < 0.001$
2. Average R-squared (ARS) = 0.490, $P < 0.001$
3. Average adjusted R-squared (AARS) = 0.483, $P < 0.001$
4. Average block VIF (AVIF) = 1.603, acceptable if ≤5, ideally ≤3.3
5. Average full collinearity VIF (AFVIF) = 2.602, acceptable if ≤5, ideally ≤3.3
6. Tenenhaus GoF (GoF) = 0.549, small ≥0.1, medium ≥0.25, large ≥0.36
7. Sympson's paradox ratio (SPR) = 1.000, acceptable if ≥0.7, ideally = 1
8. R-squared contribution ratio (RSCR) = 1.000, acceptable if ≥0.9, ideally = 1
9. Statistical suppression ratio (SSR) = 1.000, acceptable if ≥0.7
10. Nonlinear bivariate causality direction ratio (NLBCDR) = 1.000, acceptable if ≥0.7

As for APC, ARS, and AARS indexes, we can conclude that the model has sufficient predictive validity, since *P*-values in these three cases were lower than 0.05, the maximum value accepted for a 95 % confidence level. Similarly, AVIF and AFVIF indexes show values lower than 3.3 units. Thus, it is concluded that there are no problems of collinearity between latent variables analyzed. In addition, note that the goodness of fit index of Tenenhaus has is considerably higher than 0.36, the minimum recommended value. Similar interpretations can be offered for the remaining indexes, which indicates that the model is appropriate and can be interpreted.

11.4.2.2 Coefficients of Latent Variables

Indexes above indicate that the model is in general efficient. However, it is also important to analyze latent variables separately. Thus, Table 11.12 shows validity indexes for every latent variable and allows for the interpretation of these indexes.

First, all latent variables show R-squared and adjusted R-squared values higher than 0.02, that is, the minimum acceptable value. Therefore, from a parametric point of view, all latent variables have enough predictive validity. Similarly, all

Table 11.12 Coefficient of latent variables—model 3

To	From			
	Capacity	Production organization and material flow	Material handling	Economic performance
R-squared	0.37		0.733	0.366
Adj. R-squared	0.366		0.727	0.357
Composite reliability	0.836	0.886	0.894	0.922
Cronbach's alpha	0.739	0.838	0.851	0.898
AVE	0.562	0.609	0.628	0.664
AFVIF	1.516	1.751	3.457	3.686
Q-squared	0.367		0.712	0.37

latent variables report Q-squared values higher than 0 and similar to R-squared values, which indicates that they have predictive validity from a nonparametric point of view.

Also, composite reliability and Cronbach's Alpha indexes in all cases show values higher than 0.7, the minimum acceptable value. Therefore, all latent variables of this model have enough internal reliability. Furthermore, AVE index in all cases is higher than 0.5. Thus, every latent variable has sufficient convergent validity. Finally, AVIF index shows values lower than 3.3, which indicates that there are no collinearity problems among variables analyzed.

11.4.2.3 Direct Effects

Figure 11.6 depicts the model proposed after its analysis and enables to interpret it as it follows:

- H1: There is enough statistical evidence to declare that production organization and material flow have a direct positive effect on production capacity and inventory management within a JIT environment. When the first latent variable increases its standard deviation by one unit, the standard deviation of the second latent variable rises by 0.61 units.
- H2: There is sufficient statistical evidence to declare the production capacity and inventory management have a direct and positive effect on benefits of material handling in a successful JIT system, since when the first latent variable increases its standard deviation by one unit, the standard deviation of the second latent variable increases by 0.21 units.
- H3: There is not enough statistical evidence to declare that production capacity and inventory management have a direct and positive impact on the economic performance of companies in a successful JIT environment. The *P*-value obtained from the statistical significance test over the hypothesis was higher than 0.05, the maximum accepted value for a 95 % confidence level.
- H4: There is not enough statistical evidence to state that production organization and material flow have a direct and positive impact on the indexes of economic performance indexes of companies within a successful JIT implementation. The *P*-value obtained from the statistical significance test over the hypothesis exceeds 0.05, the maximum value allowed for a 95 % confidence level.
- H5: There is enough statistical evidence to point out that production organization and material flow have a direct and positive effect on the following aspects of Production Capacity and Inventory Management: reducing inventories and accidents, strengthening relationships between manufacturers and suppliers, and increasing inventory rotation. When the first latent variable increases its standard deviation by one unit, the standard deviation of the second latent variable rises by 0.46 units.
- H6: There is enough statistical evidence to state that benefits of material handling in a JIT environment have a direct and positive impact on the economic

Table 11.13 Effect sizes of model 3

	Capacity	Production organization and material flow	Material handling
Capacity		0.37	
Economic performance	0.011	0.05	0.672
Material handling	0.1	0.266	

performance of companies, since when the first latent variable increases its standard deviation by one unit, the standard deviation of the second latent variable increases by 0.80 units.

Table 11.13 shows the effect sizes or amount of variability of dependent latent variables that is explained by independent variables. The table also allows for the interpretation of data.

- Latent variable called Production Capacity and Inventory Management is explained in 37 % by latent variable Production Organization and Material Flow.
- Latent variable Material Handling is explained in 37 % by two variables. Thus, Production Capacity and Inventory Maintenance can explain 10 %, while Production Organization and Material Flow explains 27 % of this variability.
- Latent variable Economic Performance can be 73 % explained by three latent variables. Thus, Production Capacity and Inventory Management explain 1.1 % of variability, while 5 % originates from Production Organization and Material Flow, and latent variable Material Handling can explain 67.2 %. Therefore, it can be stated that Material Handling has the most significant impact on the economic performance.

11.4.2.4 Sum of Indirect Effects of Model 3

Table 11.14 shows the sum of indirect effects of Model 3. All the estimated parameters are statistically significant, since the *P*-value in all cases is lower than 0.05, the maximum allowed value for a 95 % confidence level. The table also allows for the following conclusions:

Table 11.14 Sum of indirect effects of model 3

To	From	
	Capacity	Production organization and material flow
Economic performance	0.167 ($P = 0.002$) ES = 0.065	0.485 ($P = 0.004$) ES = 0.262
Material handling		0.127 ($P < 0.001$) ES = 0.074

- The largest indirect effect is caused by latent variable Production Organization and Material Flow over latent variable Economic Performance. When the first increases its standard deviation by one unit, the standard deviation of the second increases by 0.485 units. Moreover, the former latent variable explains up to 26.2 % of the variability of the latter.
- Production Organization and Material Flow has an indirect positive effect on latent variable Material Handling, because when the first latent variable increases its standard deviation by one unit, the standard deviation of the second latent variable increases by 0.127 units. Also, the first latent variable explains up to 7.4 % of variability of the second latent variable.
- Production Capacity and Inventory Management has an indirect and positive effect on Economic Performance, since when the first latent variable increases its standard deviation by one unit, the standard deviation of the second latent variable rises by 0.167 units. First latent variable can also explain up to 6.5 % of variability of the second latent variable.

11.4.2.5 Total Effects of Model 3

Table 11.15 shows the total effects from relations between variables and, according to it, all total effects are statistically significant, since the *P*-value of all estimated parameters was lower than 0.05, the maximum value accepted for a 95 % confidence level. The table also allows for the following interpretation of data:

- The largest total effect is also a direct effect caused by latent variable Material Handling over latent variable Economic Performance, since when the first latent variable increases its standard deviation by one unit, the standard deviation of the second variable increases by 0.8 units. This is also one of the highest values in the analyzed models, and the first latent variable explains 67 % of variability of the second latent variable.
- The second largest total effect is also a direct effect occurring from latent variable Production Organization and Material Flow over latent variable Production Capacity and Inventory Management, since when the former increases its standard deviation by one unit, the standard deviation of the latter

Table 11.15 Total effects of model 3

TO	From			
	Capacity	Production organization and material flow	Economic performance	Material handling
Capacity		0.608		
Economic performance	0.195	0.578		0.8
Material handling	0.209	0.585		

increases 0.608 units. Moreover, first latent variable explains 37 % of variability of second latent variable.

- The third largest total effect is caused by latent variable Production Organization and Material Flow over Economic Performance of companies, since when the first latent variable increases its standard deviation by one unit, the standard deviation of the second variable increases by 0.578 units. Moreover, the former latent variable accounts for 31.2 % of variability of the latter latent variable.
- The fourth total effect, according to its size, is caused by latent variable Production Organization and Material Flow over latent variable Material Handling, since when the first latent variable increases its standard deviation by one unit, the standard deviation of the second latent variable increases by 0.585 units. Moreover, the former latent variable explains up to 34 % of variability of the latter. Similar interpretations can be offered for the remaining relations between variables.

11.4.2.6 Conclusions of Model 3

This model associated four latent variables; two concerned JIT elements for the production process and two referred to benefits obtained from a successful JIT implementation. The results from the evaluation of the model, its hypotheses, and the relation among latent variables allow for the following final conclusions:

- Both latent variables Production Organization and Material Flow and Production Capacity and Inventory Management have no direct and positive effect on the financial performance of the company. The impacts are rather indirect through benefits associated with handling of materials, such as reduced material handling, close relationships between suppliers and manufacturers, increased inventory turnovers, and reduced lot sizes.
- Companies should always seek for benefits associated with material handling, since they have a direct and positive effect on their economic performance in a JIT environment.

References

Absi N, Dauzère-Pérès S, Kedad-Sidhoum S (2012) Just-in-time planning and lot-sizing. In: Ríos-Mercado RZ, Ríos-Solís YA (eds) Just-in-time systems. Springer New York, pp 191–207

Ahmad MF, Zakuan N, Jusoh A, Takala J (2012) Relationship of TQM and business performance with mediators of SPC, lean production and TPM. Procedia Soc Behav Sci 65:186–191. doi:10.1016/j.sbspro.2012.11.109

Al-Tahat MD, Mukattash AM (2006) Design and analysis of production control scheme for Kanban-based JIT environment. J Franklin Inst 343(4–5):521–531. doi:10.1016/j.jfranklin.2006.01.001

Alcaraz JLG, Maldonado AA, Iniesta AA, Robles GC, Hernández GA (2014) A systematic review/survey for JIT implementation: Mexican maquiladoras as case study. Comput Ind 65 (4):761–773. doi:http://dx.doi.org/10.1016/j.compind.2014.02.013

Almomani MA, Aladeemy M, Abdelhadi A, Mumani A (2013) A proposed approach for setup time reduction through integrating conventional SMED method with multiple criteria decision-making techniques. Comput Ind Eng 66(2):461–469. doi:10.1016/j.cie.2013.07.011

Amasaka K (2007) Applying New JIT—Toyota's global production strategy: epoch-making innovation of the work environment. Robot Comput Integr Manuf 23(3):285–293. doi:10.1016/j.rcim.2006.02.001

Amit RK, Mehta P, Tripathi RR (2015) Optimal shelf-space stocking policy using stochastic dominance under supply-driven demand uncertainty. Eur J Oper Res 246(1):339–342. doi:10.1016/j.ejor.2015.04.031

Argilés-Bosch JM, Martí J, Monllau T, Garcia-Blandón J, Urgell T (2014) Empirical analysis of the incidence of accidents in the workplace on firms' financial performance. Saf Sci 70:123–132. doi:10.1016/j.ssci.2014.05.012

Arıkan E, Fichtinger J, Ries JM (2014) Impact of transportation lead-time variability on the economic and environmental performance of inventory systems. Int J Prod Econ 157:279–288. doi:10.1016/j.ijpe.2013.06.005

Avelar-Sosa L, García-Alcaraz J, Vergara-Villegas OO, Maldonado-Macías AA, Alor-Hernández G (2015) Impact of traditional and international logistic policies in supply chain performance. Int J Adv Manuf Technol 76(5–8):913–925. doi:10.1007/s00170-014-6308-3

Bettayeb B, Bassetto SJ, Sahnoun M (2014) Quality control planning to prevent excessive scrap production. J Manuf Syst 33(3):400–411. doi:10.1016/j.jmsy.2014.01.001

Blaga P, Jozsef B (2014a) Human resources, quality circles and innovation. Procedia Econ Finan 15:1458–1462. doi:10.1016/S2212-5671(14)00611-X

Blaga P, Jozsef B (2014b) Increasing human resource efficiency in the production process. Procedia Technol 12:469–475. doi:10.1016/j.protcy.2013.12.516

Caggiano A, Teti R (2012) Digital manufacturing cell design for performance increase. Procedia CIRP 2:64–69. doi:10.1016/j.procir.2012.05.041

Calvo-Mora A, Ruiz-Moreno C, Picón-Berjoyo A, Cauzo-Bottala L (2014) Mediation effect of TQM technical factors in excellence management systems. J Bus Res 67(5):769–774. doi:10.1016/j.jbusres.2013.11.042

Chakravorty SS (2009) Improving distribution operations: implementation of material handling systems. Int J Prod Econ 122(1):89–106. doi:10.1016/j.ijpe.2008.12.026

Chan FTS (2001) Effect of Kanban size on just-in-time manufacturing systems. J Mater Process Technol 116(2–3):146–160. doi:10.1016/S0924-0136(01)01022-6

Chen H, Jeter D (2008) The role of auditing in buyer-supplier relations. J Contemp Account Econ 4(1):1–17. doi:10.1016/S1815-5669(10)70027-8

Chen Y, Xiao Q, Tang X (2011) Product layout optimization and simulation model in a multi-level distribution center. Syst Eng Procedia 2:300–307. doi:10.1016/j.sepro.2011.10.042

Chiarini A (2013) The main methods of lean organization: Kanban, cellular manufacturing, SMED and TPM. In: Lean organization: from the tools of the toyota production system to lean office. Springer, Milan, pp 81–116

Chinniah Y (2015) Analysis and prevention of serious and fatal accidents related to moving parts of machinery. Saf Sci 75:163–173. doi:10.1016/j.ssci.2015.02.004

Chlivickas E (2014) International cooperation and innovations for developing human resources system. Procedia Soc Behav Sci 110:276–283. doi:10.1016/j.sbspro.2013.12.871

Crumpton MA (2015) 3—Human resources as a strategic partner. In: Crumpton MA (ed) Strategic human resources planning for academic libraries. Chandos Publishing, Oxford, pp 23–29

Cua KO, McKone KE, Schroeder RG (2001) Relationships between implementation of TQM, JIT, and TPM and manufacturing performance. J Oper Manage 19(6):675–694. doi:10.1016/S0272-6963(01)00066-3

David I, Eben-Chaime M (2003) How far should JIT vendor–buyer relationships go? Int J Prod Econ 81–82:361–368. doi:10.1016/S0925-5273(02)00274-8

De Toni A, Nassimbeni G (2000) Just-in-time purchasing: an empirical study of operational practices, supplier development and performance. Omega 28(6):631–651. doi:http://dx.doi.org/10.1016/S0305-0483(00)00016-5

Deep K, Singh PK (2015) Design of robust cellular manufacturing system for dynamic part population considering multiple processing routes using genetic algorithm. J Manuf Syst 35:155–163. doi:10.1016/j.jmsy.2014.09.008

Dong Y, Carter CR, Dresner ME (2001) JIT purchasing and performance: an exploratory analysis of buyer and supplier perspectives. J Oper Manage 19(4):471–483. doi:10.1016/S0272-6963(00)00066-8

Dong JX, Lee CY, Song DP (2015) Joint service capacity planning and dynamic container routing in shipping network with uncertain demands. Transp Res Part B Methodol 78:404–421. doi:10.1016/j.trb.2015.05.005

Ertay T, Ruan D, Tuzkaya UR (2006) Integrating data envelopment analysis and analytic hierarchy for the facility layout design in manufacturing systems. Inf Sci 176(3):237–262. doi:10.1016/j.ins.2004.12.001

Farris JA, Van Aken EM, Doolen TL, Worley J (2009) Critical success factors for human resource outcomes in Kaizen events: an empirical study. Int J Prod Econ 117(1):42–65. doi:10.1016/j.ijpe.2008.08.051

Fernández FB, Pérez MÁS (2015) Analysis and modeling of new and emerging occupational risks in the context of advanced manufacturing processes. Procedia Eng 100:1150–1159. doi:10.1016/j.proeng.2015.01.478

Fiscus, JE (1987) Kanban and just-in-time at Toyota: Edited by the Japanese management association. Eng Manage Int 4(2):157. (trans: David JL. Productivity Press, Cambridge, MA, U.S.A., 1986. 192 pages, 40 figures, hardcover. ISBN 0-915299-08-0. List price: $29.95). doi:http://dx.doi.org/10.1016/0167-5419(87)90022-6

Fukukawa T, Hong SC (1993) The determination of the optimal number of Kanbans in a just-in-time production system. Comput Ind Eng 24(4):551–559. doi:10.1016/0360-8352(93)90197-6

Gilbert FW, Young JA, O'Neal CR (1994) Buyer-seller relationships in just-in-time purchasing environments. J Bus Res 29(2):111–120. doi:10.1016/0148-2963(94)90015-9

Glover WJ, Farris JA, Van Aken EM, Doolen TL (2011) Critical success factors for the sustainability of Kaizen event human resource outcomes: an empirical study. Int J Prod Econ 132(2):197–213. doi:10.1016/j.ijpe.2011.04.005

Gyulai D, Kádár B, Monostori L (2014) Capacity planning and resource allocation in assembly systems consisting of dedicated and reconfigurable lines. Procedia CIRP 25:185–191. doi:10.1016/j.procir.2014.10.028

Hajakbari MS, Minaei-Bidgoli B (2014) A new scoring system for assessing the risk of occupational accidents: a case study using data mining techniques with Iran's Ministry of Labor data. J Loss Prev Process Ind 32:443–453. doi:10.1016/j.jlp.2014.10.013

Hong YS, Huh WT, Kang C (2015) Sourcing assemble-to-order inventories under supplier risk uncertainty. Omega. doi:http://dx.doi.org/10.1016/j.omega.2015.06.011

Hou TH, Hu WC (2011) An integrated MOGA approach to determine the Pareto-optimal Kanban number and size for a JIT system. Expert Syst Appl 38(5):5912–5918. doi:10.1016/j.eswa.2010.11.032

Huang CH, Lin PY (2014) Influence of spatial layout on airflow field and particle distribution on the workspace of a factory. Building Environ 71:212–222. doi:10.1016/j.buildenv.2013.09.014

Huang CY, Huang TS, Chen WL (2009) Communication protocols for order management in collaborative manufacturing. Int J Prod Econ 122(1):257–268. doi:10.1016/j.ijpe.2009.06.004

Humphreys P, Shiu WK, Lo VHY (2003) Buyer–supplier relationship: perspectives between Hong Kong and the United Kingdom. J Mater Process Technol 138(1–3):236–242. doi:10.1016/S0924-0136(03)00078-5

Huq Z (1999) Conventional shop control procedures to approximate JIT inventory performance in a job shop. J Manuf Syst 18(3):161–174. doi:10.1016/S0278-6125(99)80029-6

Huq Z, Huq F (1994) Embedding JIT in MRP: the case of job shops. J Manuf Syst 13(3):153–164. doi:10.1016/0278-6125(94)90001-9

Huumonen J (2011) Conceptualizing agility of enterprises. Hum Factors Ergon Manuf 21(2):132–146. doi:10.1002/hfm.20255

Inman R, Anthony RS, Sale KW, Jr G, Whitten D (2011) Agile manufacturing: relation to JIT, operational performance and firm performance. J Oper Manage 29(4):343–355. doi:10.1016/j.jom.2010.06.001

Jayaram J, Droge C, Vickery SK (1999) The impact of human resource management practices on manufacturing performance. J Oper Manage 18(1):1–20. doi:10.1016/S0272-6963(99)00013-3

Jensen JB, Malhotra MK, Philipoom PR (1996) Machine dedication and process flexibility in a group technology environment. J Oper Manage 14(1):19–39. doi:10.1016/0272-6963(95)00030-5

Ji M, Guo W, Zhu H, Yang Y (2015) Optimization of loading sequence and rehandling strategy for multi-quay crane operations in container terminals. Transp Res Part E Logistics Transp Rev 80:1–19. doi:10.1016/j.tre.2015.05.004

Jiang Y, Seidmann A (2014) Capacity planning and performance contracting for service facilities. Decis Support Syst 58:31–42. doi:10.1016/j.dss.2013.01.010

Kassicieh SK, Yourstone SA (1998) Training, performance evaluation, rewards, and TQM implementation success. J Qual Manage 3(1):25–38. doi:10.1016/S1084-8568(99)80102-3

Kemal Karasu M, Cakmakci M, Cakiroglu MB, Ayva E, Demirel-Ortabas N (2014) Improvement of changeover times via Taguchi empowered SMED/case study on injection molding production. Measurement 47:741–748. doi:10.1016/j.measurement.2013.09.035

Kia R, Khaksar-Haghani F, Javadian N, Tavakkoli-Moghaddam R (2014) Solving a multi-floor layout design model of a dynamic cellular manufacturing system by an efficient genetic algorithm. J Manuf Syst 33(1):218–232. doi:10.1016/j.jmsy.2013.12.005

Kim I, Tang CS (1997) Lead time and response time in a pull production control system. Eur J Oper Res 101(3):474–485. doi:10.1016/S0377-2217(96)00174-9

Kim SL, Ha D (2003) A JIT lot-splitting model for supply chain management: enhancing buyer–supplier linkage. Int J Prod Econ 86(1):1–10. doi:10.1016/S0925-5273(03)00006-9

Knechtges P, Decker MC (2014) Application of kaizen methodology to foster departmental engagement in quality improvement. J Am Coll Radiol 11(12, Part A):1126–1130. doi:http://dx.doi.org/10.1016/j.jacr.2014.08.027

Konecny PA, Thun JH (2011) Do it separately or simultaneously—An empirical analysis of a conjoint implementation of TQM and TPM on plant performance. Int J Prod Econ 133(2):496–507. doi:10.1016/j.ijpe.2010.12.009

Kouki C, Jouini O (2015) On the effect of lifetime variability on the performance of inventory systems. Int J Prod Econ 167:23–34. doi:10.1016/j.ijpe.2015.05.007

Kumar SV, Mani VGS, Devraj N (2014) Production planning and process improvement in an impeller manufacturing using scheduling and OEE techniques. Procedia Mater Sci 5(0):1710–1715. doi:http://dx.doi.org/10.1016/j.mspro.2014.07.360

Kumiega A, Van Vliet B (2008) 30—Kaizen: continuous improvement. In: Kumiega A, Van Vliet B (eds) Quality money management. Academic Press, Burlington, pp 271–277

Kwak TC, Kim JS, Moon C (2006) Supplier–buyer models for the bargaining process over a long-term replenishment contract. Comput Ind Eng 51(2):219–228. doi:10.1016/j.cie.2006.02.004

Lage JM, Filho G (2010) Variations of the Kanban system: literature review and classification. Int J Prod Econ 125(1):13–21. doi:http://dx.doi.org/10.1016/j.ijpe.2010.01.009

Lagemann H, Meier H (2014) Robust capacity planning for the delivery of industrial product-service systems. Procedia CIRP 19:99–104. doi:10.1016/j.procir.2014.05.021

Lee CY, Johnson AL (2012) Two-dimensional efficiency decomposition to measure the demand effect in productivity analysis. Eur J Oper Res 216(3):584–593. doi:10.1016/j.ejor.2011.08.004

Lemak DJ, Reed R (1997) Commitment to total quality management: Is there a relationship with firm performance? J Qual Manage 2(1):67

Lengnick-Hall ML, Lengnick-Hall CA, Rigsbee CM (2013) Strategic human resource management and supply chain orientation. Hum Resour Manage Rev 23(4):366–377. doi:10.1016/j.hrmr.2012.07.002

Liberopoulos G, Koukoumialos S (2005) Tradeoffs between base stock levels, numbers of Kanbans, and planned supply lead times in production/inventory systems with advance demand information. Int J Prod Econ 96(2):213–232. doi:10.1016/j.ijpe.2004.04.006

Liu H, Hwang SL, Liu TH (2009) Economic assessment of human errors in manufacturing environment. Saf Sci 47(2):170–182. doi:10.1016/j.ssci.2008.04.006

Lortie M (2012) Analysis of the circumstances of accidents and impact of transformations on the accidents in a beverage delivery company. Saf Sci 50(9):1792–1800. doi:10.1016/j.ssci.2012.04.008

Lovell MC (2003) Optimal lot size, inventories, prices and JIT under monopolistic competition. Int J Prod Econ 81–82:59–66. doi:10.1016/S0925-5273(02)00349-3

Lummus RR (1995) A simulation analysis of sequencing alternatives for JIT lines using Kanbans. J Oper Manage 13(3):183–191. doi:10.1016/0272-6963(95)00026-O

Macbeth DK, Baxter LF, Farguson N, Neil GC (1988) Buyer-Vendor Relationship with Just-in-Time: lessons from US multinationals. Ind Eng 20(9):38–41

Marodin GA, Saurin TA, Tortorella GL, Denicol J (2015) How context factors influence lean production practices in manufacturing cells. Int J Adv Manuf Technol 79(5–8):1389–1399. doi:10.1007/s00170-015-6944-2

Martínez-Jurado PJ, Moyano-Fuentes J, Jerez-Gómez P (2014) Human resource management in lean production adoption and implementation processes: success factors in the aeronautics industry. BRQ Bus Res Q 17(1):47–68. doi:10.1016/j.cede.2013.06.004

Myung YS, Moon I (2014) A network flow model for the optimal allocation of both foldable and standard containers. Oper Res Lett 42(6–7):484–488. doi:10.1016/j.orl.2014.08.004

Nen M (2015) Professional insertion of highly qualified human resources. Case study—Romania. Procedia Soc Behav Sci 191:1891–1896. doi:10.1016/j.sbspro.2015.04.615

Nenonen S (2011) Fatal workplace accidents in outsourced operations in the manufacturing industry. Saf Sci 49(10):1394–1403. doi:10.1016/j.ssci.2011.06.004

Niskanen T, Lauttalammi J (1989) Accidents in materials handling at building construction sites. J Occup Accidents 11(1):1–17. doi:10.1016/0376-6349(89)90002-3

Ohno K (2011) The optimal control of just-in-time-based production and distribution systems and performance comparisons with optimized pull systems. Eur J Oper Res 213(1):124–133. doi:10.1016/j.ejor.2011.03.005

Ohno H, Matsubae K, Nakajima K, Kondo Y, Nakamura S, Nagasaka T (2015) Toward the efficient recycling of alloying elements from end of life vehicle steel scrap. Resour Conserv Recycl 100:11–20. doi:10.1016/j.resconrec.2015.04.001

Oropesa-Vento M, García-Alcaraz JL, Rivera L, Manotas DF (2015) Effects of management commitment and organization of work teams on the benefits of Kaizen: Planning stage. DYNA 82(191):76–84

Pan JN, Nguyen HTN (2015) Achieving customer satisfaction through product–service systems. Eur J Oper Res 247(1):179–190. doi:10.1016/j.ejor.2015.05.018

Panayiotou CG, Cassandras CG (1999) Optimization of Kanban-based manufacturing systems. Automatica 35(9):1521–1533. doi:10.1016/S0005-1098(99)00074-6

Pattanaik LN, Sharma BP (2009) Implementing lean manufacturing with cellular layout: a case study. Int J Adv Manuf Technol 42(7–8):772–779. doi:10.1007/s00170-008-1629-8

Pourbabai B (1988) A just-in-time/group technology based scheduling model. In: Radharamanan R (ed) Robotics and factories of the future '87. Springer, Berlin Heidelberg, pp 105–110

Power D, Sohal AS (2000) Human resource management strategies and practices in just-in-time environments: Australian case study evidence. Technovation 20(7):373–387. doi:10.1016/S0166-4972(99)00151-0

Pragman CH (1996) JIT II: a purchasing concept for reducing lead times in time-based competition. Bus Horiz 39(4):54–58. doi:10.1016/S0007-6813(96)90052-X

Radharamanan R, Godoy LP, Watanabe KI (1996) Quality and productivity improvement in a custom-made furniture industry using Kaizen. Comput Ind Eng 31(1–2):471–474. doi:10.1016/0360-8352(96)00177-5

Ragin-Skorecka K (2014) Agile enterprise: a human factors perspective. Hum Factors Ergon Manuf. doi:10.1002/hfm.20610

Rahman NAA, Sharif SM, Esa MM (2013a) Lean manufacturing case study with Kanban system implementation. Procedia Econ Finan 7:174–180. doi:http://dx.doi.org/10.1016/S2212-5671(13)00232-3

Rahman NAA, Sharif SM, Esa MM (2013b) Lean manufacturing case study with Kanban system implementation. Procedia Econ Finan 7(0):174–180. doi:http://dx.doi.org/10.1016/S2212-5671(13)00232-3

Recht R, Wilderom C (1998) Kaizen and culture: on the transferability of Japanese suggestion systems. Int Bus Rev 7(1):7–22. doi:10.1016/S0969-5931(97)00048-6

Reda HM (1987) A review of "Kanban"—the Japanese "just-in-time" production system. Eng Manage Int 4(2):143–150. doi:10.1016/0167-5419(87)90019-6

Ruckart PZ, Burgess PA (2007) Human error and time of occurrence in hazardous material events in mining and manufacturing. J Hazardous Mater 142(3):747–753. doi:10.1016/j.jhazmat.2006.06.117

Rusu G, Avasilcai S (2014) Linking human resources motivation to organizational climate. Procedia Soc Behav Sci 124:51–58. doi:10.1016/j.sbspro.2014.02.459

Sargent J, Matthews L (2009) China versus Mexico in the global EPZ industry: Maquiladoras, FDI quality, and plant mortality. World Dev 37(6):1069–1082. doi:10.1016/j.worlddev.2008.10.002

Schaefer B, Konur D (2015) Economic and environmental considerations in a continuous review inventory control system with integrated transportation decisions. Transp Res Part E Logistics Transp Rev 80:142–165. doi:10.1016/j.tre.2015.05.006

Selçuk B (2013) Adaptive lead time quotation in a pull production system with lead time responsive demand. J Manuf Syst 32(1):138–146. doi:10.1016/j.jmsy.2012.07.017

Spencer MS (1998) Group technology at John Deere: production planning and control issues. In: Suresh NC, Kay JM (eds) Group technology and cellular manufacturing. Springer, US, pp 501–516

Špicar R, Januška M (2015) Use of Monte Carlo modified Markov Chains in capacity planning. Procedia Eng 100:953–959. doi:10.1016/j.proeng.2015.01.454

Sundar R, Balaji AN, Satheesh Kumar RM (2014) A review on lean manufacturing implementation techniques. Procedia Eng 97:1875–1885. doi:10.1016/j.proeng.2014.12.341

Sylvain L, Duguay R (1997) Integrating MRP, KANBAN and bar coding system to achieve JIT procurement. Prod Inventory Manage J First Q:18–26

Tabassi AA, Abu Bakar AH (2009) Training, motivation, and performance: the case of human resource management in construction projects in Mashhad, Iran. Int J Project Manage 27(5):471–480. doi:10.1016/j.ijproman.2008.08.002

Teagarden MB, Butler MC, Von Glinow MA (1992) Mexico's maquiladora industry: Where strategic human resource management makes a difference. Organ Dyn 20(3):34–47. doi:10.1016/0090-2616(92)90023-G

Thomas AJ, Jones GR, Vidales P (2006) An integrated approach to TPM and six sigma development in the castings industry. In: Pham DT, Eldukhri EE, Soroka AJ (eds) Intelligent production machines and systems. Elsevier Science Ltd., Oxford, pp 620–625

van der Laan E, Salomon M, Dekker R (1999) An investigation of lead-time effects in manufacturing/remanufacturing systems under simple PUSH and PULL control strategies. Eur J Oper Res 115(1):195–214. doi:10.1016/S0377-2217(98)00108-8

Villa A, Taurino T (2013) From JIT to seru, for a production as lean as possible. Procedia Eng 63:956–965. doi:10.1016/j.proeng.2013.08.172

Williams DJ (1991) 1—An introduction to the manufacturing cell and its integration and control. In: WilliamsPaul Rogers DJ (eds) Manufacturing cells. Newnes, pp 1–9

Wu Z, Jianghua W (2015) Price discount and capacity planning under demand postponement with opaque selling. Decis Support Syst 76:24–34. doi:10.1016/j.dss.2015.02.002

Yalcin A (2004) Supervisory control of automated manufacturing cells with resource failures. Robot Comput Integr Manuf 20(2):111–119. doi:10.1016/j.rcim.2003.10.001

Chapter 12
Causal Models JIT Elements Associated with Product and Obtained Benefits

This chapter introduces two structural equation models to assess the impact of JIT elements related to the final product over JIT benefits obtained. The 13 items or JIT elements associated with the product were comprised into two major categories by means of a factor analysis, which will be described below. This factor analysis distributed all items into the two categories or also called latent variables or factors. These latent variables were named and will be proposed and tested in these models. Finally, readers can consult the methodology addressed in Chap. 5 and previous chapters for a full comprehension of procedures concerning the evaluation and interpretation of these models.

12.1 Factor Analysis of JIT Elements Associated with the Product

The first step to perform a factor analysis is to determine its feasibility. Table 12.1 illustrates two indexes that determined the factorability of items. KMO index to measure sampling adequacy shows a value higher than 0.8, which is the minimum acceptable value for this research. Therefore, factor analysis of variables is feasible. Similarly, Bartlett's sphericity test shows a high Chi-squared value and has 78° of freedom, which gives this test a significance value of 0.000. This is a value lower than 0.5, the maximum value allowed for a 95 % confidence level. Therefore, according to these two tests, factor analysis is a viable technique to reduce dimensions in data.

Table 12.2 shows the results from the factor analysis. The table illustrates the explained variance and cumulative variance for every factor or latent variable. It can be observed that two factors are enough to explain 65.139 % of the variance contained in the 13 items. The first factor explains 56.981 % of the whole variance, while the second one explains 8.158 %.

Similarly, Table 12.3 shows items with their corresponding factor loadings and the latent variable to which they belong. The first factor or latent variable was named Quality Planning and comprises eight items or variables. The second factor

J.L. García-Alcaraz and A.A. Maldonado-Macías, *Just-in-Time Elements and Benefits*,
Management and Industrial Engineering, DOI 10.1007/978-3-319-25919-2_12

Table 12.1 KMO and Bartlett's test

Kaiser-Meyer-Olkin measure of sampling adequacy		0.929
Bartlett's test of sphericity	Approx. Chi-Square	1210.251
	df	78
	Sig.	0.000

Table 12.2 Total variance explained

Component	Initial Eigen values			Extraction sums of squared loadings			Rotation sums of squared loadings
	Total	Percentage of variance	Cumulative percentage	Total	Percentage of variance	Cumulative percentage	Total
1	7.408	56.981	56.981	7.408	56.981	56.981	7.096
2	1.060	8.158	65.139	1.060	8.158	65.139	5.147

Table 12.3 Items, factor loadings, and factors of model 1

Items	Factor loadings	Factor
Continuous quality improvement	0.926	Quality planning
Statistical quality control	0.893	
Quality-oriented training	0.853	
Quality circles	0.800	
Zero defects	0.793	
Quality development programs	0.764	
Total quality control	0.738	
Quality culture	0.632	
100 % quality inspection	0.992	Quality management
High visibility of quality control	0.863	
Long-term commitment to quality	0.834	
Simplifying the process of total quality	0.698	
Regulate quality and reliability of audits	0.695	

or latent variable is composed of five items or variables and is named Quality Management.

Although both latent variables refer to quality, they have a different approach. While Planning Quality refers to the actions concerning anticipation, Quality Management concerns organization and execution of plans and programs designed by Quality Planning.

12.2 Model 1. JIT Elements: Quality Planning and Quality Management. Benefits: Human Resources, Quality, and Economic Benefits

Models in this chapter seek to associate five latent variables. Two latent variables concern JIT elements for the product and three latent variables concern benefits obtained from an effective JIT implementation, although these benefits vary in every model. However, Model 1 is integrated as follows:

- JIT elements: Quality Planning and Quality Management
- Benefits of JIT: Human Resources, Quality, and Economic Benefits

Model 1 considers Quality Planning as the independent latent variable upon which all other latent variables depend. On the other hand, Economic Benefits were set as the final dependent latent variable, which can be affected directly and indirectly by all other latent variables. That is, economic benefits of companies depend on established Quality Planning and Human Resources who follow these quality plans and seek to achieve product quality. Finally, ten working hypotheses were proposed to demonstrate relations between latent variables. They were statistically tested to validate their significance according to the methodology described in Chap. 5.

12.2.1 *Hypotheses of Model 1*

Figure 12.1 illustrates the ten working hypotheses that relate one variable with another. Discussions to support them are provided below.

The way quality plans and programs are developed in companies depends on two aspects: their conception and the knowledge of tools necessary to their execution (Corredor and Goñi 2011; Ooi 2014; Valmohammadi and Roshanzamir 2015). This means that managers and administrators are likely to plan activities based on what they know and are aware of concerning techniques to manage quality. However, when managers are not familiar with certain techniques that are necessarily required, companies usually rely on the recruitment of specialists. One may thus not be surprised by the great number of many firms which nowadays provide consulting and advice to companies (Jayaram et al. 2010).

Similarly, quality plans and programs are based on the commitment of companies to their costumers (Shi et al. 2014), which is why the assessment of certain variables and characteristics of product quality is the result of costumer demands. Therefore, effective Quality Management requires appropriate planning, which should rely on useful techniques, such as statistical process control, quality circles, zero-defects philosophy, diffusion of the quality control along the company and, of course, training programs provided to workers (Ahmad et al. 2012; Vinodh and Joy 2011). These techniques could provide companies with useful information to monitor and assess quality, which should be to some extent visible through several

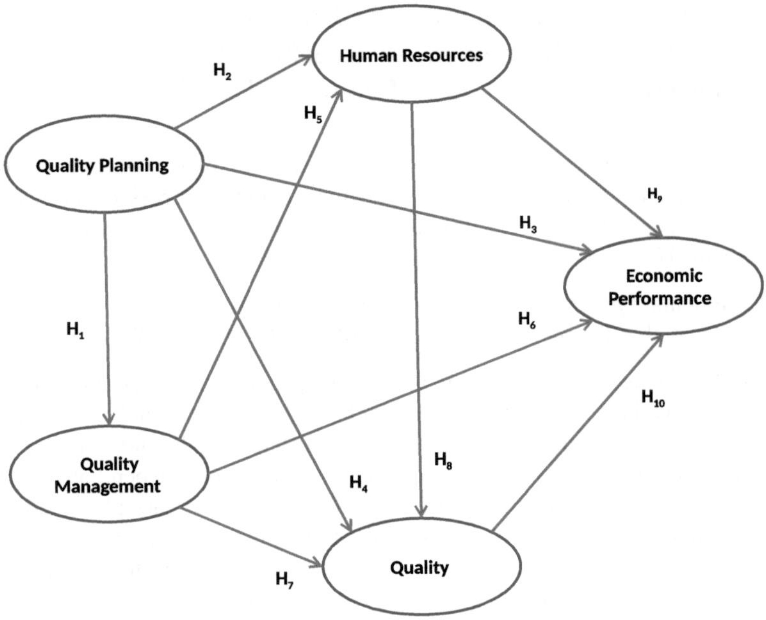

Fig. 12.1 Working hypotheses of model 1

means of communication. For instance, companies can make use of small notes in visible places and simplify the process of total quality so that all workers become acquainted with the quality reached and required. This would allow them to track their possible mistakes and even improve their performance. Therefore, as a means to contribute to this discussion, the following hypotheses are proposed.

H1: *There is a direct and positive relation between quality planning and quality management during the implementation of a JIT program*
H2: *There is a direct and positive relation between quality planning and benefits for human resources during the implementation of a JIT program.*

Although it is often stated that quality plans and programs should be targeted at consumer satisfaction (Shadur 1995), it should not be forgotten that quality is only a means to achieve high levels of billing, which can become the greatest economical benefit for companies (Molina-Azorín et al. 2015; O'Neill et al. 2015). A company that offers poor quality products will sooner or later be unable to sell them. Thus, the significance of quality plans and programs is that they must increase wealth in companies and to shareholders. Again, quality is a means to achieve the ultimate goal.

However, quality plans cannot be implemented by themselves; they require the support of human resources at all levels of the company, whether they be managers or operators. Thus, quality plans should focus on meeting not only the demands of shareholders and customers, but also the needs of these human resources. Quality plans that are fully or partially executed can actually improve workers motivation. They become satisfied with their achievements (Ahmad et al. 2012; Corredor and Goñi 2011), although most of these require a great amount of collaboration and communication among members (Wiengarten and Pagell 2012; Lester 2007; Arbogast 1997). Hence, in order to contribute to this discussion, the following hypotheses can be formulated.

H3: *There is a direct and positive relation between quality planning in a JIT environment and the economic performance reached by companies.*
H4: *There is a direct and positive relation between quality planning in a JIT environment and the indexes achieved of product quality.*

Quality Management and how companies diffuse it has an impact on human resources (Kannan and Tan 2005; Ahmed and Rafiq 2002). Companies must rely on a highly motivated workforce (Taskov and Mitreva 2015; Diefendorff and Seaton 2015; Anderman and Gray 2015), which can be achieved when these workers become familiar with the situation of quality plans and programs. Therefore, companies diffuse this important information through different means, such as statistical reports. Thus, when employees are aware of the quality achieved and are provided with such statistical information, their attitude might change and a more motivated employee can be more efficient (Anderman and Gray 2015; Zámečník 2014; Rusu and Avasilcai 2014).

Quality audits, both internal and external, have become a traditional technique to inform of the status of planned and expected quality (Maroun 2015; Ball et al. 2015). For instance, internal quality audits can be run by the quality department of the company, since employees in this department are familiar with the production process. However, when workers are used to observe the process that often, sometimes they may be unable to recognize or detect problems

Therefore, external audits are very supportive, since specialized personnel from outside companies are hired to perform this assessment and may be able to quickly identify inconsistencies (Miko and Kamardin 2015; Kouaib and Jarboui 2014).

As previously mentioned, these reports on the state of quality systems can be a source of motivation and satisfaction for operators. They may be able to set their own personal goals and challenge themselves. Moreover, this system of quality management is expected to be as simple as possible, so that all employees can be able to interpret it (Calvo-Mora et al. 2014; Corredor and Goñi 2011). Finally, when all levels of the company have become familiar and understood the state of quality in their plants, communication increases, which also implies higher performance indices especially economic indexes. In order to contribute to this research, the following hypotheses are proposed.

H5: *There is a direct and positive relation between quality management and benefits for human resources in a JIT environment.*
H6: *There is a direct and positive relation between quality management and the economic performance reached by companies during JIT implementation.*

Quality Management mainly depends on the techniques that top management can identify to achieve the planned targets and the techniques that guarantee companies their desired results (Lemak and Reed 1997; Zairi 1993). Thus, establishing a program for visibility of quality control should reflect increased quality in the production process and the final product. However, similar effects can be obtained, for example, by applying an inspection at 100 % (Myklebust 2013; Ferretti et al. 2013), although in this globalization era it may imply high costs of quality assurance. Yet, many human errors will still be likely to happen (Love et al. 1995; Oropesa-Vento et al. 2015). This means that product quality could be ensured, although its low cost may not always be guaranteed. Nevertheless, when companies are highly and long-term commited to the planning and execution of these quality plans and programs, they ensure increased quality in their production processes and products. Promoting education and providing training among operators and the other employees is a suitable option to assure such commitment. (Karatepe and Karadas 2012; Howard and Thomas Foster 1999; Xie et al. 2014; Gyulai et al. 2014; Wang et al. 2013).

Quality process should be simplified as much as it is possible, and this concerns crucial JIT element. Simplifying quality process could ensure quality and consequently reduce administrative costs related to it. However, this may seem antagonistic to a 100 % quality inspection of raw materials, since such simplification implies a high level of integration between suppliers and manufacturers and thus high levels of trust, while with 100 % quality inspection companies precisely communicate that such level of trust in raw materials has not been reached. As a means to support the following discussion, the following hypotheses were proposed.

H7: *There is a direct and positive relation between quality management and product quality features in a JIT environment.*

The hypothesis that associates human resources with product quality was proposed in model 2 from Chap. 11. Readers may consult the chapter once more. However, such a relationship is included in this model to depict and understand the indirect effects between two variables through this segment.

H8: *There is a direct and positive relation between the benefits associated with human resources and product quality.*

It has been widely mentioned in this research that human resources is the most valuable capital for companies, since their knowledge can generate wealth. New theories and deep lines of research have focused on managing corporate knowledge, and companies have also suggested that corporate strategy should be implemented according to the general skills and abilities of human resources (Çınar and Karcıoğlu 2013). Similarly, experts in location processes have recommended that before deciding the location of a new subsidiary, companies must consider the capabilities of human resources of the area, the cost of labor, and trade union protection, among others (Xie et al. 2014; Gyulai et al. 2014; Wang et al. 2013; Fullerton and McWatters 2002).

Similarly, motivation can be promoted by means of incentive programs that reward the abilities of workers and their effort, although companies must make sure these are not misinterpreted and a greater amount of products is generated but quality becomes compromised. Furthermore, companies should provide training emphasized on collaborative work, since individualism often occurs when incentives or rewards systems exists. They may be due to feelings of competition or inherent attitudes of the human being. However, human resources should always be satisfied, able to work in collaboration, and hold efficient communication among them, since can bring a number of benefits for them and the company (Fullerton and McWatters 2002; Armstrong et al. 2015; Veldman and Gaalman 2014). In order to contribute to this line of research, the next working hypothesis is proposed.

H9: *There is a direct and positive relation the between benefits associated with human resources and the economic performance of companies in a JIT environment.*

Hypothesis 10 (H_{10}) was previously proposed and tested by the second model included in Chap. 10. It is thus unnecessary to perform its validation once more. However, the hypothesis is illustrated in the model to prove the indirect effects that may occur between two latent variables through this segment.

H10: *Quality features and metrics in a finished product during the process of implementing a program or philosophy JIT have a direct and positive impact with the economic performance that the company reaches.*

12.2.2 Results of Model 1

This model, as well as all previous models, was evaluated based on methodology initially described. The results of such evaluation are illustrated graphically in Fig. 12.2, which shows the results of the direct effects among latent variables through the beta parameter and a P-value that measures the statistical significance of every segment or effect. Blue relations are statistically significant, while the red segments were found statistically not significant.

12.2.2.1 Efficiency Indexes of Model 1

According to the APC, ARS, and AARS indexes, the model is efficient from a statistical point of view. The P-value to measure the statistical significance of the parameter was in all cases lower than 0.05, which is the maximum cutoff value for a 95 % confidence level of 95 %.

Similarly, AVF and AFVIF indexes show values lower than 3.3, the maximum cutoff value. Therefore, no problems of collinearity exist in the model. Finally, it is worth mentioning the index of Tenenhaus that equals 0.628 and far exceeds the recommended minimum cutoff value, 0.36 units. Similar interpretations can offered

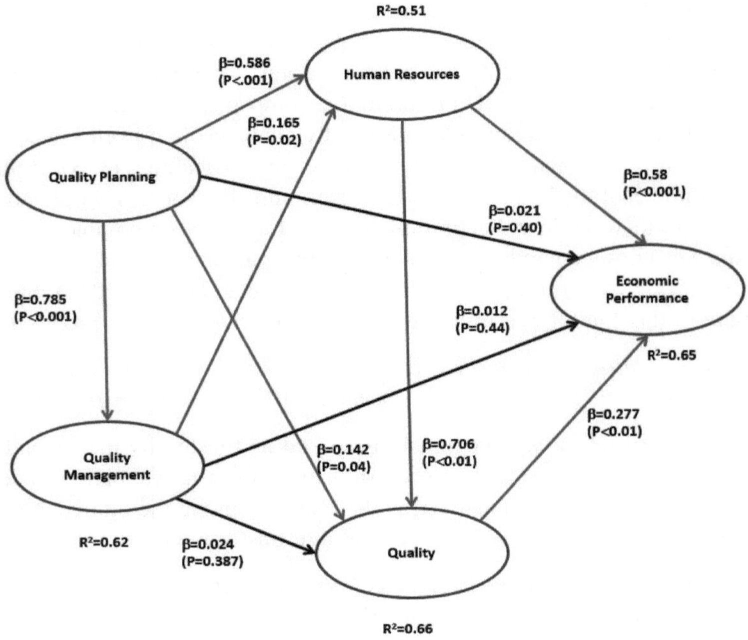

Fig. 12.2 Evaluation of hypotheses of model 1

for the other estimated indices, which lead to the conclusion that the model can be evaluated.

1. Average path coefficient (APC) = 0.330, $P < 0.001$
2. Average R-squared (ARS) = 0.609, $P < 0.001$
3. Average adjusted R-squared (AARS) = 0.602, $P < 0.001$
4. Average block VIF (AVIF) = 2.534, acceptable if ≤5, ideally ≤3.3
5. Average full collinearity VIF (AFVIF = 3.252, acceptable if ≤5, ideally ≤3.3
6. TenenhausGoF (GoF) = 0.628, small ≥ 0.1, medium ≥ 0.25, large ≥ 0.36
7. Simpson's paradox ratio (SPR) = 0.700, acceptable if ≥0.7, ideally = 1
8. R-squared contribution ratio (RSCR) = 0.988, acceptable if ≥0.9, ideally = 1
9. Statistical suppression ratio (SSR) = 1.000, acceptable if ≥0.7
10. Nonlinear bivariate causality direction ratio (NLBCDR) = 1.000, acceptable if ≥0.7.

12.2.2.2 Coefficients of Latent Variables

Indices reported above proved that the model was in general efficient and could be used to interpret the relations between variables. However, it is also important to

Table 12.4 Coefficient of the latent variables of model 1

To	Quality planning	Quality management	Quality	Human resources	Economic performance
R-squared		0.617	0.659	0.509	0.65
Adjusted R-squared		0.614	0.651	0.502	0.64
Composite reliability	0.932	0.883	0.885	0.887	0.899
Cronbach alpha	0.915	0.834	0.805	0.841	0.859
AVE	0.664	0.603	0.72	0.612	0.64
AVIF	3.022	2.5	3.261	3.177	2.898
Q-squared		0.62	0.686	0.504	0.66

analyze every latent variable of this model separately. Table 12.4 depicts validity indexes for each of these variables.

First of all, based on R-square and Adjusted R-squared coefficients, the model and all latent variables have acceptable predictive validity from a parametric approach, since all values of both coefficients are higher than 0.02 that is the minimum acceptable value. Moreover, all Q-squared values for these variables are positive and higher than 0 and are close R-squared and adjusted R-squared values. Therefore, the model and its latent variables have predictive validity from a non-parametric perspective.

As for internal validity, Table 12.4 shows that both composite reliability and Cronbach's Alpha indexes have values higher than 0.7, the minimum value allowed in this research. This indicates that every latent variable of this model has sufficient internal validity or is adequately expressed.

Also, according to AVE values, all latent variables have sufficient convergent validity, since in all cases values are higher than 0.5, the minimum acceptable value in this study. Finally, Table 12.4 also indicates that there are no collinearity problems in latent variables, since all values are lower than 3.3 units.

12.2.2.3 Direct Effects—Validation of Hypotheses of Model 1

After the general evaluation of the model and the analysis of all its latent variables, it is possible to state that the following conclusions can be drawn.

- H1: There is enough statistical evidence to state that Quality Planning has a direct and positive effect on quality management during the implementation of JIT philosophy, since when the first latent variable increases its standard deviation by one unit, the standard deviation of the second latent variable increases by 0.785 units.
- H2: There is sufficient statistical evidence to declare that quality planning in a JIT environment has a direct and positive effect on benefits associated with human resources, since when the first latent variable increases its standard

deviation by one unit, the standard deviation of the second latent variable increases by 0.586 units.

- H3: There is not enough statistical evidence to state that quality planning has a direct and positive effect on the economic performance of companies in a JIT environment, since the P-value of statistical significance for this hypothesis was 0.40 and exceeds the maximum value accepted for a 95 % confidence level. From an industrial point of view, one may have considered this relationship as significant; however, indirect and total effects between these variables should not be taken for granted. They will be discussed later in this section.
- H4: There is sufficient statistical evidence to state that quality planning has a direct and positive effect on quality product in a JIT environment, since when the first latent variable increases its standard deviation by one unit, the standard deviation of the second latent variable rises 0.142 units.
- H5: There is enough statistical evidence to declare that quality management has a direct and positive impact on benefits that associated with human resources during the implementation of JIT philosophy. When the first latent variable increases its standard deviation by one unit, the standard deviation of the second latent variable rises by 0.165 units.
- H6: There is not enough statistical evidence to point out that quality management in JIT environments has a direct and positive effect on the economic performance of companies, since the statistical significance test on this hypothesis showed a P-value equal to 0.44, while 0.05 is the maximum cutoff value set in this research for a 95 % confidence level.
- H7: There is not enough statistical evidence to declare that quality management has a direct and positive impact on product quality in JIT environments. The P-value of this hypothesis equaled 0.387 units, which exceeds the maximum allowed value for statistical inferences with a 95 % confidence level.
- H8: There is sufficient statistical evidence to declare that benefits associated with human resources in a JIT implementation have a direct and positive impact on product quality. When the first latent variable increases its standard deviation by one unit, the standard deviation of the second latent variable increases by 0.706 units.
- H9: There is sufficient statistical evidence to state that benefits associated with human resources in JIT environments have a direct and positive impact on the economic performance of companies, since when the first latent variable increases its standard deviation by one unit, the standard deviation of the second latent variable rises by 0.58 units.
- H10: Enough statistical evidence indicates that quality product in JIT systems has a direct and positive impact on the economic performance of companies because when the former latent variable increases its standard deviation by one unit, the standard deviation of the latter latent variable increases by 0.277 units.

For latent-dependent variables depending on independent variables, it is important to apportion the value of the R-squared and discuss the percentage of direct effect that is explained through the different independent latent variables.

Table 12.5 Effect size of direct effects of model 1

To	From			
	Quality planning	Quality management	Quality	Human resources
Quality management	0.62			
Quality	0.093	0.013		0.578
Human resources	0.412	0.096		
Economic performance	0.012	0.006	0.205	0.463

Table 12.5 shows the effect sizes that independent latent variables have on dependent latent variables.

- The highest effect size corresponds to the relation between quality planning and quality management, which is due to the effect of a single variable.
- The latent variable Quality is explained by three latent variables in 66 %, since the R-squared value is 0.66. However, the strongest effect on this variable is caused by Human Resources, since the effect size is 0.578. This demonstrates that workers are the keys to obtain the established product quality.
- The latent variable Economic Performance of the company is explained by four latent variables in 65 %, since R-square has a value of 0.65, However two of these relations were statistically not significant. Once more, Human Resources is the latent variable that explains the highest percentage of this dependent variable, with an effect size of 0.463. Human Resources thus seem to be the most valuable asset for companies; moreover, they can bring economic benefits.
- The dependent latent variable Human Resources is explained by two independent latent variables in 51 %, since the R-squared value is 0.51. Latent variable Quality Planning is the most significant, since its effect size is 0.42 units.

Latent variables in relations that were statistically not significant have extremely small effect sizes, some of which are less than 0.02. For instance, the effect of Quality Management over Quality can seem illogical; however, it will be addressed once more in the analysis of indirect effects.

12.2.2.4 Sum of Indirect Effects of Model 1

As Fig. 12.2 shows, latent variables can have indirect effects over other variables, which become essential to explain why their direct relationship was statistically not significant. Moreover, it should be remembered that a same latent variable can have different effects on another variable through different segments. Table 12.6 shows the sum of indirect effects that one latent variable may have on other latent variables.

Table 12.6 Sum of indirect effects of model 1

To	From		
	Quality planning	Quality management	Human resources
Quality	0.486 ($P < 0.001$) ES = 0.318	0.116 ($P < 0.022$) ES = 0.061	
Human resources	0.129 ($P < 0.013$) ES = 0.091		
Economic performance	0.579 ($P < 0.001$) ES = 0.342	0.121 ($P = 0.009$) ES = 0.056	0.195 ($P < 0.001$) ES = 0.156

- The direct effect caused by the latent variable Quality Planning over Quality of the final product had a low value of 0.142. However, the sum of indirect effects that may occur through latent variables Human Resources and Quality Management is of 0.486 and it can explain for up to 31.8 % of that variable.
- The direct effect that the latent variable Quality Management caused on Quality of the product turned out to be statistically not significant. However, an indirect effect occurs between them and can explain 6.1 % of its variability.
- The direct effect that the latent variable Quality Planning caused on Economic Performance was statistically not significant. However, the sum of indirect effects that occur between them has a high value of 0.579 and the independent latent variable can explain up to 34.2 % of variability in dependent latent variable, which can occur through other latent variables, including Human Resources. This demonstrates again the importance of workforce for a company.
- The direct effect caused by the latent variable Quality Management over the latent variable Economic Performance was statistically not significant. Yet, the sum of indirect effects between them showed a value of 0.121. These indirect effects occur through Human Resources and Quality. It seems that Human Resources are crucial for the economic performance of companies.

12.2.2.5 Total Effects of Model 1

The sum of direct and indirect effects equals the total effects that one latent variable can have over another. Total effects can equal direct effects when indirect effects do not occur between latent variables. Table 12.7 illustrates these effects and allows for their following interpretations.

- According to their size, four total effects show values higher than 0.7, which indicates strong relation between these variables concerned.
- The latent variable Quality Planning causes large effects on all other latent variables of this model. All its effect sizes are higher than 0.5, which means that major changes occur in subsequent latent variables when Quality Planing increases its standard deviation by one unit. This points out the significance of quality planning for the survival of companies.

Table 12.7 Total effects of model 1

To	From			
	Quality planning	Quality management	Quality	Human resources
Quality management	0.785 ($P < 0.001$) ES = 0.617			
Quality	0.628 ($P < 0.001$) ES = 0.411	0.092 ($P = 0.013$) ES = 0.048		0.706 ($P < 0.001$) ES = 0.578
Human resources	0.715 ($P < 0.001$) ES = 0.503	0.165 ($P = 0.021$) ES = 0.096		
Economic performance	0.558 ($P < 0.001$) ES = 0.330	0.109 ($P < 0.031$) ES = 0.050	0.277 ($P < 0.001$) ES = 0.205	0.776 ($P < 0.001$) ES = 0.619

- The latent variable Quality Management causes direct effects over three latent variables: Quality, Economic Performance, and Quality Planning. Although all these were statistically not significant, the sum of the total effect resulted in significant relations for the three of them. Indirect effects occur through Human Resources and Quality. Therefore, this can demonstrate that Quality is the result of skills and abilities of human resources.

12.2.3 *Conclusions of Model 1*

This model provided with relevant information since certain relationships or hypotheses tested were initially found not significant but became significant when all types of effects were analyzed. However, the relevance is not due to their sudden significant, but to the latent variables that were involved in making these relations meaningful. It may be impossible for managers to conceive that quality management could not have a direct and positive impact on the economic performance of companies, or that quality planning does not have a direct and positive effect on product quality.

Moreover, based on the results found from the evaluation of the model, the following conclusions were drawn.

- Benefits obtained for human resources from the implementation of JIT philosophy are variables that companies should strive to achieve and never underestimate, since in this analysis, workforce allowed for the existence of total and indirect effects between variables whose direct effect were statistically not significant.
- Companies must invest in training and motivation techniques for human resources, since they represent the most valuable asset an organization can rely on; moreover, they allow for the generation of wealth.

- Product quality is the means by which companies can benefit from financial resources for their shareholders. However, they should also strive to obtain benefits for human resources.
- Relations of Quality Planning—Quality Management and Human Resources—Economic Performance were reported as the two largest relations, according to their total effect sizes.

12.3 Model 2: JIT Elements: Quality Planning and Quality Management. Benefits: Engineering Process, Production Process, and Material Handling

The benefits obtained from the implementation of a JIT program or philosophy was divided into six categories or latent variables in Chap. 3. Three of these were considered in the previous model, while the remaining three will be addressed in this section.

Therefore, this second model integrates 5 latent variables and analyzes 30 items or variables comprised in these latent variables and distributed as follows:

- Elements of JIT associated with the product (13 items or variables included)
  - Quality Planning (8 items or variables included)
  - Quality Management (5 items or variables included)
- Benefits obtained after JIT implementation (17 items or variables included)
  - Engineering processes (4 item or variables included)
  - Production process (7 items or variables included)
  - Material handling (6 items or variables included)

The 10 analyzed relationships between latent variables were proposed as working hypotheses and illustrated in Fig. 12.3. Three relations have been tested in previous models and are thus briefly addressed in this section. Yet, they were not removed from this model in order to assess the possible indirect effects that exist through these or relations.

12.3.1 Hypotheses of Model 2

This model introduces 10 hypotheses generated based on believed relationships among latent variables. Three of them were previously proposed, tested, and validated. Thus, they are mentioned here but not discussed or tested once more.

The first model of this chapter shown in Fig. 12.1 proposed the first hypothesis of this model. Results of its analysis are illustrated graphically in Fig. 12.2; thus, the

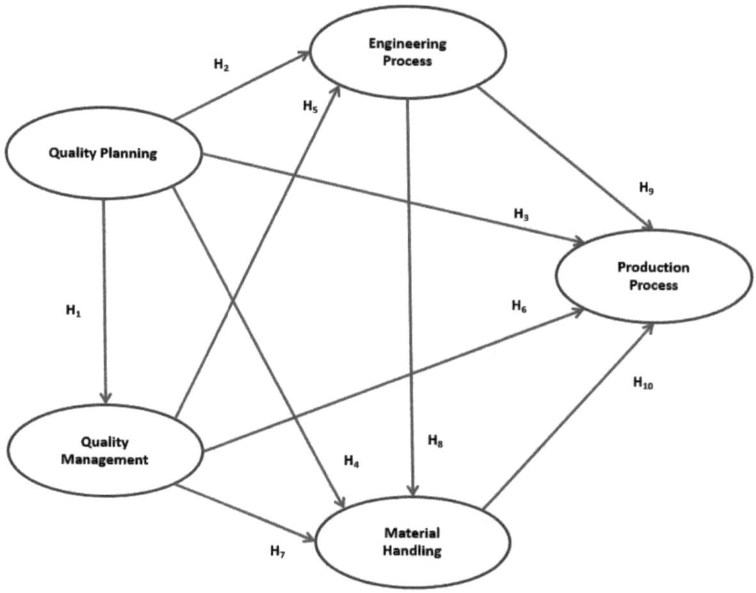

Fig. 12.3 Proposed model 2

relationship is mentioned but not discussed once more. However, the hypothesis is present in this model to illustrate the indirect effects that might exist through this segment.

H1: *There is a direct and positive relation between plans and programs generated through product quality planning and the benefits obtained in the engineering process in JIT environments.*

It is traditionally believed that continuous improvement of a production process is merely reflected on product quality. However, although it is a major goal (Kumiega and Van Vliet 2008), increased efficiency in the processes of production can also bring a wide range of benefits, such as better productive maintenance, quicker changes between product designs, and a reduction of space requirements that result in a more adequate distribution of machines and equipment (Glover et al. 2011; Farris et al. 2009). Moreover, one may recall quality circles, whose aim is to solve different kinds of issues or situations throughout the production process. Proposals to solve these situations may involve different areas in the company (Knechtges and Michael 2014).

Thus, it is likely that many elements associated with product quality have an impact on engineering processes. For instance, the implementation of a philosophy of zero defects implies little or no rework in production lines. Therefore, the number of defective parts will be minimal which could result in the reduce duction of costs

(Myklebust 2013; Ferretti et al. 2013). Similarly, although they focus on quality, training programs have direct consequences on the increasing innovation that emerges from quality groups or circles and the so-called Kaizen events. This indicates that an improved production process brings benefits in numerous aspects and especially favors a more appropriate use of resources (Oropesa-Vento et al. 2015; Radharamanan et al. 1996).

Similarly, Quality Planning or programs must clearly state and specify the personnel responsible for every task to perform, the time set to perform this task, and the way it must be accomplished. (Pragman 1996; Arikan et al. 2014b). This will allow companies to know how a product will be manufactured and what characteristics it must meet. As a result, fewer delays in the production line may occur and delivery times would be reduced (Tersine and Hummingbird 1995). Nowadays, it may be impossible to find a company that does not rely on a certain plan or program to achieve quality in its generated products.

In order to contribute to this research, the following hypotheses are proposed.

H2: *There is a direct and positive relation between plans and programs generated through product quality planning with the benefits obtained in the engineering process in successful JIT environments.*
H3: *There is a direct and positive relation between plans and programs generated through the product quality planning with the benefits obtained in a production process within a JIT environment.*

It was previously mentioned that an appropriate quality plan could not guarantee product quality on its own. Production process can be to some extent efficient and follow quality programs and plans, but it might experience issues of materials handling that may result in their deterioration (Arikan et al. 2014a). Research used to emphasize greatly on the fact that materials handling represented the major source of accidents, and the movement of materials did not generate any value for the company. However, financial outlays in companies also concern the losses of material that may occur during its management. Thus, this is an additional cost that should be analyzed (Bendul and Skorna 2015).

Properly adjusted and managed quality programs and plans must integrate aspects to reduce inventories (Huq 1999; Ramasesh 1990), minimize transportation and movement of raw materials (Azevedo et al. 2014; Chakravorty 2009); an increase in inventory turnover, since companies produce the amount of product required to fulfill a production order (Lee et al. 2014), they seek to purchase lot sizes that comprise the exact amount of raw material needed for production and to backup possible errors (Lovell 2003). All this encourages a close relation with suppliers of raw materials and promotes greater integration among all members of the supply chain and departments of companies. Similarly, some manufacturing companies organize working groups or quality circles that propose forms of reducing the number of defects in products caused by material handling. Studies have indicated that these dynamics do bring expected benefits (Kumiega and Van Vliet 2008; Moore 2007; Taghizadegan 2006).

Thus, in order to contribute to this research, the following hypothesis is proposed.

H4: *There is a direct and positive relation between plans and programs generated through product quality planning and the benefits obtained in materials handling within a JIT environment.*

The effectiveness of quality plans and programs also depends on how they are executed in production lines with the participation of human resources (Göleç 2015; Mourtzis and Doukas 2014; Ahsan et al. 2015). It is expected that the way in which plans are implemented impact on the reduction of space required and movement distances and the ability of companies to quickly adjust to possible engineering changes between production runs. Certain authors call these rapid changes in design as concurrent engineering (Zidane et al. 2015; Sapuan and Mansor 2014; Demoly et al. 2013), which refers to how manufacturing companies adjust the production line running for the introduction of new changes in design without stopping the flow of materials.

Quality plans and programs must be implemented according to the installed capacity of the production process and ought to be focused on increasing productivity in the entire production system. Quality implies zero errors and thus involves producing more with the same amount (Myklebust 2013; Westkämper and Warnecke 1994). Thus, companies that generate large amounts of waste may not be considered productive or efficient due to poor quality programs and inadequately trained operators (Ohno et al. 2015; Bettayeb et al. 2014). Therefore, an efficient production system is likely the result of appropriate quality plans and clearly defined techniques to achieve quality. Besides, knowledge and understanding of many of these techniques increases flexibility in the process. For instance, companies with more than one piece of machinery or equipment to perform a same production operation may become more flexible than those manufacturing companies relying on a single machine to perform the same production operation. (Zhang et al. 2009; Wahab et al. 2008; Moattar Husseini et al. 2006). However, flexibility is often a synonym of universality of machines and tools that may be impossible to achieve with high-tech and specialized equipments.

The execution and implementation of quality plans also impact on how material is handled, since it is likely that companies do not generate finished products in order to store them for long periods of time in their warehouses. They rather produce what has been requested by a production order, which increases inventory and asset turnover (Hong et al. 1992). The following hypotheses support this discussion.

H5: *There is a direct and positive relation between the quality management of a product with the benefits obtained in the engineering process within a JIT environment.*

H6: *There is a direct and positive relation between quality management of a product and the benefits obtained in the production process when JIT program or philosophy is implemented.*

H7: *There is a direct and positive relation between quality management of a product and the benefits obtained in materials handling within a JIT environment.*

Hypothesis 8 (H8) for this model was proposed by Model 3 in the tenth chapter of this book. The results from its evaluation are similarly shown in Fig. 10.6 in the same chapter.

H8: *There is a direct and positive relation between the benefits obtained in engineering process and those gained in materials handling when JIT program or philosophy is implemented.*

Hypothesis 9 (H9) was first introduced in the previous chapter by Model 1 in Fig. 11.1. The results of its evaluation were illustrated in Fig. 11.2. This hypothesis is mentioned but not discussed once more. It is also present in this model to illustrate the indirect effects that might exist through this segment.

H9: *There is a direct and positive relation between the benefits obtained in engineering process and those obtained in the production process when a JIT program or philosophy is implemented.*

Inventory and material handling reduction, as well as an increasing inventory turnover can surge levels of productivity and efficiency in the productive process (Lee et al. 2015). However, inventory turnover implies shorter delivery times, since raw materials rarely stop running along the succession process and some finished products are not stored in the warehouses (Demeter and Matyusz 2011; Alfaro and Tribó 2003). Similarly, the relationship that manufacturing companies hold with their suppliers of raw materials may be significant for short deliveries, since both parts can collaborate to cope with the uncertainty in demand (Heydari 2014). However, this requires high levels of integration from all participants in the supply chain, which often becomes a challenge (David and Eben-Chaime 2003; O' Neal 1987). Therefore, since it is considered that benefits obtained in the material handling can impact on efficiency indexes of the production process in JIT environment, the following hypothesis can be formulated.

H10: *There is a direct and positive relation between the benefits of materials handling and those gained in the production process in a JIT environment.*

12.3.2 Results of Model 2

Figure 12.4 shows the assessment of this model. The three red segments or relations indicate that due to their statistical evaluation, they were found statistically not significant, since their *P*-value was lower than 0.05 units. Thus, only seven hypotheses were statistically significant, three of which were previously evaluated in other models.

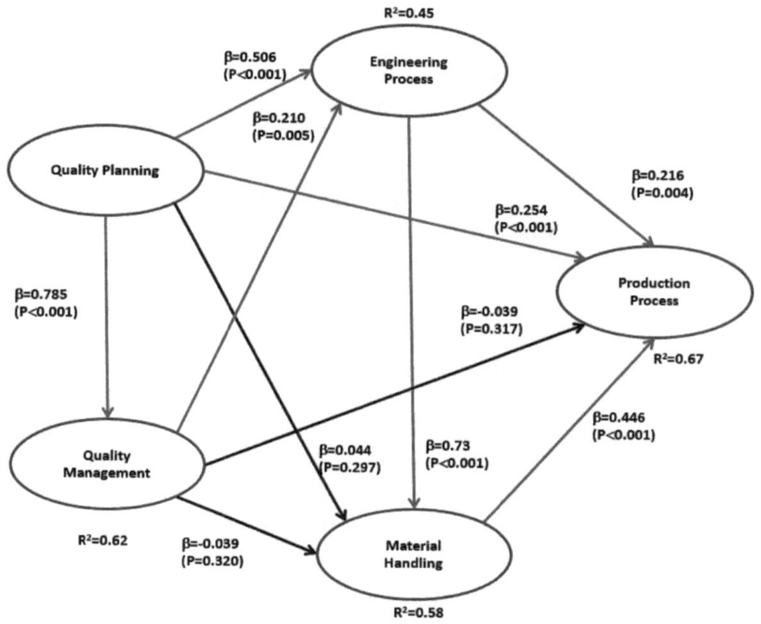

Fig. 12.4 Validation of the hypotheses of model 2

12.3.2.1 Efficiency Indexes of Model 2

Before its interpretation, the model must be tested through efficiency and adjustment indexes to prove with a high confidence level that it is efficient and predictive. The efficiency indexes are listed below.

1. Average path coefficient (APC) = 0.327, $P < 0.001$
2. Average R-squared (ARS) = 0.580, $P < 0.001$
3. Average adjusted R-squared (AARS) = 0.572, $P < 0.001$
4. Average block VIF (AVIF) = 2.463, acceptable if ≤5, ideally ≤3.3
5. Average full collinearity VIF (AFVIF) = 3.199, acceptable if ≤5, ideally ≤3.3
6. TenenhausGoF (GoF) = 0.613, small ≥ 0.1, medium ≥ 0.25, large ≥ 0.36
7. Simpson's paradox ratio (SPR) = 0.800, acceptable if ≥0.7, ideally = 1
8. R-squared contribution ratio (RSCR) = 0.984, acceptable if ≥0.9, ideally = 1
9. Statistical suppression ratio (SSR) = 1.000, acceptable if ≥0.7
10. Nonlinear bivariate causality direction ratio (NLBCDR) = 1.000, acceptable if ≥0.7.

According to the previous list, it is possible to propose the following interpretations of these indexes.

- According to ARS, APC, and AARS indexes, the model is in general efficient and has acceptable goodness of fit. The *P*-value in all three cases is lower than 0.05, the maximum cutoff value for a 95 % confidence level.
- As for collinearity, there are no problems of collinearity among variables according to AVIF and AFVIF indexes, whose values are lower than 3.3
- The remaining indexes are also suitable to this model; however, Tenenhaus index shows a highly favorable result. The minimum value acceptable for this index is 0.36 units, while it reported a value of 0.613 units in this model. Therefore, it is again concluded that the model is statistically viable.

12.3.2.2 Coefficients of Latent Variables of Model 2

Previous assessment indexes demonstrated generic efficiency of the model. However, all latent variables must be assessed through additional efficiency indexes to determine their efficiency. Table 12.8 presents the coefficients of the latent variables in Model 2. Based on this data, it is possible to state the following conclusions:

- According to R-squared and Adjusted R-square values, all dependent latent variables have sufficient predictive validity from a parametric view, since in all cases the value obtained exceeds 0.02, the minimum value accepted in this research.
- The Q-squared value in all cases is higher than 0 and similar to both R-squared and Adjusted R-squared indexes. Thus, dependent latent variables have predictive validity from a parametric point of view.
- As for internal validity, the third and fourth lines of Table 12.8 show that all variables show values higher than 0.7, the minimum acceptable value in this research for composite validity and the Cronbach's alpha index.
- AVE index is higher than 0.5 in all latent variables of this model. Therefore, they all have convergent validity.

Table 12.8 Coefficients of latent variables of model 2

	Quality planning	Quality management	Material handling	Engineering process	Production process
R-squared		0.617	0.578	0.453	0.67
Adjusted R-squared		0.614	0.569	0.446	0.661
Composite reliability	0.932	0.883	0.889	0.857	0.915
Cronbach's alpha	0.915	0.834	0.834	0.749	0.888
AVE	0.664	0.603	0.668	0.666	0.643
AVIF	3.081	2.574	3.162	3.206	3.257
Q-squared		0.62	0.612	0.456	0.712

- AVIF index for very latent variables is lower than 3.3. Hence, there are no collinearity issues in any latent variable.

12.3.2.3 Direct Effects of Model 2—Validation of Hypotheses

After the evaluation of the model according to the methodology proposed and described in previous chapters, it is possible to state the following conclusions from hypotheses initially proposed.

- H1: There is sufficient statistical evidence to declare that plans and programs generated through product Quality Planning have a direct and positive impact on Quality Management, since when JIT has been successfully implemented in manufacturing companies. When the first latent variable increases its standard deviation by one unit, the standard deviation of the second latent variable increases by 0.785 units.
- H2: There is enough statistical evidence to state that plans and programs developed through product Quality Planning have a direct and positive impact on the benefits obtained in Engineering Process in a successful JIT environment, since when the first latent variable increases its standard deviation by one unit, the standard deviation of the second latent variable increases by 0.506 units.
- H3: Enough statistical evidence enables to declare that plans and programs developed through product Quality Planning have a direct and positive effect on benefits obtained for the Production Process in a successful JIT environment, since when the first latent variable increases its standard deviation in one unit, the standard deviation of the second latent variable increases by 0.254 units.
- H4: There is not enough statistical evidence to state that plans and programs generated through product Quality Planning have a direct and positive effect on JIT benefits for Materials Handling, since the P-value obtained from the statistical significance test for this parameter was 0.297, a value highly above 0.05 units, the maximum allowed value for inferences with a 95 % confidence level. This result may be surprising since it might seem illogical. However, by analyzing the indirect effects between these variables, it will be noted that the intervention of other latent variables act as mediators in this relationship that is certainly meaningful.
- H5: Enough statistical evidence can state that product Quality Management has a direct and positive effect on the benefits obtained in the Engineering Process when a JIT program or philosophy of JIT is successfully implemented, since when the first latent variable increases its standard deviation by one unit, the standard deviation of the second latent variable rises 0.210 units.
- H6: There is not enough statistical evidence to declare that product Quality Management has a direct and positive impact on the benefits obtained in the Production Process in a successful JIT environment. The *P*-value obtained for the estimated parameter is 0.317 and thus exceeds the maximum cutoff value 0.05, for statistical inferences with 95 % confidence level. This result could

seem contradictory to the industrial practice; however, the analysis of indirect effects between variables reported that efficient materials handling and adequate indexes of the engineering process are required for the improvement of production processes.

- H7: There is not enough statistical evidence to declare that Quality Management has a direct and positive effect on the JIT benefits obtained in material handling. The *P*-value obtained from the statistical significance test of this parameter is 0.320 and exceeds the maximum allowed value for statistical inferences made with 95 % confidence level. However, the analysis of indirect effects between these latent variables demonstrated that the mediating effect of the variable benefits in the Engineering Process is required for these two variables to have a relation.
- H8: Sufficient statistical evidence enables to declare that JIT benefits obtained in the engineering process have a direct and positive impact on those JIT benefits gained in Material Handling. When the first latent variable increases its standard deviation by one unit, the standard deviation of the second latent variable increases by 0.73 units.
- H9: There is sufficient statistical evidence to state that JIT benefits obtained in the engineering process have a direct and positive effect on those JIT benefits gained in the production process, since when the first latent variable increases its standard deviation by one unit, the standard deviation of the second latent variable rises by 0.216 units.
- H10: There is enough statistical evidence to state that benefits obtained in Material Handling have a direct and positive impact on those benefits gained in the production process when a JIT program or philosophy is successfully implemented. When the first latent variable increases its standard deviation by one unit, standard deviation of the second latent variable also increases by 0.446 units.

Dependent latent variables may be affected by more than one latent variable. Thus, the value of R-squared for these dependent latent variables should be decomposed into its components. Table 12.9 shows the sizes of the effects that certain latent variables have on others.

Table 12.9 Direct effects sizes of model 2

To	From			
	Quality planning	Quality management	Material handling	Engineering process
Quality management	0.62			
Material handling	0.025	0.018		0.571
Engineering process	0.333	0.121		
Production process	0.175	0.021	0.349	0.166

Data contained in this table also allows for the following interpretation of data.

- The latent variable Quality Planning shows the largest effect, according to its size, over latent variable Quality Management, and it is due to a direct effect between both variables with a value of 0.62 units.
- Dependent latent variable Material Handling is explained by three independent latent variables in 58 %, since the R-squared equals 0.58 units. However, the highest size effect over this variable is caused by latent variable Engineering Process, which has an impact of 0.571 units.
- Latent variable Engineering Process is explained by two variables in 45 %, since the R-squared value is 0.45. However, Quality Planning has the largest impact over Engineering Process, since the size of the effect is 0.333 units.
- The latent variable Production Process is explained by four variables. However, only two of the relationships with these variables are statistically significant. The variable is explained in 67 %, since the value of the R-squared is 0.67 units. Moreover, the latent variable Material Handling seems to cause the most significant effect, since it explains 34.9 % of its variability.
- The latent variables involved in relations that were statistically not significant have considerably small size effects.

12.3.2.4 Sum of Indirect Effects of Model 2

As mentioned above, one independent latent variable can have an indirect effect on another latent variable through a mediator latent variable, which can be given by two or more segments. Table 12.10 depicts the sum of the indirect effects of latent variables over other latent variables. The table also shows the *P*-values from the statistical significance test for the estimated parameters and the size of the effects. Table 12.11 allows for the following data interpretations.

- Six indirect effects between latent variables analyzed in this model, five of which are statistically significant because the *P*-values obtained from the statistical significant was less than 0.05 units. However, one of these indirect effects is statistically not significant since the *P* value of the estimated parameter is higher than 0.05.

Table 12.10 Sum of indirect effects of model 2

To	From		
	Quality planning	Quality management	Engineering process
Material handling	0.459 ($P < 0.001$) ES = 0.262	0.153 ($P = 0.004$) ES = 0.072	
Engineering process	0.165 ($P = 0.002$) ES = 0.108		
Production process	0.338 ($P < 0.001$) ES = 0.233	0.096 ($P = 0.120$) ES = 0.050	0.326 ($P < 0.001$) ES = 0.251

- According to its size, the indirect effect that the latent variable Quality Planning causes on the latent variable Material Handling is the largest reported in this model. When the first latent variable increases its standard deviation by one unit, the standard deviation of the second latent variable rises 0.459 units. Moreover, the former explains up to 26.2 % of the variability of the later. Note that the direct relation between these two variables was found statistically not significant, which indicates that the latent variables Quality Management and Engineering Process are mediator variables in this relation. That is, Quality Planning has an effect on material handling only if Quality Management is successfully achieved and benefits in engineering process are obtained.
- Another significant indirect effect is by the latent variables Quality Planning and Production Process, since when the first latent variable increases its standard deviation by one unit, the standard deviation of the second latent variable increases 0.338 units. Moreover, the former can explain up to 23.3 % of the variability of the latter. These indirect effects occur through other three variables included in the model.
- As for the relationship between the latent variables Quality Management and Production Process, both the direct and indirect effects are statistically not significant.
- Another indirect effect of great interest from an industrial point of view is the one caused by Engineering Processes on latent variable Production Process, since when the former increases its standard deviation by one unit, the standard deviation of the latter increases by 0.251 units. Moreover, the first latent variable can explain up to 25.1 % of the variability of the second latent variable.

12.3.2.5 Total Effects of Model 2

Table 12.11 shows the total effects between the variables of this model. Note that the total effects concern the sum of both direct and indirect effects. Data provided in this table also allows for the following interpretations.

- Ten 10 relationships were analyzed as working hypothesis in this model, three of which were statistically not significant when direct effects were analyzed. However, according to Table 12.11, merely two of these were statistically not significant, since their total effects were also statistically not significant. This concerns the relationships of Quality Management with Materials Handling and Engineering Process.
- The highest total effect in this model is a direct effect. It is caused by Quality Planning over Quality Management. The second highest total effect is also a direct effect caused by Engineering Process and Material Handling.
- The third largest total effect, according to its size, is caused by Quality Planning on Production Process, since when the first latent variable increases its standard deviation by one unit, the standard deviation of the second latent variable rises by 0.592 units. Moreover, the former can explain up to 40.8 % of the variability of the latter.

Table 12.11 Sum of total effects of model 2

To	From			
	Quality planning	Quality management	Material handling	Engineering process
Quality management	0.785 ($P < 0.001$) ES = 0.617			
Material handling	0.503 ($P < 0.001$) ES = 0.287	0.114 ($P = 0.08$) ES = 0.054		0.73 ($P < 0.001$) ES = 0.571
Engineering process	0.67 ($P < 0.001$) ES = 0.441	0.21 ($P = 0.005$) ES = 0.121		
Production process	0.592 ($P < 0.001$) ES = 0.408	0.057 ($P = 0.245$) ES = 0.030	0.446 ($P < 0.001$) ES = 0.349	0.541 ($P < 0.001$) ES = 0.417

- Another relation that must be carefully analyzed is the one existing between Engineering Process and the Production Process, since when the first latent variable increases its standard deviation by one unit, the standard deviation of the second latent variable also rises by 0.541 units. Moreover, the former can explain up 41.7 % of the variability of the latter. Remaining relationships between variables can be similarly interpreted.

12.3.2.6 Conclusions of Model 2

This model integrated five latent variables and tested 10 hypotheses. Based on the results of the evaluation, it is possible to conclude the following.

- The direct, indirect, and total relationship between the latent variables Quality Management and Production Process was proved to be statistically not significant. This indicates that other latent variables but Engineering Process and Material Handling may be the mediators. A suitable alternative would be to test the mediating effect of Human Resources, since they execute these quality plans and programs. It is also worth noting that the indirect effect that is reported is the sum of all possible effects that exist, since the effect of the segments is through the engineering process variable is statistically significant.
- Although direct effect in the relationship between the latent variables Quality Planning and Material Handling turned out to be statistically significant, it was found that the sum of their indirect effects is one of the largest effects sizes, and is given through the latent variables Engineering Process and Quality Management.
- The largest total effects in this model were also direct effects. They concerned two relationships between, Quality Planning and Quality Management, on the one hand, and Engineering Process and Materials Handling, on the other hand.

References

Ahmed PK, Rafiq M (2002) Chapter six—total quality management (TQM) and internal marketing. In: Ahmed PK, Rafiq M (eds) Internal marketing. Butterworth-Heinemann, Oxford, pp 91–124

Ahmad MF, Zakuan N, Jusoh A, Takala J (2012) Relationship of TQM and business performance with mediators of SPC, lean production and TPM. Procedia Soc Behav Sci 65:186–191. doi:10.1016/j.sbspro.2012.11.109

Ahsan AMMN, Habib MA, Khoda B (2015) Resource based process planning for additive manufacturing. Comput-Aided Des. 69:112–125. doi:http://dx.doi.org/10.1016/j.cad.2015.03.006

Alfaro JA, Tribó JA (2003) Operations manager turnover and inventory fluctuations. Int J Prod Econ 81–82:51–58. doi:10.1016/S0925-5273(02)00268-2

Anderman EM, Gray D (2015) Motivation, learning, and instruction. In: Wright JD (ed) International encyclopedia of the social and behavioral sciences, 2nd edn. Elsevier, Oxford, pp 928–935

Arbogast GW (1997) A case study: Statistical analysis in a production quality improvement project. J Qual Manage 2(2):267–277. doi:10.1016/S1084-8568(97)90007-9

Arıkan E, Fichtinger J, Ries JM (2014a) Impact of transportation lead-time variability on the economic and environmental performance of inventory systems. Int J Prod Econ 157:279–288. doi:10.1016/j.ijpe.2013.06.005

Arıkan E, Fichtinger J, Ries JM (2014b) Impact of transportation lead-time variability on the economic and environmental performance of inventory systems. Int J Prod Econ 157:279–288. doi:10.1016/j.ijpe.2013.06.005

Armstrong CS, Blouin JL, Jagolinzer AD, Larcker DF (2015) Corporate governance, incentives, and tax avoidance. J Acc Econ 60(1):1–17. doi:10.1016/j.jacceco.2015.02.003

Azevedo R, Martins C, Teixeira JC, Barroso M (2014) Obstacle clearance while performing manual material handling tasks in construction sites. Saf Sci 62:205–213. doi:10.1016/j.ssci.2013.08.016

Ball F, Tyler J, Wells P (2015) Is audit quality impacted by auditor relationships? J Contemp Acc Econ 11(2):166–181. doi:10.1016/j.jcae.2015.05.002

Bendul JC, Skorna ACH (2015) Exploring impact factors of shippers' risk prevention activities: a European survey in transportation. Transp Res Part E Logistics Transp Rev (0). doi:http://dx.doi.org/10.1016/j.tre.2015.05.008. Available online 29 June 2015 (in Press)

Bettayeb B, Bassetto SJ, Sahnoun M (2014) Quality control planning to prevent excessive scrap production. J Manuf Syst 33(3):400–411. doi:10.1016/j.jmsy.2014.01.001

Calvo-Mora A, Ruiz-Moreno C, Picón-Berjoyo A, Cauzo-Bottala L (2014) Mediation effect of TQM technical factors in excellence management systems. J Bus Res 67(5):769–774. doi:10.1016/j.jbusres.2013.11.042

Çınar O, Karcıoğlu F (2013) The relationship between strategic management, institutionalization and human resource management: a survey study with family businesses located in the Northeast Anatolia Sub Economic Region of Turkey. Procedia Soc Behav Sci 99:835–842. doi:10.1016/j.sbspro.2013.10.555

Corredor P, Goñi S (2011) TQM and performance: is the relationship so obvious? J Bus Res 64 (8):830–838. doi:10.1016/j.jbusres.2010.10.002

Chakravorty SS (2009) Improving distribution operations: implementation of material handling systems. Int J Prod Econ 122(1):89–106. doi:10.1016/j.ijpe.2008.12.026

David I, Eben-Chaime M (2003) How far should JIT vendor–buyer relationships go? Int J Prod Econ 81–82:361–368. doi:10.1016/S0925-5273(02)00274-8

Demeter K, Matyusz Z (2011) The impact of lean practices on inventory turnover. Int J Prod Econ 133(1):154–163. doi:10.1016/j.ijpe.2009.10.031

Demoly F, Dutartre O, Yan XT, Eynard B, Kiritsis D, Gomes S (2013) Product relationships management enabler for concurrent engineering and product lifecycle management. Comput Ind 64(7):833–848. doi:10.1016/j.compind.2013.05.004

Diefendorff JM, Seaton GA (2015) Work motivation. In: Wright JD (ed) International encyclopedia of the social and behavioral sciences, 2nd edn. Elsevier, Oxford, pp 680–686

Farris JA, Van Aken EM, Doolen TL, Worley J (2009) Critical success factors for human resource outcomes in Kaizen events: an empirical study. Int J Prod Econ 117(1):42–65. doi:10.1016/j.ijpe.2008.08.051

Ferretti S, Caputo D, Penza M, D'Addona DM (2013) Monitoring systems for zero defect manufacturing. Procedia CIR 12:258–263. doi:10.1016/j.procir.2013.09.045

Fullerton RR, McWatters CS (2002) The role of performance measures and incentive systems in relation to the degree of JIT implementation. Acc Organ Soc 27(8):711–735. doi:10.1016/S0361-3682(02)00012-0

Glover WJ, Farris JA, Van Aken EM, Doolen TL (2011) Critical success factors for the sustainability of Kaizen event human resource outcomes: an empirical study. Int J Prod Econ 132(2):197–213. doi:10.1016/j.ijpe.2011.04.005

Göleç A (2015) A relationship framework and application in between strategy and operational plans for manufacturing industry. Comput Ind Eng 86:83–94. doi:10.1016/j.cie.2014.10.007

Gyulai D, Kádár B, Monostori L (2014) Capacity planning and resource allocation in assembly systems consisting of dedicated and reconfigurable lines. Procedia CIRP 25:185–191. doi:10.1016/j.procir.2014.10.028

Heydari J (2014) Coordinating supplier's reorder point: A coordination mechanism for supply chains with long supplier lead time. Comput Oper Res 48:89–101. doi:10.1016/j.cor.2014.03.011

Hong JD, Hayya JC, Kim SL (1992) JIT purchasing and setup reduction in an integrated inventory model. Int J Prod Res 30(2):255–266. doi:10.1080/00207549208942893

Howard LW, Thomas Foster S (1999) The influence of human resource practices on empowerment and employee perceptions of management commitment to quality. J Qual Manage 4(1):5–22. doi:10.1016/S1084-8568(99)80093-5

Huq Z (1999) Conventional shop control procedures to approximate JIT inventory performance in a job shop. J Manuf Syst 18(3):161–174. doi:10.1016/S0278-6125(99)80029-6

Jayaram J, Ahire SL, Dreyfus P (2010) Contingency relationships of firm size, TQM duration, unionization, and industry context on TQM implementation—a focus on total effects. J Oper Manage 28(4):345–356. doi:10.1016/j.jom.2009.11.009

Kannan VR, Tan KC (2005) JIT, total quality management, and supply chain management: understanding their linkages and impact on business performance. Omega 33(2):153–162. doi:10.1016/j.omega.2004.03.012

Karatepe OM, Karadas G (2012) The effect of management commitment to service quality on job embeddedness and performance outcomes. J Bus Econ Manage 13(4):614–636. doi:10.3846/16111699.2011.620159

Knechtges P, Michael CD (2014) Application of Kaizen methodology to foster departmental engagement in quality improvement. J Am Coll Radiol 11(12, Part A):1126–1130. doi:http://dx.doi.org/10.1016/j.jacr.2014.08.027

Kouaib A, Jarboui A (2014) External audit quality and ownership structure: interaction and impact on earnings management of industrial and commercial Tunisian sectors. J Econ Finan Adm Sci 19(37):78–89. doi:10.1016/j.jefas.2014.10.001

Kumiega A, Van Vliet B (2008) 30—Kaizen: continuous improvement. In: Kumiega A, Van Vliet B (eds) Quality money management. Academic Press, Burlington, pp 271–277

Lee, HH, Zhou J, Hsu PH (2014) The role of innovation in inventory turnover performance. Decis Support Syst (0). doi:http://dx.doi.org/10.1016/j.dss.2015.02.010

Lee HH, Zhou J, Hsu PH (2015) The role of innovation in inventory turnover performance. Decis Support Syst 76:35–44. doi:10.1016/j.dss.2015.02.010

Lemak DJ, Reed R (1997) Commitment to total quality management: Is there a relationship with firm performance? J Qual Manage 2(1):67

Lester A (2007) 17—Quality management. In: Lester A (ed) Project management, planning and control, 5th edn. Butterworth-Heinemann, Oxford, pp 73–83

Love CE, Guo R, Irwin KH (1995) Acceptable quality level versus zero-defects: some empirical evidence. Comput Oper Res 22(4):403–417. doi:10.1016/0305-0548(94)00049-E

Lovell MC (2003) Optimal lot size, inventories, prices and JIT under monopolistic competition. Int J Prod Econ 81–82:59–66. doi:10.1016/S0925-5273(02)00349-3

Maroun W (2015) Reportable irregularities and audit quality: insights from South Africa. Acc Forum 39(1):19–33. doi:10.1016/j.accfor.2014.03.002

Miko NU, Kamardin H (2015) Impact of audit committee and audit quality on preventing earnings management in the pre- and post- Nigerian corporate governance code 2011. Procedia Soc Behav Sci 172:651–657. doi:10.1016/j.sbspro.2015.01.415

Moattar Husseini SM, O'Brien C, Hosseini ST (2006) A method to enhance volume flexibility in JIT production control. Int J Prod Econ 104(2):653–665. doi:10.1016/j.ijpe.2005.05.025

Molina-Azorín JF, Tarí JJ, Pereira-Moliner J, López-Gamero MD, Pertusa-Ortega EM (2015) The effects of quality and environmental management on competitive advantage: a mixed methods study in the hotel industry. Tourism Manage 50:41–54. doi:10.1016/j.tourman.2015.01.008

Moore R (2007) 8—Kaizen. In: Moore R (ed) Selecting the right manufacturing improvement tools. Butterworth-Heinemann, Burlington, pp 159–172

Mourtzis D, Doukas M (2014) Design and planning of manufacturing networks for mass customisation and personalisation: challenges and outlook. Procedia CIRP 19:1–13. doi:10.1016/j.procir.2014.05.004

Myklebust O (2013) Zero defect manufacturing: a product and plant oriented lifecycle approach. Procedia CIRP 12:246–251. doi:10.1016/j.procir.2013.09.043

O' Neal C (1987) The buyer seller linkage in a JIT environment. J Purchasing Mater Manage 7–13

O'Neill P, Sohal A, Teng CW (2015) Quality management approaches and their impact on firms' financial performance—An Australian Study. Int J Prod Econ. doi:http://dx.doi.org/10.1016/j.ijpe.2015.07.015. Available online 21 July 2015 (in Press)

Ohno H, Matsubae K, Nakajima K, Kondo Y, Nakamura S, Nagasaka T (2015) Toward the efficient recycling of alloying elements from end of life vehicle steel scrap. Resour Conserv Recycl 100:11–20. doi:10.1016/j.resconrec.2015.04.001

Ooi KB (2014) TQM: a facilitator to enhance knowledge management? a structural analysis. Expert Syst Appl 41(11):5167–5179. doi:10.1016/j.eswa.2014.03.013

Oropesa-Vento M, García-Alcaraz JL, Rivera L, Manotas DF (2015) Effects of management commitment and organization of work teams on the benefits of Kaizen: planning stage. DYNA 82(191):76–84

Pragman CH (1996) JIT II: a purchasing concept for reducing lead times in time-based competition. Bus Horiz 39(4):54–58. doi:10.1016/S0007-6813(96)90052-X

Radharamanan R, Godoy LP, Watanabe KI (1996) Quality and productivity improvement in a custom-made furniture industry using Kaizen. Comput Ind Eng 31(1–2):471–474. doi:10.1016/0360-8352(96)00177-5

Ramasesh RV (1990) Recasting the traditional inventory model to implement just-in time purchasing. Prod Inventory Manage 31:71–75

Rusu G, Avasilcai S (2014) Linking human resources motivation to organizational climate. Procedia Soc Behav Sci 124:51–58. doi:10.1016/j.sbspro.2014.02.459

Sapuan SM, Mansor MR (2014) Concurrent engineering approach in the development of composite products: a review. Mater Des 58:161–167. doi:10.1016/j.matdes.2014.01.059

Shadur MA (1995) Total quality—Systems survive, cultures change. Long Range Plan 28(2):115–125. doi:10.1016/0024-6301(95)98594-I

Shi Y, Prentice C, He W (2014) Linking service quality, customer satisfaction and loyalty in casinos, does membership matter? Int J Hospitality Manage 40:81–91. doi:10.1016/j.ijhm.2014.03.013

Taghizadegan S (2006) Chapter 6—design for lean/Kaizen Six sigma. In: Taghizadegan Salman (ed) Essentials of lean six sigma. Butterworth-Heinemann, Burlington, pp 59–101

Taskov N, Mitreva E (2015) The motivation and the efficient communication both are the essential pillar within the building of the TQM (total quality management) system within the

Macedonian Higher Education Institutions. Procedia Soc Behav Sci 180:227–234. doi:10.1016/j.sbspro.2015.02.109

Tersine RJ, Hummingbird EA (1995) Lead time reduction: the search for competitive advantage. Int J Oper Prod Manage 15(2):8–18

Valmohammadi C, Roshanzamir S (2015) The guidelines of improvement: Relations among organizational culture, TQM and performance. Int J Prod Econ 164:167–178. doi:10.1016/j.ijpe.2014.12.028

Veldman J, Gaalman G (2014) A model of strategic product quality and process improvement incentives. Int J Prod Econ 149:202–210. doi:10.1016/j.ijpe.2013.03.002

Vinodh S, Joy Dino (2011) Structural equation modelling of lean manufacturing practices. Int J Prod Res 50(6):1598–1607. doi:10.1080/00207543.2011.560203

Wahab MIM, Desheng Wu, Lee Chi-Guhn (2008) A generic approach to measuring the machine flexibility of manufacturing systems. Eur J Oper Res 186(1):137–149. doi:10.1016/j.ejor.2007.01.052

Wang W, Plante RD, Tang J (2013) Minimum cost allocation of quality improvement targets under supplier process disruption. Eur J Oper Res 228(2):388–396. doi:10.1016/j.ejor.2013.01.048

Westkämper E, Warnecke HJ (1994) Zero-defect manufacturing by means of a learning supervision of process chains. CIRP Ann Manuf Technol 43(1):405–408. doi:10.1016/S0007-8506(07)62240-9

Wiengarten F, Pagell M (2012) The importance of quality management for the success of environmental management initiatives. Int J Prod Econ 140(1):407–415. doi:10.1016/j.ijpe.2012.06.024

Xie W, Jiang Z, Zhao Y, Hong J (2014) Capacity planning and allocation with multi-channel distribution. Int J Prod Econ 147 Part A:108–116. doi:http://dx.doi.org/10.1016/j.ijpe.2013.08.005

Zairi M (1993) Competitive manufacturing: combining total quality with advanced technology. Long Range Plan 26(3):123–132. doi:10.1016/0024-6301(93)90014-7

Zámečník R (2014) The measurement of employee motivation by using multi-factor statistical analysis. Procedia Soc Behav Sci 109:851–857. doi:10.1016/j.sbspro.2013.12.553

Zhang Q, Vonderembse MA, Cao M (2009) Product concept and prototype flexibility in manufacturing: implications for customer satisfaction. Eur J Oper Res 194(1):143–154. doi:10.1016/j.ejor.2007.12.013

Zidane YJT, Stordal KB, Johansen A, Van Raalte S (2015) Barriers and challenges in employing of concurrent engineering within the Norwegian construction projects. Procedia Econ Finan 21:494–501. doi:10.1016/S2212-5671(15)00204-X

Chapter 13
Causal Models of JIT Elements and Causes of Slow JIT Implementation

Companies frequently report a number of problems when implementing Just-in-Time (JIT) philosophy in their production systems. Literature review in this book indicated that these issues are often due to the absence of an element that is required for the successful functioning of the philosophy. Thus, the absence of these elements is what causes that companies experience a slow progress in the implementation of a JIT system. Thirteen causes were found and discussed in previous chapters based on the literature formerly consulted (Kumar 2010; Kumar and Garg 2000; Kumar et al. 2004; Singh and Garg 2011).

The aim of this chapter thus, is to associate the three previously discussed categories of JIT elements (human resources, production process, and product) with the previously analyzed causes of a slow JIT implementation to determine the elements that minimize issues that often force companies to abandon the philosophy in a critical stage. As in the previous models proposed, the first step for the proposal and analysis of the three causal models presented in this chapter was the identification of latent variables that could include the 13 causes of a slow JIT implementation. A factor analysis for this was hence carried out.

13.1 Factor Analysis

The factor analysis was performed according to methodology described in previous chapters. The aim of this procedure was to generate the latent variables in which the 13 elements causing slow implementation would be integrated. Results demonstrated that the determinant of the correlation matrix was close to 0. Table 13.1 shows the results from the factorability tests, from which the following can be concluded:

- KMO index to measure sampling adequacy shows a value higher than 0.8, which is the minimum cutoff acceptable value for this research. Therefore, factor analysis of variables is feasible.

J.L. García-Alcaraz and A.A. Maldonado-Macías, *Just-in-Time Elements and Benefits*, Management and Industrial Engineering, DOI 10.1007/978-3-319-25919-2_13

Table 13.1 Factor analysis

Kaiser–Meyer–Olkin measure of sampling adequacy		0.943
Bartlett's test of sphericity	Approx. Chi-square	1480.551
	df	78
	Sig.	0.000

Causes of slow JIT implementation

Table 13.2 Total variance explained

Component	Initial eigenvalues			Extraction sums of squared loadings		
	Total	% of variance	Cumulative %	Total	% of variance	Cumulative %
1	8.348	64.213	64.213	8.348	64.213	64.213
2	0.910	6.998	71.211			

– According to Bartlett sphericity test, the obtained the chi-squared value was 1480.551 units with 78° of freedom. Moreover, the P-value for statistical test of this parameter was 0. Therefore, the correlation matrix is different from an identity matrix and the factor analysis can be thus carried out on data.

Results of the factor analysis carried out on data are introduced in Table 13.2. The table shows that all the items analyzed were integrated in a single category of causes of a slow JIT implementation. This was decided since the second showed values highly distant from to the unit (note that, according to methodology, one the criteria to generate factors is to rely in eigenvalues higher than or equal to one unit). However, this second example remained in Table 13.2 as a justification of the decision made.

Therefore, according to Table 13.2, since one single factor explains 64.213 % of the variance contained in the dataset, all the observed variables, or items can be integrated into one latent variable. This latent variable will be named slow implementation. Table 13.3 shows the factor loadings obtained for these variables.

Table 13.3 Component matrix

Cause of slow JIT implementation	Factor loadings
Lack of teamwork	0.874
Lack of training	0.867
Poor and inadequate maintenance of machinery and equipment	0.859
Lack of communication among levels	0.844
Lack of multifunctional workers	0.827
Lack of support from research and development (R&D) departments	0.823
Lack of understanding of JIT techniques	0.820
Negative attitude of workers	0.818
Lack of management support in quality programs	0.807
Use of traditional methods for quality control	0.791
Lack of costumer awareness of quality policies	0.758
Casual and informal quality audits	0.715
High cost of the implementation	0.563

13.2 Model 1: JIT Elements of Human Resources and Causes of Slow JIT Implementation

Chapter 10 associated JIT elements from human resources with the benefits obtained from a JIT implementation. However, it has been mentioned that the absence of any of these elements may cause slow progress of JIT production systems. Figure 13.1 graphically introduces the relations proposed between four latent variables, three of which refer to JIT elements of human resources and were identified for the proposal of models in Chap. 10. The remaining latent variable corresponds to slow implementation of JIT. Thus, as some of the hypotheses proposed in this model were already established, this section will focus on justifying the newly emerged hypotheses between latent variables.

13.2.1 Proposed Working Hypotheses of Model 1

Figure 13.1 illustrates the proposed model with working hypotheses. Previously assessed and discussed relations are H1, H2, and H3. Thus, merely H4, H5, and H6 will be evaluated. However, all variables and hypotheses must be present in the model.

Education and training that companies provide to their human resources shall not be considered a cost but rather an investment. Human resources are those who really generate wealth for the company, since they perform all kinds of tasks inside

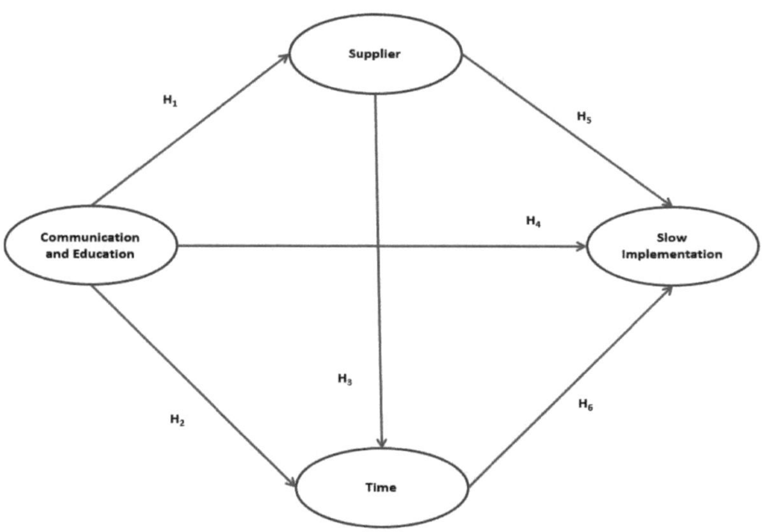

Fig. 13.1 Proposed model 1

and run operations that are performed by machines. Moreover, they are responsible for adding value to unprocessed raw materials to transform them into finished products with features requested by the customer (Globerson and Korman 2001; Jacobs 2010; Jones 2001; Morsch and Griest 1969; Ferretti et al. 2013; Myklebust 2013). However, achieving that level of perfection does not happen overnight, it is a hard-working process. Unfortunately, administrators in the course of this process many only focus on the economic investments that such training implies as well as the benefits that would be obtained (Love et al. 1995) without perhaps taking care of a more humanistic approach.

Such attitude from administrators can be easily observed by workers and all human resources in general. For instance, training courses tend to miss essential material for the sessions, or these materials are of poor quality. Similarly, training courses might happen during rest periods or days of human resources involved, or these are requested to attend those sessions during extra hours that will not be economically rewarded (Howard and Foster 1999; Lemak et al. 1997). Such attitudes may lead to a lack of motivation of the workforce (Barron and Hulleman 2015; Diefendorff and Seaton 2015).

However, opposite situations may happen, where operators adopt negative attitudes toward themselves and their professional development and will be likely to refuse to attend training courses offered. Such operators can quickly fall behind technological changes that occur continuously within the company; therefore, they might represent a constant source of errors in the production line, which can involve economic losses for companies (Macheka et al. 2013; Singh and Garg 2011). Hence, trained operators are likely to become more efficient and could be able to minimize their mistakes. However, if these occurred, trained workers could manage to detect them and would be empowered to correct them at the right time, preventing an entire production order from being poorly manufactured or without completely meeting the design specifications. On the other hand, it might be difficult for supervisors and managers to empower unwilling operators in the production process. As a result, communication between management departments and these would be probably challenged (Cheung et al. 2012; Howard and Foster 1999; Paul et al. 2000). In order to support this discussion, the following hypothesis is proposed.

H4: *Communication and education offered by companies to workers during JIT implementation have a direct and negative effect on the causes of its slow implementation.*

JIT philosophy is an essential part of the supply chain, which begins precisely with suppliers of the raw materials to be transformed into finished products (Amasaka 2002; Fullerton and McWatters 2001; Machuca 2002). Therefore, these suppliers must be properly integrated into the manufacturing processes of the producer companies; otherwise, these might need large or numerous warehouses to store their raw material due to the lack of trust in and integration with their suppliers. As a result, administrative and labor costs could increase (David and Eben-Chaime 2003; Humphreys et al. 2003; Kwak et al. 2006). Moreover, these companies would experience greater movement of materials. Consequently, the number of accidents could surge as well those damages that could occur to raw material (Chinniah 2015; Niskanen and Lauttalammi 1989).

Similarly, when suppliers are not properly integrated into the production process, manufacturing companies could frequently face shortage of raw material, which might cause the stoppage of the entire factory or production line, and companies would have to assume the respective costs. Integration between suppliers and producers, however, includes an optimal level of communication and thus a close relationship between them. Thus, both parts must strive to work in arduous teamwork and be responsible for the quality of the product they send to market (Chai et al. 2013).

In order to properly select their suppliers, manufacturing companies are nowadays able to rely on a wide range of selection techniques, which from various perspectives can ease that selection process (Lee et al. 2013; Rajesh and Malliga 2013). This process needs a group of people to perform the evaluation and make the appropriate decisions; however, certain attributes in the assessment must be assigned with a greater weight, which often can suppose challenges due to the different perspectives of the members of the evaluation group concerning those aspects that should be prioritized when evaluating and selecting suppliers (Igarashi et al. 2013). For instance, an accounting manager might emphasize on economic attributes, while a production manager would certainly focus on technical attributes. Despite the many issues that may arise from this perspective, evaluation methods are generally accepted (De Boer et al. 2001; Ho et al. 2010). In order to contribute to this discussion, the following hypothesis is proposed.

H5: *Proper evaluation and selection of certified suppliers has a direct and negative impact on the causes of a slow JIT implementation.*

One essential element of any JIT system is short delivery times to customers. (Kim and Tang 1997; Pahl et al. 2005; Pragman 1996) This forces manufacturing companies to implement a pull or push production system to meet the required demands of existing production orders (Selçuk 2013). However, achieving short deliveries implies the integration of several factors, such as a high training of workers and high investment for the proper implementation of such system (Baumann et al. 2014; Morsch and Griest 1969; Taskov and Mitreva 2015). If this cannot be considered an investment instead of a source of cost, companies might not be able to obtain their desired benefits.

Short delivery times aim to improve the flow of material and thus avoid large and long storage of raw materials, accumulation of inventories in process, and storage of finished products in warehouses (Tersine and Hummingbird 1995). As the previous JIT element, it requires highly trained human resources able to perform activities required by the new product design and respond quickly to changes in the production line from one design to another. However, this training should also emphasize on quality (Chiarini 2013; Ferradás and Salonitis 2013). Moreover, to ensure a continuous flow of raw materials and short delivery times, companies must ensure high quality of product. This involves a constant monitoring of the production system by making use of new information and communication technologies for quality control. Traditional techniques for quality control, on the other hand, have been reported as one of the causes of slow progress of JIT in production companies, although it is well known that plans and programs are not enough to ensure product quality. It is usually suitable that companies run audits to be aware of the status of quality and be informed of the corrective actions that they must take if necessary to guarantee such short delivery times to customers (Madu et al. 1995; Roderick and English 1990; Shturtz 1992).

Similarly, human resources are the main quality generators (Crumpton 2015; Rusu and Avasilcai 2014). Therefore, companies should make an effort to promote long-term contracts with their employees as much as it is possible for both parts. This improves worker motivation and increases their morale. On the other hand, employees who do not feel recognized or integrated permanently may find few reasons to become more efficient and outstand in the activities they perform (Rusu and Avasilcai 2014). Also, it has been stated that motivation among workers could be improved through approaches that oppose to traditional forms of motivation. That is, it is usually sought to integrate human resources into the company by integrating them into senior management departments, as they represent the company. Nevertheless, some authors state that this integration should happen in the other way around, by the involvement of senior managers with the employees, as it may be easier for managers to understand that it could be for employees to become acquainted with their managers. Quality circles to resolve problems can be an opportunity for this new type of integration. Thus, the hypothesis below can be formulated as a means to support this discussion.

H6: *Short delivery times with customers and long-term contracts with employees have a direct and negative impact on the causes of a slow implementation of a JIT program or philosophy.*

13.2.2 Evaluation of Model 1

Figure 13.2 illustrates the model initially proposed in Fig. 13.1. All parameters or hypotheses were tested, and, according to their P-values, they all are statistically significant. The model can thus be evaluated.

13.2.2.1 Efficiency Indexes of Model 1

After testing the statistical significance of the segments or hypothesis, the model was tested by using the following efficiency indexes:

1. Average path coefficient (APC) = 0.249, $P < 0.001$
2. Average R-squared (ARS) = 0.192, $P = 0.004$

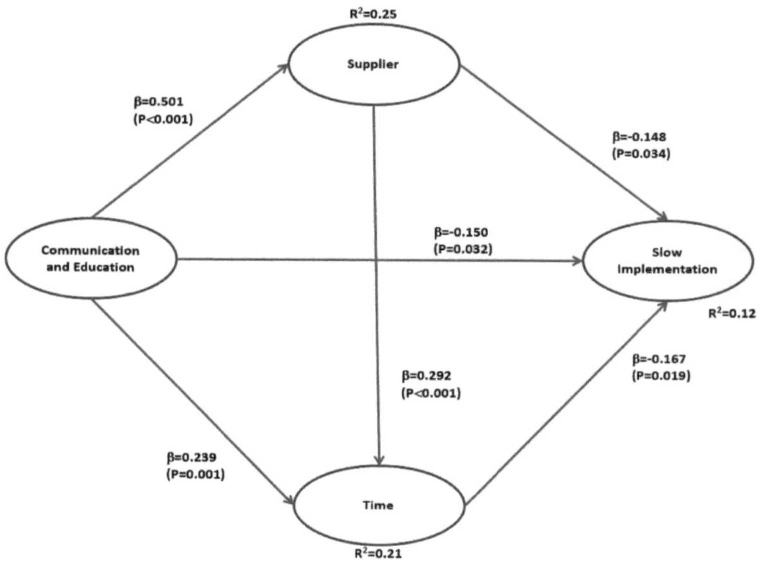

Fig. 13.2 Hypotheses validation of model 1

3. Average adjusted *R*-squared (AARS) = 0.180, *P* = 0.006
4. Average block VIF (AVIF) = 1.230, acceptable if ≤5, ideally ≤3.3
5. Average full collinearity VIF (AFVIF) = 1.297, acceptable if ≤5, ideally ≤3.3
6. Tenenhaus GoF (GoF) = 0.350, small ≥ 0.1, medium ≥ 0.25, large ≥ 0.36
7. Sympson's paradox ratio (SPR) = 1.000, acceptable if ≥0.7, ideally = 1
8. *R*-squared contribution ratio (RSCR) = 1.000, acceptable if ≥0.9, ideally = 1
9. Statistical suppression ratio (SSR) = 1.000, acceptable if ≥0.7
10. Nonlinear bivariate causality direction ratio (NLBCDR) = 0.917, acceptable if ≥0.7.

The results showed above allow for the following interpretations:

- APC, AARS, and ARS indexes show a *P*-value below 0.05 for the statistical significance test. Therefore it can be concluded that, in general, all relationships between latent variables analyzed are statistically significant as well as the predictive validity of the model.
- According to AVIF and AVIF indexes, there exist no collinearity problems among latent variables, as both indexes shows values lower than 3.3 units.
- Similarly, the Tenenhaus index (GoF) has a value of 0.350 units that is close to 0.36, the idealized value for this index.

  Every remaining efficiency index from the table can be provided with similar interpretations.

13.2.2.2 Coefficients of Latent Variables

Indexes above provide a generic analysis of the model. This section, however, introduces the efficiency levels reached by every latent variable. Based on data in Table 13.4, it is possible to conclude the following:

- All dependent latent variables show *R*-squared and adjusted *R*-squared values higher than 0.02 units. Therefore, they all have predictive validity from a parametric point of view. Likewise, same dependent latent variables show *Q*-square values greater than 0, which are also similar to their *R*-squared values. Thus, dependent latent variables have predictive validity from a nonparametric point of view.
- Third and fourth lines of the table concern measures of reliability and Cronbach's alpha. Both values are higher than 0.7, the minimum acceptable value. Thus, all latent variables in the model have internal validity.
- As for convergent validity, it is observed that the average variance extracted (AVE) exceeds 0.5, the acceptable minimum value and therefore efficiency index is also met.
- PVIF index shows values lower than 3.3 in every latent variable. Thus, this model does not experience collinearity problems.

13.2.2.3 Direct Effects—Validation of Hypotheses of Model 1

This model analyzes six working hypotheses, three of which were not previously tested in any other model. Based on results shown in Fig. 13.2, it is possible to infer the following conclusions for the three hypotheses raised.

- H4. There is enough statistical evidence to declare that communication and education in companies when implementing a JIT philosophy have a direct and negative effect on the causes of slow implementation of such philosophy. When the first latent variable increases its standard deviation by one unit, the standard deviation of the second latent variable decreases in 0.150 units.

Table 13.4 Coefficients of latent variables of model 1

Parameter	Communication and education	Supplier	Time	Slow implementation
R-squared		0.251	0.207	0.118
Adj. *R*-squared		0.245	0.196	0.099
Composite reliability	0.899	0.871	0.791	0.958
Cronbach's alpha	0.869	0.778	0.706	0.952
AVE	0.561	0.693	0.654	0.642
FVIF	1.407	1.416	1.267	1.101
Q-squared		0.252	0.211	0.128

- H5. There is sufficient statistical evidence to state that appropriate evaluation and selection of suppliers and their certification have a direct and negative impact on the causes of slow JIT implementation, since when the first latent variable increases its standard deviation by one unit, the standard deviation of the second latent variable decreases by 0.148 units.
- H6. There is enough statistical evidence to declare that short delivery times with customers and long-term contracts with employees have a direct and negative impact on the causes of slow implementation of a JIT program or philosophy, because when the first latent variable increases its standard deviation by one unit, the standard deviation of the second latent variable decreases by 0.167 units.

13.2.2.4 Sum of Indirect Effects of Model 1

Indirect effects between latent variables through other latent variables must also be analyzed. These indirect effects can occur across latent variables identified in previous chapters. For this reason, they had to be illustrated in the model.

According to the data included in Table 13.5, it is possible to conclude the following:

- The indirect effect between latent variable Communication and Education and latent variable Slow Implementation of JIT is statistically significant, since when the first latent variable increases its standard deviation by one unit, the standard deviation of the second latent variable decreases by 0.138 units. Moreover, the former explains 3.6 % of the variability of the latter.
- The indirect effect between latent variable Suppliers and latent variable Slow Implementation of JIT is statistically not significant, as the *P*-value for the statistical significance test of the parameter is 0.202. It is thus higher than 0.05, the maximum accepted value in this research for inferences with 95 % confidence level.

Table 13.5 Sum of indirect effects of model 1

To	From	
	Communication and education	Supplier
Time	0.146 ($P = 0.006$) ES = 0.055	
Slow implementation	−0.138 ($P = 0.044$) ES = 0.036	−0.049 ($P = 0.202$) ES = 0.012

13.2.2.5 Total Effects of Model 1

Total effects are the sum of direct and indirect effects. This section will merely focus on the impact received by latent variable Slow Implementation of JIT from other latent variables. All effects are statistically significant at a 95 % confidence level. Therefore, results of this latent variable can be summarized as it follows:

- Communication and Education have a direct and negative impact on Slow Implementation JIT, since when the first latent variable increases its standard deviation by one unit, the standard deviation of the second decreases by 0.289 units. Moreover, the former can explain up to 7.6 % the variability of the latter.
- Latent variable Supplier has a direct and negative effect on latent variable Slow Implementation of JIT philosophy, because when the former increases its standard deviation by one unit, the standard deviation of the latter decreases by 0.197 units. Moreover first latent variable can explain up to 4.8 % of the variability of second latent variable.
- Latent variable Time has a direct and negative effect on latent variable Slow Implementation of JIT, since when the first latent variable increases its standard deviation by one unit, the standard deviation of the second latent variable decreases by 0.167 units. Moreover, the former explains up to 4.2 % of the variability of the latter.

13.2.2.6 Conclusions of Model 1

Six hypotheses were proposed in this model, three of which were previously addressed and thus were not discussed one more. However, the section focused on demonstrating and testing the remaining three. Based on the results, it is possible to offer the following conclusions:

- The three latent variables that integrate the category of JIT elements of human resources (Communication and Education, Suppliers, and Time) have a direct and negative impact on the slow implementation of this philosophy. However, the most important negative effect reported concerns latent variable Time, since it causes the largest effects, even though all relations showed similar values.
- Only one of the indirect effects on latent variable Slow Implementation of JIT was statistically significant and concerned the indirect effect caused by latent variable Communication and Education.
- The total effects of the three latent variables associated with JIT elements of human resources on slow implementation of JIT were statistically significant. However, largest total effect is caused by latent variable Communication and Education. Thus, it may be suitable for companies to focus management efforts on this aspect to ensure a successful continuity of JIT philosophy in their production systems.

13.3 Model 2: JIT Elements of Production Process and Causes of Slow JIT Implementation

This model integrates four latent variables in total, three of which refer to JIT elements of the production process (Production Organization and Material Flow, Lean Techniques, and Capacity) that were identified for models proposed in Chap. 11. The remaining latent variable corresponds to Slow Implementation of JIT and comprises the causes of a slow progress of such philosophy in companies.

Some of the hypotheses proposed in this model were already studied in previous models. Thus, this section will focus on justifying the newly emerged hypotheses between latent variables. Previously proposed and analyzed hypotheses refer to H1, H2, and H3 in this model, while hypotheses to be tested concern H4, H5, and H6. Figure 13.3 graphically represents the variables and hypotheses.

13.3.1 Proposed Working Hypotheses of Model 2

Readers may address Chap. 11 as a reminder of those hypotheses proposed there that have an impact on this model. The following section will addressed newly proposed hypotheses H4, H5, and H6, according to Fig. 13.3.

Companies organize their machinery and equipment according to certain plant distribution in mind with benefit from this layout (Chen et al. 2011; Pattanaik and Sharma 2009). For instance, manufacturing cells seeks to integrate in a single space

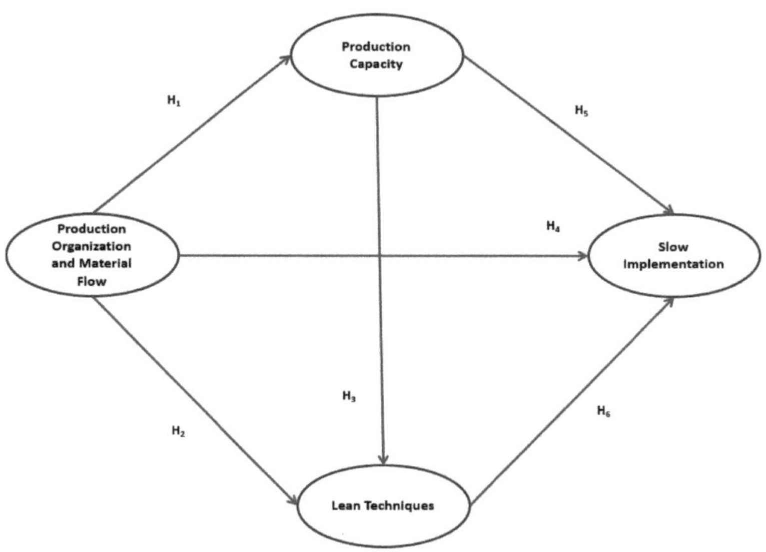

Fig. 13.3 Raised hypotheses of model 2

all machines and equipment used to produce a family of products from start to finish, which brings autonomy to managers and administrators (Caggiano and Teti 2012; Marodin et al. 2015). Group technology is another form of plant layout, which organizes machines and equipment that perform similar activities of different products in one place. The purpose of group technology is to achieve high levels of specialization (Jensen et al. 1996; Johnson and Manoochehri 1990; Pourbabai 1988). These forms of physical distribution in plants require multifunctional workers who also maintain extensive communication among them and with higher levels of the organization hierarchy, which are considered essential elements of JIT.

Along with these approaches, companies can implement many other techniques that facilitate the flow of materials inside the production process. For instance, one of the most common techniques is Kanban, also known as the cards system. It allows operators to be familiar with the exact activities that should be performed on every type of raw material to be processed, although it does imply training and understanding from thee operators who must become acquainted with this internal information system (Chan 2001; Lage et al. 2010; Rahman et al. 2013).

As many other techniques, the cards system emerged from a system of improvements proposed by workers and which is often called Kaizen. Kaizen is defined as a system of suggestions where employees propose improvements to the production system (Glover et al. 2011; Oropesa-Vento et al. 2015; Ortiz 2010). These improvement systems are significant since they seek to identify genuine knowledge from operators concerning the tasks they perform. Thus, Kaizen brings new forms of teamwork and better maintenance of machines used, among other benefits. However, workers integrate into and feel part of companies when they are motivated and observe great commitment to improvement plans and programs from senior management (Recht and Wilderom 1998). That is, a suggestion system is likely to become useless if it collects, for instance, one thousand proposals and merely considers 1 % of these received. Workers become more efficient and motivated when companies implement their proposed improvements and recognize their work. Thus, in order to contribute to this discussion, the following hypothesis is proposed.

H4: *Production organization and material flow has a direct and negative effect on the causes of a slow implementation of JIT in production systems.*

Companies that schedule production above their installed capacity may be unable to deliver orders on time. Thus, they may either demand their customers for extra time or rely on outsourcing techniques (Amirteimoori and Kordrostami 2012; Kumar et al. 2014), although the latter would imply that the outsourced companies would be responsible for the quality of the final product of companies that employed them (Gonzalez et al. 2006; Yilmaz and Bedük 2014). Thus, accepting production orders above installed capacity will always be an obstacle to achieve commitments with the customer.

Some companies with well-identified customers, who often demand production orders of a same product, might increase the amount of product to generate in small production processes. This allows them to rely on a safety stock and deliver the quantity of parts demanded by customers (Chiquoine and Hjalmarsson 2009). Safety stock involves a number of costs and expenses for companies, since they must store generated products in their own warehouses at their own expense, hoping to eventually receive a production order for that product. However, these safety stocks help manufacturing companies cope with uncertainty in demand uncertainty, which is generally due to poor communication between the sales department and the production department (Zhao et al. 2015).

A constant complaint by companies is the lack of awareness of a costumer regarding their production capacity and quality indices, since their demands generally lead to an excess of production in the system and changes in product features or quality specifications (Largoni et al. 2015). To handle these situations and make suitable changes, companies need the support of the department of research and development, even though it can be challenging to achieve. Most of these changes imply great amount of work for members of this department, not to mention the different opinions from the engineering and production department can emerge, which implies to reprogram and calibrate all or several of the machines involved in a production order (Kumari et al. 2015).

Moreover, these changes in machines and equipment will not only delay delivery times of the requested order, they will also represent several other associated costs, which in the end will be reflected on the cost of production of the parts. Similarly, workers might adopt an unsuitable attitude toward these constant changes in a production order, since they might need to learn new activities and methods to adequately perform their tasks. Thus, the ideal scenario would be one where demand is always fully controlled, which will not happen (Karatop et al. 2015; Siemieniuch et al. 2015).

When these modifications frequently occur in production orders, managers must monitor the new product features and run regular audits to ensure these features are guaranteed (Velthuis and van Asseldonk 2011). Moreover, new product characteristics may require new skills and abilities from operators which they may have not yet developed. Thus training may be needed, although it increases the cost of production. According to this discussion, it is hence possible to propose the following hypothesis as contribution.

H5: *Scheduling below the installed production capacity and inventory management has a direct and negative impact on the causes of a slow JIT implementation.*

JIT as a philosophy is strongly integrated as a lean manufacturing technique in production systems. Hence, this technique or philosophy requires the support of many other lean techniques (Abdulmalek and Rajgopal 2007), which are designed to generate higher quality products in a much more standardized process (Rahani and al-Ashraf 2012). Then the inappropriate integration of these elements with JIT can be the cause of its slow progress. For instance, one of the basic JIT elements is total and suitable preventive maintenance for machines and tools used to process raw material. Thus, a lack of such equipment care causes a slow implementation of JIT in companies, since the production process can stop or generate poor-quality parts (Cua et al. 2001; Chiarini 2013; Willmott and McCarthy 2001).

A flexible production system that is able to accept changes in orders placed is an essential JIT element. Otherwise, inflexible production systems might be unable to follow a change in product design. They would not be capable of generating such new product, the production order would be lost, and companies would assume the implied economic costs (Merschmann and Thonemann 2011; Moattar Husseini et al. 2006; Wahab et al. 2008). This refers to another tool of lean manufacturing, which refers to the ability of companies to make rapid changes in the production line. Manufacturing companies with these skills will be hence more flexible than those without this ability.

Pull production system is another crucial JIT element. Companies that implement this system seek to produce the exact amount that must be produced under the explicit order of a contract (Kim and Tang 1997; Ohno 2011; van der Laan et al. 1999). Producing in order to keep products in stock would merely indicate that the relationship between these companies and their customers is not effective, they are not adequately familiar with the product in demand, and they may not be acquainted with the sector to where belong.

Finally, JIT philosophy begins with the purchase of raw material from suppliers (Gilbert et al. 1994; Macbeth et al. 1988; Mohan and Singh 1995). Therefore, companies must ensure that such raw material is preferably purchased in small lots. This would reduce costs associated with their storage. Therefore, as a contribution regarding this discussion concerning the role of lean techniques in a JIT implementation, it is possible to formulate the following hypothesis.

H6: *The lean techniques properly applied in the production system under a JIT environment will have a direct and negative impact on the causes of the implementation of such philosophy.*

13.3.2 Evaluation of Model 2

As all previous models, Model 2 was evaluated according to methodology described. Results from this evaluation are shown in Fig. 13.4. Some items included in latent variables concerning the category of JIT elements for production process were eliminated since they caused collinearity problems with the items shown in latent variable Slow Implementation of JIT. Thus, results reported in this section concerning the hypotheses proposed in Chap. 11 may differ from those found in the chapter, where all latent variables included their corresponding items. However, hypotheses remain statistically significant even with the elimination of some of these items.

On the one hand, Fig. 13.4 shows that five of the initially proposed hypotheses were statistically significant. On the other hand, the red segment was reported as statistically not significant. This hypothesis was rejected. Efficiency indexes of this model are as follows.

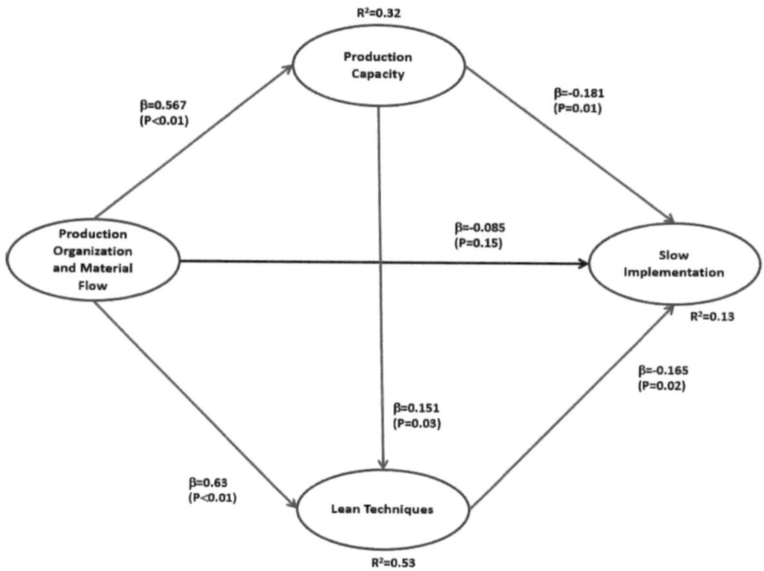

Fig. 13.4 Hypothesis validation of model 2

13.3.2.1 Efficiency Indexes of Model 2

Efficiency indexes obtained from the Model 2 are listed below.

1. Average path coefficient (APC) = 0.296, $P < 0.001$
2. Average R-squared (ARS) = 0.326, $P < 0.001$
3. Average adjusted R-squared (AARS) = 0.316, $P < 0.001$
4. Average block VIF (AVIF) = 1.604, acceptable if ≤5, ideally ≤ 3.3
5. Average full collinearity VIF (AFVIF) = 1.723, acceptable if ≤5, ideally ≤ 3.3
6. Tenenhaus GoF (GoF) = 0.459, small ≥ 0.1, medium ≥ 0.25, large ≥ 0.36
7. Sympson's paradox ratio (SPR) = 1.000, acceptable if ≥0.7, ideally = 1
8. R-squared contribution ratio (RSCR) = 1.000, acceptable if ≥0.9, ideally = 1
9. Statistical suppression ratio (SSR) = 1.000, acceptable if ≥0.7
10. Nonlinear bivariate causality direction ratio (NLBCDR) = 1.000, acceptable if ≥0.7

Based on the results, the following conclusions can be inferred concerning the general efficiency of the model.

- The APC, ARS, and AAR S parameters show statistically significant P-values, which indicates that hypotheses raised in the model are in general valid and the model has predictive validity.

- According to AVIF and AFVIF indexes, collinearity has been resolved. Both indexes show average values lower than 3.3 that is the maximum acceptable value.
- The Tenenhaus index stands out once more. The value of this parameter for this model is 0.459 units, which is visibly higher than the required minimum value; that is 0.36.
- Similar interpretations can be offered for values of remaining efficiency indexes. All of them indicate that the model is efficient.

13.3.2.2 Coefficients of Latent Variables of Model 2

Indexes above provide a generic analysis of the model. Table 13.6, however, shows the efficiency levels reached by latent variables analyzed in this model.

Based on data provided by Table 13.6, the following interpretations can be provided:

- According *R*-squared and adjusted *R*-squared indexes, all dependent latent variables have sufficient predictive validity from a parametric view, since all *P*-values obtained in these indexes are higher than 0.02. Also, as for *Q*-square index, all dependent latent variables show values higher than 0 and similar to those *R*-squared values. This indicates once more that of dependent latent variables have predictive validity from a nonparametric point of view.
- Based on values obtained in both composite reliability and Cronbach's alpha index, all latent variables analyzed have internal validity. Values obtained for these indexes were higher than 0.7 units, the minimum acceptable value.
- AVE values for all latent variables were higher than 0.5, the minimum acceptable value.

Table 13.6 Coefficients of latent variables of model 2

Parameter	Lean techniques	Production organization and inventory management	Slow implementation	Capacity
R-squared	0.529		0.127	0.321
Adj. *R*-squared	0.523		0.108	0.316
Composite reliability	0.913	0.89	0.961	0.846
Cronbach's alpha	0.888	0.835	0.955	0.728
AVE	0.601	0.67	0.671	0.648
FVIF	2.105	2.215	1.122	1.451
Q-squared	0.535		0.141	0.321

- As for collinearity, all indexes of variance inflation are less than 3.3, the maximum value allowed. As previously stated, items were removed from some latent variables to solve problems of collinearity.

13.3.2.3 Direct Effects of Model 2—Validation of Hypotheses

Since three of the initial hypotheses were of interest for their description, results are presented in tabular form below based on information illustrated in Fig. 13.4.

- H4. There is not enough statistical evidence to declare that Production Organization and Material Flow has a direct and negative effect on the causes of Slow Implementation of JIT, since the *P*-value for the significance test of the estimated parameter was 0.15 and higher than 0.05, the maximum value allowed for inferences made with a 95 confidence level. This may seem antagonistic to industrial practice. However, after the analysis of indirect effects, this relationship between latent variables will be noted through a mediator variable.
- H5. There is sufficient statistical evidence to state that scheduling production below installed Capacity and Inventory Management have a direct and negative impact on the causes of Slow Implementation of JIT, since when the first latent variable increases its standard deviation by one unit, the standard deviation of the second latent variable decreases by 0.181 units.
- H6. There is enough statistical evidence to declare that Lean Techniques properly applied in the production system will have a direct and negative impact on the Slow Implementation of implementation of JIT. When the first latent variable increases its standard deviation by one unit, the standard deviation of the latent variable decreases by 0.165 units.

13.3.2.4 Sum of Indirect Effects of Model 2

Hypothesis 4 (H4) that was found statistically not significant remained in the illustrative description of this model in Fig. 13.4 in order to analyze the possible indirect effects that may occur between two latent variables through this segment. The sum of indirect effects that independent latent variables have on the dependent latent variables is shown in Table 13.7.

Data on Table 13.7 allows for the following interpretations:

- Two latent variables associated with JIT elements of production processes have indirect impacts on the slow process of implementation of JIT.
- One of these two effects is statistically significant. That is, latent variable Production Organization and Material Flow has an indirect and negative impact on Slow Implementation of JIT, since when the former increases its standard deviation by one unit, the standard deviation of the latter increases by 0.221 units.

Table 13.7 Sum of indirect effects of model 2

To	From	
	Production organization and inventory management	Capacity
Lean techniques	0.085 ($P = 0.071$) ES = 0.061	
Slow implementation	−0.221 ($P = 0.003$) ES = 0.063	−0.025 ($P = 0.336$) ES = 0.007

- The indirect effect caused by latent variable Capacity of production over a Slow Implementation of JIT is statistically not significant. The *P*-value obtained for the estimated parameter was 0.336. It is a value higher than 0.05, the maximum value allowed for inferences at 95 % confidence level.

13.3.2.5 Total Effects of Model 2

Total effects concerning the latent variables discussed in Chap. 11 will not be discussed in this section. However, total effects on latent variable Slow Implementation of JIT will be addressed, since this latent variable is considered the focal point of this chapter.

- All total effects on latent variable Slow Implementation of JIT are statistically significant.
- Latent variable Lean Technique has a negative total effect on latent variable Slow Implementation of JIT, since when the first latent variable increases its standard deviation by one unit, the standard deviation of the second latent variable decreases second decreases by 0.165 units. Moreover, the former can explain up to 4.9 % of the variability of the latter.
- Latent variable Production Organization and Material Flow has a negative total effect on latent variable Slow Implementation of JIT, since when the first latent variable increases its standard deviation by one unit, the standard deviation of the second decreases by 0.306 units. Moreover, the former explains 8.8 % of the variability of the latter.
- Latent variable Capacity of production has a negative total effect on latent variable Slow Implementation, since when the former increases its standard deviation by one unit, the standard deviation of the latter decreases by 0.206 units. Moreover, the first latent variable explains 6.1 % of the variability of the second latent variable.

13.3.2.6 Conclusions of Model 2

After the evaluation of the model with methodology previously described, it is possible to reach the following conclusions.

- The direct effect caused by Production Organization and Material Flow over Slow Implementation of JIT turned out to be statistically not significant. However, both direct relations between latent variables Lean Techniques and Slow Implementation, and Capacity and Slow Implementation were statistically significant.
- All existing total effects between latent variables from the category of production processes and latent variable Slow Implementation of JIT proved to be statistically significant.

Therefore, it is recommended that managers take special care of production schedules, and focus on those that are below the installed capacity of companies. This helps significantly reduce problems associated with the slow implementation of just-in-time systems. However, lean techniques integrated into the manufacturing process are crucial. As mediator effects, they may reduce a slow JIT implementation.

13.4 Model 3: Product Characteristics and Causes of Slow Implementation of JIT

This last model of the book integrates three latent variables in total, two of which refer to JIT elements of the quality of the product (Quality Planning and Quality Management), which were identified for models proposed in Chap. 12. The remaining latent variable corresponds to Slow Implementation of JIT and comprises the causes of a slow progress of such philosophy in companies. Thus, one hypothesis proposed in this model was analyzed and tested in the previous chapter, it will not be addressed again.

13.4.1 Proposed Working Hypotheses of Model 3

The three latent variables identified for this model are illustrated in Fig. 13.5. This is the simplest model presented in the book due to the number of latent variables involved and the few effects that they generate.

Three working hypotheses emerged from the proposed model. The first hypothesis of this model is initially addressed in Chap. 12 but is still present in this model to analyze the possible effects that may occur between two variables through this relation.

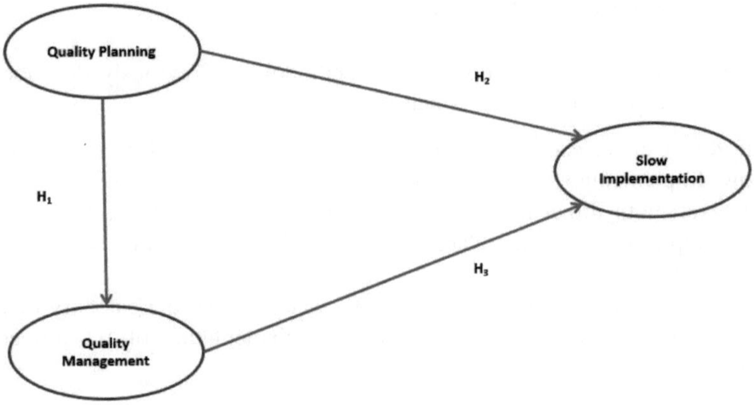

Fig. 13.5 Working hypotheses proposed of model 3

H1: *Product quality planning has a direct and positive impact on quality management and execution in a JIT environment.*

Quality plans and programs should go along with educational programs to help workers, and human resources in general, develop skills. This plans must specify the objective of the program, the skills to develop, the needs for the skills, and the activities performed that would benefit from the development of these skills (Madu et al. 1995; Shturtz 1992). Otherwise, a lack of such education plan will result in operators unfamiliar with the techniques to perform their tasks or the working methods of the company in general (Tabassi and Bakar 2009; Trad and Kalpić 2014).

However, when companies do not count on enough educational or training programs, it might also denote a lack of senior management commitment to quality, education of the workforce, and shareholders. Similarly, such lack of commitment from senior management would make it difficult to continue production plans based on 0 defects (Ferretti et al. 2013; Myklebust 2013), workers could lose their motivation to produce high quality products (Barron and Hulleman 2015; Diefendorff and Seaton 2015; Singh and Garg 2011), and improvements gained from quality circles would be largely reduced. It can be stated that senior management commitment to quality programs, the maintenance of such programs, Kaizen, and Kanban, among others, is one the most crucial factors that companies must ensure to favor a successful JIT implementation. Hence, in order to contribute to this line of research, the following hypothesis is proposed.

H2: *Product quality planning has a direct and negative impact on the causes of slow JIT implementation.*

Quality results from the effort made by senior management and employees to achieve the product characteristics required (Ahmed and Rafiq 2002; Taskov and Mitreva 2015). Therefore, when companies generate a quality plan, they must also study how it will be secured. That is, companies must know those actions that will be performed to maintain such desired quality from raw material to the finished product (Shturtz 1992; Vora 1992). On the one hand, some authors claim that is important to start with a 100 % inspection of all raw materials. Unfortunately, this approach might be highly expensive and is an indication that suppliers are not properly integrated with the production process of the manufacturing companies (Miltenbirg 1990; Steele 2001).

On the other hand, it is preferred that manufacturing companies train suppliers in the best way possible and integrate them into their production processes. This can be a form of ensuring quality programs for long terms by basing on trust. Such integration from suppliers could also allow for the simplification of the quality process quality throughout the company and would promote collaborative effort between both parts (Dong et al. 2001; Kwak et al. 2006).

Finally, it is also important that managers continuously monitor quality through periodic, not sporadic, audits, since rare and informal inspections are one of the factors that compromise a successful JIT implementation (Ball et al. 2015; Lee and Park 2013). Thus, quality is not the result of meeting standards. It is ensured by the work and relationships established among members involved in the generation of a product. Thus, according to this discussion, it is possible to formulate the following working hypothesis.

H3: *Quality management and execution have a direct and negative impact on the causes of a slow JIT implementation.*

13.4.2 Evaluation of Model 3

The model was evaluated using previously described methodology. Results are illustrated in Fig. 13.6. As in previous models, *P*-values for the statistical significance test and beta values were obtained for each parameter. Based on these *P*-values, although they are low, it can be stated that all relationships between latent variables are statistically significant.

13.4.2.1 Efficiency Indexes of Model 3

Figure 13.6 illustrates results from the evaluation of Model 3 depicted in Fig. 13.5. Efficiency indexes employed for these evaluation are describes below.

1. Average path coefficient (APC) = 0.359, $P < 0.001$
2. Average R-squared (ARS) = 0.301, $P < 0.001$
3. Average adjusted R-squared (AARS) = 0.293, $P < 0.001$
4. Average block VIF (AVIF) = 1.656, acceptable if ≤ 5, ideally ≤ 3.3
5. Average full collinearity VIF (AFVIF) = 1.653, acceptable if ≤ 5, ideally ≤ 3.3
6. Tenenhaus GoF (GoF) = 0.425, small ≥ 0.1, medium ≥ 0.25, large ≥ 0.36

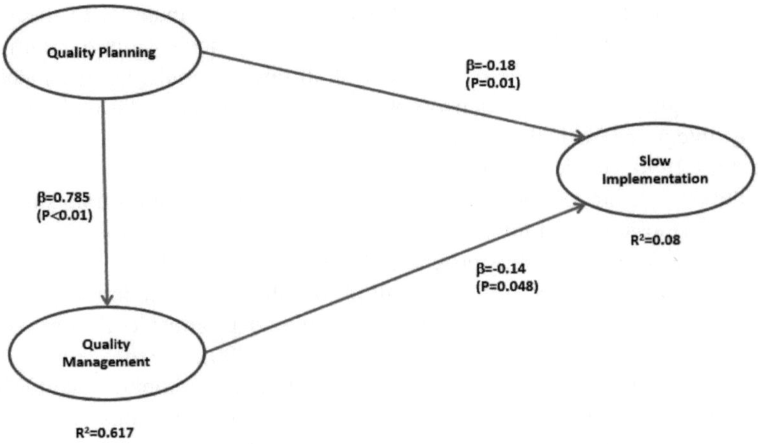

Fig. 13.6 Validation of hypotheses of model 3

7. Sympson's paradox ratio (SPR) = 1.000, acceptable if ≥0.7, ideally = 1
8. *R*-squared contribution ratio (RSCR) = 1.000, acceptable if ≥0.9, ideally = 1
9. Statistical suppression ratio (SSR) = 1.000, acceptable if ≥0.7
10. Nonlinear bivariate causality direction ratio (NLBCDR) = 1.000, acceptable if ≥0.7.

The following interpretations to those indexes can be offered.

- ARS, APC, and AARS indexes all show a *P*-value that is below 0.05 units. Thus, it can be concluded that the model has predictive validity and the estimated parameters are statistically significant in general terms.
- Based on the value of AVIF and AFVIF indexes, it is concluded that there is no collinearity in variables analyzed.
- The Tenenhaus index is once more remarkable. It shows a value of 0.425, which is outstandingly higher than 0.36, the minimum value accepted for this index.
- Similar interpretations can be proposed for the remaining indexes, which also indicate that this model is efficient and valid.

13.4.2.2 Coefficients of Latent Variables of Model 3

Table 13.8 reports efficiency indexes for every latent variable contained in this model, since indexes above merely provide a generic evaluation of the model.

Based on data contained in Table 13.8, it is possible to conclude the following.

- The two dependent latent variables have predictive validity from a parametric perspective, since *R*-squared and adjusted *R*-square indexes show values higher than 0.02, in all cases. Also, the value of the *Q*-square for these same dependent

Table 13.8 Coefficient of latent variables of model 3

Parameter	Quality planning	Quality management	Slow implementation
R-squared		0.617	0.080
Adj. *R*-squared		0.614	0.067
Composite reliability	0.932	0.883	0.958
Cronbach's alpha	0.915	0.834	0.952
AVE	0.664	0.603	0.642
FVIF	2.421	2.434	1.058
Q-squared		0.620	0.088

latent variables is higher than 0 and very similar to *R*-square and adjusted *R*-square values. Therefore, all dependent latent variables have predictive validity from a nonparametric perspective.

- According to Cronbach's Alpha and composite reliability indexes, all latent variables have internal validity, since in all cases values are higher than 0.7.
- AVE index, located in the fifth line of the table, shows values lower than 0.5 in each of the variables analyzed. Therefore, variables have convergent validity.
- As for collinearity, the AVIF index shows values lower than 3.3 for all latent variables. Thus, there are no collinearity problems within these variables.

13.4.2.3 Direct Effects—Validation of Hypotheses of Model 3

Based on Figs. 13.5 and 13.6, the following conclusions can be offered regarding the direct effects that exist among latent variables of this model.

- H1. There is enough statistical evidence to state that product quality planning has a direct and positive impact on quality management and execution in a JIT environment, since when the first latent variable increases its standard deviation by one unit, the standard deviation of the second latent variable increases by 0.785 units.
- H2. There is sufficient statistical evidence to declare that product quality planning has a direct and negative impact on the causes of a slow JIT implementation. When the first latent variable increases its standard deviation by one unit, the standard deviation of the second latent variable decreases by 0.18 units.
- H3. There is enough statistical evidence to point out that product quality management and execution have a direct and negative impact on the slow implementation of JIT, because when the first latent variable increases its standard deviation by one unit, the standard deviation of the second latent variable decreases by 0.14 units.

In the case of dependent latent variables, direct effects can be decomposed as it follows.

- Latent variable Quality Management is explained by latent variable Quality Planning. The total effect is direct and first latent variable explains 61.7 % of the variability of the second latent variable.
- Latent variable Slow Implementation is explained in 8 % by two latent variables. Quality Planning explains 4.7 % of this variability, while quality management explains 3.3 %.

13.4.3 *Sum of Indirect Effects of Model 3*

Due to its simplicity, this model merely reported one indirect effect caused by latent variable Quality Planning over latent variable Slow Implementation of JIT philosophy. The effect occurs through latent variable Quality Management. Therefore, it is possible to conclude the following.

- Quality Planning has an indirect, negative, and significant effect on Slow Implementation of JIT, since when the former increases its standard deviation by one unit, the standard deviation of the latter decreases by 0.106 units ($P = 0.034$). Moreover, the first latent variable can explain up to 2.8 % of the variability of the second latent variable.

13.4.3.1 Total Effects of Model 3

The sum of the total effects includes direct and indirect effects. For this model, total effects are outlined in Table 13.9.

Based on data in Table 13.9, it is possible to state the following conclusions regarding total effects among latent variables proposed in this model.

- All total effects are statistically significant since the *P*-value obtained in al cases is lower than 0.05.
- Latent variable Quality Planning has a negative total effect on latent variable Slow Implementation of JIT. When the first latent variable increases its standard deviation by one unit, the standard deviation of the second latent variable decreases by 0.288 units. Moreover, the former can explain 7.5 % of the variability of the latter.
- Latent variable Quality Management has a negative total effect on latent variable Slow Implementation of JIT, since when the first latent variable increases its standard deviation by one unit, the second decreases by 0.14 units.

Table 13.9 Sum of total effects of model 3

To	From	
	Quality planning	Quality management
Quality management	0.785 ($P < 0.001$) ES = 0.617	
Slow implementation	−0.288 ($P < 0.001$) ES = −0.075	−0.140 ($P = 0.048$) ES = −0.033

13.4.3.2 Conclusions of Model 3

Model 3 proposed three working hypothesis to relate JIT elements of product quality to the causes of a slow JIT implementation. Previous evaluation and analysis of the model allowed for the following final conclusions regarding these relations.

- The direct effects that latent variables Quality Planning and Quality Management cause over a slow implementation of JIT is statistically significant. All total effects are also statistically significant.
- The generation of quality products ensures a continuous flow of material. Therefore, the benefits sought by JIT are reached.
- Product quality in the production process is based on the collaboration between JIT and human resources.

References

Abdulmalek FA, Rajgopal J (2007) Analyzing the benefits of lean manufacturing and value stream mapping via simulation: a process sector case study. Int J Prod Econ 107(1):223–236. doi:10.1016/j.ijpe.2006.09.009

Ahmed PK, Rafiq M (2002) Chapter six—total quality management (TQM) and internal marketing. In: Ahmed PK, Rafiq M (eds) Internal marketing. Butterworth-Heinemann, Oxford, pp 91–124

Amasaka K (2002) "New JIT": A new management technology principle at Toyota. Int J Prod Econ 80(2):135–144. doi:10.1016/S0925-5273(02)00313-4

Amirteimoori A, Kordrostami S (2012) Production planning in data envelopment analysis. Int J Prod Econ 140(1):212–218. doi:10.1016/j.ijpe.2011.09.025

Ball F, Tyler J, Wells P (2015) Is audit quality impacted by auditor relationships? J Contemp Acc Econ 11(2):166–181. doi:10.1016/j.jcae.2015.05.002

Barron KE, Hulleman CS (2015) Expectancy-value-cost model of motivation. In: Wright JD (ed) International encyclopedia of the social & behavioral sciences, 2nd edn. Elsevier, Oxford, pp 503–509

Baumann T, Harfst S, Swanger A, Saganski G, Alwerfalli D, Cell A (2014) Developing competency-based, Industry-driven manufacturing education in the USA: bringing together industry, government and education sectors. Procedia—Soc Behav Sci 119:30–39. doi:10.1016/j.sbspro.2014.03.006

Caggiano A, Teti R (2012) Digital manufacturing cell design for performance increase. Procedia CIRP 2:64–69. doi:10.1016/j.procir.2012.05.041

Chai J, Liu JNK, Ngai EWT (2013) Application of decision-making techniques in supplier selection: a systematic review of literature. Expert Syst Appl 40(10):3872–3885. doi:10.1016/j.eswa.2012.12.040

Chan FTS (2001) Effect of kanban size on just-in-time manufacturing systems. J Mater Process Technol 116(2–3):146–160. doi:10.1016/S0924-0136(01)01022-6

Chen Y, Xiao Q, Tang X (2011) Product layout optimization and simulation model in a multi-level distribution center. Syst Eng Proc 2:300–307. doi:10.1016/j.sepro.2011.10.042

Cheung C, Baum T, Wong A (2012) Relocating empowerment as a management concept for Asia. J Bus Res 65(1):36–41. doi:10.1016/j.jbusres.2011.07.012

Chiarini A (2013) The main methods of lean organization: Kanban, Cellular Manufacturing, SMED and TPM lean organization: from the tools of the toyota production system to lean office. Springer, Milan, vol 3, pp 81–116

Chinniah Y (2015) Analysis and prevention of serious and fatal accidents related to moving parts of machinery. Saf Sci 75:163–173. doi:10.1016/j.ssci.2015.02.004

Chiquoine B, Hjalmarsson E (2009) Jackknifing stock return predictions. J Empir Finance 16 (5):793–803

Crumpton MA (2015) 3—Human resources as a strategic partner. In: Crumpton MA (ed) Strategic human resources planning for academic libraries. Chandos Publishing, Oxford, pp 23–29

Cua KO, McKone KE, Schroeder RG (2001) Relationships between implementation of TQM, JIT, and TPM and manufacturing performance. J Oper Manag 19(6):675–694. doi:10.1016/S0272-6963(01)00066-3

David I, Eben-Chaime M (2003) How far should JIT vendor–buyer relationships go? Int J Prod Econ 81–82:361–368. doi:10.1016/S0925-5273(02)00274-8

De Boer L, Labro E, Morlacchi P (2001) A review of methods supporting supplier selection. Eur J Purch Suppl Manage 7(2):75–89

Diefendorff JM, Seaton GA (2015) Work Motivation. In: Wright JD (ed) International encyclopedia of the social & behavioral sciences, 2nd edn. Elsevier, Oxford, pp 680–686

Dong Y, Carter CR, Dresner ME (2001) JIT purchasing and performance: an exploratory analysis of buyer and supplier perspectives. J Oper Manag 19(4):471–483. doi:10.1016/S0272-6963(00)00066-8

Ferradás PG, Salonitis K (2013) Improving changeover time: a tailored SMED approach for welding cells. Procedia CIRP 7:598–603. doi:10.1016/j.procir.2013.06.039

Ferretti S, Caputo D, Penza M, D'Addona DM (2013) Monitoring systems for zero defect manufacturing. Procedia CIRP 12:258–263. doi:10.1016/j.procir.2013.09.045

Fullerton RR, McWatters CS (2001) The production performance benefits from JIT implementation. J Oper Manag 19(1):81–96. doi:10.1016/S0272-6963(00)00051-6

Gilbert FW, Young JA, O'Neal CR (1994) Buyer-seller relationships in just-in-time purchasing environments. J Bus Res 29(2):111–120. doi:10.1016/0148-2963(94)90015-9

Globerson S, Korman A (2001) The use of just-in-time training in a project environment. Int J Project Manage 19(5):279–285. doi:10.1016/S0263-7863(00)00012-0

Glover WJ, Farris JA, Van Aken EM, Doolen TL (2011) Critical success factors for the sustainability of Kaizen event human resource outcomes: An empirical study. Int J Prod Econ 132(2):197–213. doi:10.1016/j.ijpe.2011.04.005

Gonzalez R, Gasco J, Llopis J (2006) Information systems outsourcing: A literature analysis. Inf Manag 43(7):821–834. doi:10.1016/j.im.2006.07.002

Ho W, Xu X, Dey PK (2010) Multi-criteria decision making approaches for supplier evaluation and selection: a literature review. Eur J Oper Res 202(1):16–24. doi:10.1016/j.ejor.2009.05.009

Howard LW, Foster ST (1999) The influence of human resource practices on empowerment and employee perceptions of management commitment to quality. J Qual Manag 4(1):5–22. doi:10.1016/S1084-8568(99)80093-5

Humphreys P, Shiu WK, Lo VHY (2003) Buyer–supplier relationship: perspectives between Hong Kong and the United Kingdom. J Mater Process Technol 138(1–3):236–242. doi:10.1016/S0924-0136(03)00078-5

Igarashi M, de Boer L, Fet AM (2013) What is required for greener supplier selection? A literature review and conceptual model development. J Purch Suppl Manag 19(4):247–263. doi:10.1016/j.pursup.2013.06.001

Jacobs RL (2010) Training and learning in the workplace. In: McGaw PPB (ed) International encyclopedia of education, 3rd edn. Elsevier, Oxford, pp 328–336

Jensen JB, Malhotra MK, Philipoom PR (1996) Machine dedication and process flexibility in a group technology environment. J Oper Manag 14(1):19–39. doi:10.1016/0272-6963(95)00030-5

Johnson TW, Manoochehri GH (1990) Adopting JIT: implications for worker roles and human resource management. Ind Manage 32(3):2

Jones MJ (2001) Just-in-time training. Adv Dev Hum Resour 3(4):480

Karatop B, Kubat C, Uygun Ö (2015) Talent management in manufacturing system using fuzzy logic approach. Comput Ind Eng 86:127–136. doi:10.1016/j.cie.2014.09.015

Kim I, Tang CS (1997) Lead time and response time in a pull production control system. Eur J Oper Res 101(3):474–485. doi:10.1016/S0377-2217(96)00174-9

Kumar V (2010) JIT based quality management: concepts and implications in Indian context. Int J Eng Sci Technol 2(1):40–50

Kumar V, Garg D (2000) JIT elements in Indian context: an analysis. Prod J 41(2):217–222

Kumar V, Garg D, Mehta N (2004) JIT elements in Indian context: an analysis. JIT Elements Indian Context: Anal 63(August):662–665

Kumar SV, Mani VGS, Devraj N (2014) Production planning and process improvement in an impeller manufacturing using scheduling and OEE techniques. Procedia Mater Sci 5:1710–1715. doi:10.1016/j.mspro.2014.07.360

Kumari S, Singh A, Mishra N, Garza-Reyes JA (2015) A multi-agent architecture for outsourcing SMEs manufacturing supply chain. Robot Comput-Integr Manuf 36:36–44. doi:10.1016/j.rcim.2014.12.009

Kwak TC, Kim JS, Moon C (2006) Supplier–buyer models for the bargaining process over a long-term replenishment contract. Comput Ind Eng 51(2):219–228. doi:10.1016/j.cie.2006.02.004

Lage Junior M, Godinho Filho M (2010) Variations of the kanban system: Literature review and classification. Int J Prod Econ 125(1):13–21. doi:10.1016/j.ijpe.2010.01.009

Largoni M, Facco P, Bernini D, Bezzo F, Barolo M (2015) Quality-by-Design approach to monitor the operation of a batch bioreactor in an industrial avian vaccine manufacturing process. J Biotechnol 211:87–96. doi:10.1016/j.jbiotec.2015.07.001

Lee C, Park MS (2013) Subjectivity in fair-value estimates, audit quality, and informativeness of other comprehensive income. Adv Acc 29(2):218–231. doi:10.1016/j.adiac.2013.05.003

Lee AHI, Kang H-Y, Lai C-M, Hong W-Y (2013) An integrated model for lot sizing with supplier selection and quantity discounts. Appl Math Model 37(7):4733–4746. doi:10.1016/j.apm.2012.09.056

Lemak DJ, Reed R, Satish PK (1997) Commitment to total quality management: Is there a relationship with firm performance? J Qual Manag 2(1):67–86. doi:10.1016/S1084-8568(97)90022-5

Love CE, Guo R, Irwin KH (1995) Acceptable quality level versus zero-defects: some empirical evidence. Comput Oper Res 22(4):403–417. doi:10.1016/0305-0548(94)00049-E

Macbeth DK, Baxter L, Farguson N, Neil G (1988) Buyer-vendor relationship with just-in-time: lessons from us multinationals. Ind Eng 20(9):38–41

Macheka L, Manditsera FA, Ngadze RT, Mubaiwa J, Nyanga LK (2013) Barriers, benefits and motivation factors for the implementation of food safety management system in the food sector in Harare Province, Zimbabwe. Food Control 34(1):126–131. doi:10.1016/j.foodcont.2013.04.019

Machuca JAD (2002) JIT facing the New Millennium. Int J Prod Econ 80(2):131–134. doi:10.1016/S0925-5273(02)00312-2
Madu CN, Kuei C-H, Winokur D (1995) Environmental quality planning: A strategic total quality management (STQM) approach. Futures 27(8):839–856. doi:10.1016/0016-3287(95)00048-2
Marodin G, Saurin T, Tortorella G, Denicol J (2015) How context factors influence lean production practices in manufacturing cells. Int J Adv Manuf Technol 79(5–8):1389–1399. doi:10.1007/s00170-015-6944-2
Merschmann U, Thonemann UW (2011) Supply chain flexibility, uncertainty and firm performance: an empirical analysis of German manufacturing firms. Int J Prod Econ 130 (1):43–53. doi:10.1016/j.ijpe.2010.10.013
Miltenbirg GJ (1990) Changing MRP's costing procedure to suit JIT. Prod Inventory Manage J 31 (2):77–83
Moattar Husseini SM, O'Brien C, Hosseini ST (2006) A method to enhance volume flexibility in JIT production control. Int J Prod Econ 104(2):653–665. doi:10.1016/j.ijpe.2005.05.025
Mohan MTR, Singh NP (1995) Buyer–supplier relationships: the case from indian diary co-operatives. Prod J 35(4):643–647
Morsch WC, Griest J (1969) A model for manpower, employment, training and education. Socio-Econ Plann Sci 2(2–4):225–250. doi:10.1016/0038-0121(69)90020-2
Myklebust O (2013) Zero defect manufacturing: a product and plant oriented lifecycle approach. Procedia CIRP 12:246–251. doi:10.1016/j.procir.2013.09.043
Niskanen T, Lauttalammi J (1989) Accidents in materials handling at building construction sites. J Occup Accid 11(1):1–17. doi:10.1016/0376-6349(89)90002-3
Ohno K (2011) The optimal control of just-in-time-based production and distribution systems and performance comparisons with optimized pull systems. Eur J Oper Res 213(1):124–133. doi:10.1016/j.ejor.2011.03.005
Oropesa-Vento M, García-Alcaraz JL, Rivera L, Manotas DF (2015) Effects of management commitment and organization of work teams on the benefits of Kaizen: planning stage. DYNA 82(191):76–84
Ortiz C (2010) Kaizen vs. lean: distinct but related. Met Finish 108(1):50–51. doi:10.1016/S0026-0576(10)80011-X
Pahl J, Voß S, Woodruff D (2005) Production planning with load dependent lead times. 4OR 3 (4):257–302. doi: 10.1007/s10288-005-0083-9
Pattanaik LN, Sharma BP (2009) Implementing lean manufacturing with cellular layout: a case study. Int J Adv Manuf Technol 42(7–8):772–779. doi:10.1007/s00170-008-1629-8
Paul RJ, Niehoff BP, Turnley WH (2000) Empowerment, expectations, and the psychological contract—managing the dilemmas and gaining the advantages. J Socio-Econ 29(5):471–485. doi:10.1016/S1053-5357(00)00083-4
Pourbabai B (1988) A just-in-time/group technology based scheduling model. In: Radharamanan R (ed) Robotics and factories of the future '87. Springer, Berlin, pp 105–110
Pragman CH (1996) JIT II: a purchasing concept for reducing lead times in time-based competition. Bus Horiz 39(4):54–58. doi:10.1016/S0007-6813(96)90052-X
Rahani AR, al-Ashraf M (2012) Production flow analysis through value stream mapping: a lean manufacturing process case study. Procedia Eng 41(0):1727–1734. doi: http://dx.doi.org/10.1016/j.proeng.2012.07.375
Rahman NAA, Sharif SM, Esa MM (2013) Lean manufacturing case study with Kanban system implementation. Procedia Econ Finance 7:174–180. doi:10.1016/S2212-5671(13)00232-3
Rajesh G, Malliga P (2013) Supplier selection based on AHP QFD methodology. Procedia Eng 64:1283–1292. doi:10.1016/j.proeng.2013.09.209
Recht R, Wilderom C (1998) Kaizen and culture: on the transferability of Japanese suggestion systems. Int Bus Rev 7(1):7–22. doi:10.1016/S0969-5931(97)00048-6
Roderick LM, English JR (1990) Laboratory for quality planning and control. Comput Ind Eng 19 (1–4):165–169. doi:10.1016/0360-8352(90)90098-7
Rusu G, Avasilcai S (2014) Linking human resources motivation to organizational climate. Procedia—Soc Behav Sci 124:51–58. doi:10.1016/j.sbspro.2014.02.459

Selçuk B (2013) Adaptive lead time quotation in a pull production system with lead time responsive demand. J Manuf Syst 32(1):138–146. doi:10.1016/j.jmsy.2012.07.017

Shturtz GT (1992) Quality planning. Met Powder Rep 47(10):57. doi:10.1016/0026-0657(92)91976-Q

Siemieniuch CE, Sinclair MA, Henshaw MJD (2015) Global drivers, sustainable manufacturing and systems ergonomics. Appl Ergon 51:104–119. doi:10.1016/j.apergo.2015.04.018

Singh S, Garg D (2011) JIT system: concepts, benefis and motivation in Indian industries. Int J Manag Bus Stud 1(1):26–30

Steele AL (2001) Cost drivers and other management issues in the JIT supply chain environment. Prod Invent Manag J 42(2):61–67

Tabassi AA, Bakar AHA (2009) Training, motivation, and performance: The case of human resource management in construction projects in Mashhad, Iran. Int J Proj Manag 27(5):471–480. doi:10.1016/j.ijproman.2008.08.002

Taskov N, Mitreva E (2015) The motivation and the efficient communication both are the essential pillar within the building of the TQM (total quality management) system within the Macedonian Higher Education Institutions. Procedia—Soc Behav Sci 180:227–234. doi:10.1016/j.sbspro.2015.02.109

Tersine RJ, Hummingbird EA (1995) Lead time reduction: the search for competitive advantage. Int J Oper Prod Manag 15(2):8–18

Trad A, Kalpić D (2014) The selection and training framework (STF) for managers in (e-)business innovation transformation projects—managerial recommendations. Procedia Technol 16:768–774. doi:10.1016/j.protcy.2014.10.026

van der Laan E, Salomon M, Dekker R (1999) An investigation of lead-time effects in manufacturing/remanufacturing systems under simple PUSH and PULL control strategies. Eur J Oper Res 115(1):195–214. doi:10.1016/S0377-2217(98)00108-8

Velthuis AGJ, van Asseldonk MAPM (2011) Process audits versus product quality monitoring of bulk milk. J Dairy Sci 94(1):235–249. doi:10.3168/jds.2010-3528

Vora JA (1992) Applying a theory of organization change to adopting JIT. Omega 20(2):193–199. doi:10.1016/0305-0483(92)90073-G

Wahab MIM, Wu D, Lee C-G (2008) A generic approach to measuring the machine flexibility of manufacturing systems. Eur J Oper Res 186(1):137–149. doi:10.1016/j.ejor.2007.01.052

Willmott P, McCarthy D (2001) 3—The top-down and bottom-up realities of TPM. In: Willmott P, McCarthy D (eds) Total productivity maintenance, 2nd edn. Butterworth-Heinemann, Oxford, pp 23–61

Yilmaz A, Bedük A (2014) Evaluation of the effect of the outsourcing on resource dependency and transaction cost approach: a research in Konya Oiz, Turkey. Procedia—Soc Behav Sci 109:737–752. doi:10.1016/j.sbspro.2013.12.538

Zhao T, Wang QJ, Bennett JC, Robertson DE, Shao Q, Zhao J (2015) Quantifying predictive uncertainty of streamflow forecasts based on a Bayesian joint probability model. J Hydrol 528:329–340. doi:10.1016/j.jhydrol.2015.06.043

MIX
Papier aus verantwortungsvollen Quellen
Paper from responsible sources
FSC® C105338

If you have any concerns about our products,
you can contact us on
ProductSafety@springernature.com

In case Publisher is established outside the EU,
the EU authorized representative is:
Springer Nature Customer Service Center GmbH
Europaplatz 3, 69115 Heidelberg, Germany

Printed by Libri Plureos GmbH
in Hamburg, Germany